Plasma Technology

B. Gross, B. Grycz and K. Miklóssy

English translation edited by

R. C. G. Leckey, B. Sc., Ph. D.
Lecturer,
School of Physical Sciences
La Trobe University
Victoria, Australia

ILIFFE BOOKS LTD LONDON

AMERICAN ELSEVIER PUBLISHING COMPANY, INC.
NEW YORK

English edition first published in 1969 by
Iliffe Books Ltd., 42 Russell Square, London, W.C.1.
in co-edition with SNTL — Publishers of Technical Literature, Prague.

American edition published in 1969 by
American Elsevier Publishing Company, Inc.
52 Vanderbilt Avenue
New York, New York 10017

Library of Congress Catalog Card Number: 68-27535

Printed in Czechoslovakia

Contents

Part 1

Introduction to the Physics of High Pressure Plasma

1.1 Properties of Plasma in Thermodynamic Equilibrium

1.1.1 Plasma and Thermodynamic Equilibrium

1.1.1.1 Plasma

Gases heated to temperatures higher than a few thousand degrees centigrade constitute a distinctive fourth state of matter – the plasma – which differs from ordinary gases particularly by the following properties:

- Apart from elastic collisions between individual molecules a considerable number of inelastic collisions occur which result in the production of molecules in excited quantum states, and sometimes cause their dissociation or ionization.
- These processes cause the transformation of the chemically uniform gas into a gaseous mixture of various particles: molecules, atoms, positive and negative ions, electrons, photons, etc.
- The plasma contains electrically charged particles – electrons and ions. It is through these particles that electric and magnetic fields can act upon the ionized gas: being conductive, the gas mixture can accept energy from an electric or magnetic field or transfer energy to it. The mixture as a whole is quasi-neutral, the concentration of positive and negative particles being roughly equal at every point ($n_+ \sim n_-$).
- Provided the density of charged particles is not too small, the properties of the gas are largely influenced by their Coulombic interaction. The characteristic feature of the Coulombic action is the relatively slow decrease in force with the distance between the particles $\left(F \sim \frac{1}{r^2}\right)$. Consequently, each particle acts simultaneously on a great number of other particles–theoret-

ically all of them ("collective interaction"). The orbits of the particles are no longer rectilinear, as they are in the ideal gas.

In this book we shall treat high pressure plasma ($p \sim 10^{-1}$ atm or more) with temperature ranging between 10^3 and 10^4 deg. K, which is almost in a state of thermodynamic equilibrium. For the purpose of our discussion, we shall consider plasma as a mixture of ideal non-degenerated chemically reacting gases.

1.1.1.2 Thermodynamic equilibrium

Let us imagine some elementary volume filled with plasma and isolated from the adjacent elements by walls perfectly impervious to all kinds of particles as well as heat, and an ideal reflector of electromagnetic radiation.

Generalizing the experience derived from the thermodynamics of closed and isolated gaseous or liquid systems, we may expect that within a certain (sufficiently long) time all "spontaneous" irreversible changes will have been completed and – except for fluctuations – a stable macroscopic physical state is established. We call this state a perfect thermodynamic equilibrium. In the simplest instances, the macroscopic physical state of the said element in thermodynamic equilibrium proves to be uniquely defined – given its mass – by any of the following pairs of independent variables of state: T, V; T, p; S, V; S, p – T being the absolute temperature, V the volume, p the pressure and S the entropy of the plasma contained in the element. All other macroscopic quantities can be computed from these independent variables with the aid of known thermodynamic relations (cf. Sec. 1.1.2.3).

The most convenient way of describing the thermodynamic state of a system is to express it by its thermodynamic potentials:

internal energy

$$E = E(S, V) \tag{1.1.1}$$

enthalpy

$$H = H(S, p) = E + pV \tag{1.1.2}$$

free energy

$$F = F(T, V) = E - TS \tag{1.1.3}$$

free enthalpy (Gibbs' potential)

$$G = G(T, p) = E - TS + pV \tag{1.1.4}$$

Most frequently these quantities are used as functions of the independent variables written in brackets. With respect to these two variables, the quantities E, H, F, G have the quality of potentials: for by partially differentiating the pertinent thermodynamic potential with respect to them, we obtain the remaining two variables. For instance

$$S = -\left(\frac{\partial F}{\partial T}\right)_V; \qquad p = -\left(\frac{\partial F}{\partial V}\right)_T. \tag{1.1.5}$$

In common forms of plasma, which are often spatially non-homogeneous and placed in external fields (gravitational, electric, etc.), the equilibrium is local, that is relating only to a small plasma element around some point. The equilibrium parameters (especially temperature) belonging to various points of the system will then differ and depend on the magnitude and distribution of the external fields.

1.1.1.3 **Statistical equilibrium**

Viewed microscopically, the state of the plasma is determined by a variety of elementary processes by which its component particles interact, e.g. elastic collisions, inelastic collisions resulting in excitation, dissociation and ionization of particles, absorption and emission of photons by compound particles (atoms, molecules, ions), etc. As a rule, these processes are accompanied by the transfer of energy, momentum, or else the formation or destruction of particles. Owing to these processes, plasma in perfect equilibrium also establishes so-called statistical equilibrium whose characteristic feature is the unique distribution of particles according to type, and within the given type according to coordinates, momenta and internal states of the particles.

A relatively general instance of such a distribution (the Gibbs distribution) is analyzed in Sec. 1.1.2.1. Here we list, without deriving them, only a few partial distributions—those most frequently encountered in plasma physics—which can all be derived from the distribution according to Eq. (1.1.11).

The probability $w(\varepsilon^{tr})\,d\varepsilon^{tr}$ that a particle selected at random will have the translational kinetic energy ε^{tr} in the interval $(\varepsilon^{tr}, \varepsilon^{tr} + d\varepsilon^{tr})$ is given by Maxwell's distribution:

$$w(\varepsilon^{tr})\,d\varepsilon^{tr} = 2(\pi kT)^{-3/2}\, e^{-\frac{\varepsilon^{tr}}{kT}} \sqrt{\varepsilon^{tr}}\, d\varepsilon^{tr} \qquad (1.1.6)$$

where $k = 1.38 \times 10^{-23}$ J deg^{-1} is Boltzmann's constant, and T the temperature of the system [°K].

Boltzmann's formula [Ref. 1.1.12] holds good for the density n_m of atoms in the mth excited state:

$$\frac{n_m}{n_0} = \frac{g_m}{g_0}\, e^{-\frac{\varepsilon_m}{kT}} \qquad (1.1.7)$$

where n_0 is the density of atoms in the ground state ($\varepsilon_0 = 0$), ε_m the excitation energy of the mth state, and g_m, g_0 are the statistical weights of the ground state and m^{th} excited state, respectively. These statistical weights describe the probability (a priori given by the number of sub-levels into which level ε_m splits due to quantum degeneracy) that some particular state can occur, given sufficiently high temperatures $\left(e^{-\frac{\varepsilon_m}{kT}} \rightarrow 1\right)$.

The density $u(\nu)$ of radiant energy (i.e. the energy of the radiation contained in 1 m^3) corresponding to electromagnetic quanta with frequency in the range $(\nu, \nu + d\nu)$ is given by Planck's formula [Ref. 1.1.12]

$$u(\nu) = \frac{8\pi h}{c^3} \frac{\nu^3}{e^{\frac{h\nu}{kT}} - 1} \qquad (1.1.8)$$

where $h = 6.625 \,.\, 10^{-34}$ Js is Planck's constant, and $c = 2.9979 \times 10^8$ m s^{-1} the velocity of light in vacuum.

For plasma in perfect thermodynamic—and therefore statistical—equilibrium, temperature T has the same value in distributions (1.1.6), (1.1.7), (1.1.8) as well as in Gibbs formula (Eq. 1.1.11) and in Saha's equation (1.1.79), irrespective of the plasma component concerned. Such a plasma is called isothermic.

The elementary processes which are conducive to equilibrium and its maintenance in the plasma may be grouped into pairs (ionization—

recombination, excitation—collision of the second kind, etc.) in which each is the inverse of the other (for the definition of an inverse process also cf. Ref. [1.1.29]). Deviations from equilibrium due to the action of some elementary process (e.g. increased ionization by collisions) are balanced directly by the corresponding reverse process, and not by some roundabout way through various intermediate processes. This assertion is the content of what is called the principle of detailed equilibrium. In this sense, every statistical equilibrium is a detailed equilibrium between direct and inverse processes. This fact is often used for examining minor deviations from equilibrium; however, especially in processes accompanied by emission and absorption of photons, much caution has to be exercised in defining the concepts of the direct and the inverse processes.

1.1.1.4 Ascertaining the properties of plasma in thermodynamic equilibrium

So far, no reliable experimental methods are available for the direct measurement of various thermodynamic properties of gases ($\sim 10^3 - 10^4$ deg. K). Such properties are the chemical composition, internal energy, enthalpy and other properties of a plasma which are essential for the macroscopic description of the flow in the nozzle of the plasma torch, the plasma rocket engine or the MHD generator. Such properties are defined theoretically—by the methods of statistical thermodynamics—and the findings are checked by various indirect methods (measurements of temperature, electron density, electrical conductivity in weak fields, etc.).

Where plasma of moderate temperature ($T = 10^4$ deg. K) and in the pressure range 10^{-3} atm $< p < 10^3$ atm is concerned, we may consider it [Ref. 1.1.2] a mixture of ideal gases behaving according to the equation of state

$$p = \frac{kT}{V} \sum_i N_i \tag{1.1.9}$$

where p, V, k, T, N_i stand for pressure, the volume of one mole of the mixture, Boltzmann's constant, temperature, and the number of i-type particles in one mole of the mixture, respectively.

If the density of charged particles is high, Coulombic action may exert a considerable influence. In Debye-Hückel's approximation, (cf. Sec. 1.1.4.4), the equation of state reads as follows:

$$p = p_0 - \Delta p = \frac{kT}{V}\sum_i N_i - \frac{e^3}{3V^{3/2}}\sqrt{\frac{\pi}{kT}}\left(\sum_i N_i q_i^2\right)^{3/2} \quad (1.1.10)$$

where e is the electron charge ($e = 1.602 \times 10^{-19}$ C) and q_i in the correction stands for the electric charge of an i-type particle expressed in e units. The rest of the symbols have the same meaning as in Eq. (1.1.9). Further approximations to the exact equation of state are found for instance in Ref. [1.1.17]. In a plasma consisting of electrons and simple ions, at a temperature of 10^4 deg. K, the correction in the equation of state begins to produce a slight effect at electron densities as low as $n_e \sim 10^{12}$ per m^3 and is comparable with p_0 at densities $n_e \sim 10^{24}$ electrons in one m^3.

In this book we shall not—with a few exceptions—allow for the correction of the equation of state due to Coulombic action. The calculation of the thermodynamic functions of plasma in equilibrium will then be reduced to three partial computations:

- calculating the thermodynamic parameters of the individual "pure" constituents, using formulae of statistical thermodynamics for the ideal gas in equilibrium;
- calculating the composition of the plasma, using the complete system of equations of chemical equilibrium (including the equations for dissociation and ionization);
- calculating the thermodynamic functions of the mixture of ideal gases from the known values of these functions for the individual components.

1.1.2 Thermodynamic Description of the Ideal Gas

1.1.2.1 The partition function (sum-over-states, Zustandssumme)

The thermodynamic properties of an ideal gas in equilibrium can be defined by methods of statistical thermodynamics, provided we know all the elementary physical states which can be assumed by individual molecules or by the gas as a whole in accordance with the laws of classical or quantum mechanics. The term "elementary physical state of the gas as a whole" will stand for every actual configuration of coordinates, momenta and internal states of all the N individual gas molecules. In the so-called semiclassical approximation (Eq. 1.1.4) we may consider discrete all elementary states and the corresponding total energies $\mathscr{E}_i$ of the gas. Some values of the energy $\mathscr{E}_i$ of the gas may be equal in several elementary states, and we shall denote by $\Omega(\mathscr{E}_i)$ the number of elementary states that have energy $\mathscr{E}_i$ in common. The probability $w(\mathscr{E}_i)$ that the gas as a whole will have the energy $\mathscr{E}_i$ is given by the Gibbs formula [Ref. 1.1.4]

$$w(\mathscr{E}_i) = \frac{1}{\mathscr{Z}} e^{\frac{\mathscr{E}_i}{kT}} \Omega(\mathscr{E}_i) \tag{1.1.11}$$

where k is the Boltzmann constant and T the temperature. The value of constant $\mathscr{Z}$ is found by relating the sum of probabilities Eq. (1.1.11) to unity

$$\mathscr{Z} = \sum_i e^{-\frac{\mathscr{E}_i}{kT}} \Omega(\mathscr{E}_i) \tag{1.1.12}$$

where the summation is carried out over all possible mutually differing states of the system of N molecules. Constant $\mathscr{Z}$ is called the partition function (sum-over-states, Zustandssumme) of the gas. Its importance is due to the fact that all thermodynamic potentials of the ideal gas can be expressed as functions of $\mathscr{Z}$ (cf. Eqs. (1.1.32) to (1.1.45)). The analogous expression

$$Z = \sum_i e^{-\frac{\varepsilon_i}{kT}} \Omega(\varepsilon_i) \tag{1.1.13}$$

where the summation is carried out over all possible elementary states of a single molecule, is termed the sum-over-states or partition function of the molecule; ε_i is the total energy of the molecule, and $\Omega(\varepsilon_i)$ the number of states of the molecule corresponding to the energy ε_i.

If the physical states of the individual molecules are mutually independent (ideal gas), the partition function of the gas as a whole can be expressed in terms of the partition functions (1.1.13) of the individual molecules. This is due to the fact that every state of the gas as a whole can be considered a combination of the states of the individual molecules. If these molecules are identical, we may write

$$\mathscr{Z} = \frac{1}{N!} Z^N \tag{1.1.14}$$

where the factor $N!$ accounts for the fact that elementary states only differing owing to the permutation of identical particles must not be considered as separate (i. e. different) states, in the expression for $\mathscr{Z}$.

The energy ε_i of each molecule consists of its translational energy ε_i^{tr} and the energy ε_i^{int} of its internal physical state:

$$\varepsilon_i = \varepsilon_i^{tr} + \varepsilon_i^{int} \tag{1.1.15}$$

Since the internal state of the molecule does not depend on its translational motion, we may write the partition function Z of the molecule (Eq. 1.1.13) as follows:

$$Z = Z^{tr} . Z^{int} \tag{1.1.16}$$

where

$$Z^{tr} = \sum_i e^{-\frac{\varepsilon_i^{tr}}{kT}} \Omega(\varepsilon_i^{tr}); \qquad Z^{int} = \sum_i e^{-\frac{\varepsilon_i^{int}}{kT}} \Omega(\varepsilon_i^{int}). \tag{1.1.17}$$

The individual states of the translational motion differ only slightly from each other (their difference is given by the uncertainty principle of quantum mechanics); we may therefore consider them continuous and in the formula for Z^{tr} substitute integration over the phase space of the molecule for the summation over i. We thus obtain (1.1.4):

$$Z^{tr} = \frac{V}{h^3} (2\pi mkT)^{3/2} \tag{1.1.18}$$

where V is the volume of the gas, h Planck's constant, m the mass of the molecule, k the Boltzmann constant, and T the temperature [°K].

1.1.2.2 Internal partition function

Atom. The internal state of an atom is given by the state of its nucleus and that of its electron cloud. Let us denote the energy of the ith excited state of the nucleus and electron cloud ε_i^{nuc} and ε_i^{el}, respectively (referred to the non-excited ground states $\varepsilon_0^{nuc} = \varepsilon_0^{el} = 0$). Let g_0^{nuc}, g_i^{nuc}, g_0^{el}, g_i^{el}

be the statistical weights of these states, which indicate how many elementary states of the nucleus or electron cloud have the energy value $\varepsilon_i^{\text{nuc}}$ or $\varepsilon_i^{\text{el}}$, respectively. To the internal partition function there applies then the formula

$$Z^{\text{int}} = Z^{\text{nuc}} \,.\, Z^{\text{el}} = \left(\sum_i g_i^{\text{nuc}} \mathrm{e}^{-\frac{\varepsilon_i^{\text{nuc}}}{kT}}\right)\left(\sum_i g_i^{\text{el}} \mathrm{e}^{-\frac{\varepsilon_i^{\text{el}}}{kT}}\right) \qquad (1.1.19)$$

where we sum over all possible energy levels of the nucleus and the electron cloud. Since at the temperatures which interest us $\mathrm{e}^{-\frac{\varepsilon_i^{\text{nuc}}}{kT}} \sim 0$ for $i \neq 0$ (and therefore $Z^{\text{nuc}} = g_0^{\text{nuc}}$), we can write

$$Z^{\text{int}} = g_0^{\text{nuc}} \left(g_0^{\text{el}} + g_1^{\text{el}} \mathrm{e}^{-\frac{\varepsilon_1^{\text{el}}}{kT}} + g_2^{\text{el}} \mathrm{e}^{-\frac{\varepsilon_2^{\text{el}}}{kT}} + \ldots\right) \qquad (1.1.20)$$

In actual computations we insert into Eq. (1.1.20) energy levels and statistical weights obtained from spectroscopic data. For some elements these values are tabulated (cf. for instance Ref. [1.1.12]). The energy levels are for the most part expressed in cm^{-1} corresponding to 1.2390×10^{-4} eV.

A so-called characteristic temperature may be assigned to the energy levels by the relationships

$$\varepsilon_i^{\text{el}} = kT_i^{\text{el}} \qquad \text{and} \qquad \varepsilon_i^{\text{nuc}} = kT_i^{\text{nuc}}.$$

In the partition functions, we may then at gas temperature T neglect all levels $\varepsilon_i^{\text{el}}$ for which $T \ll T_i^{\text{el}}$. This conveys that at temperature T these levels take no part in the formation of the total macroscopic state; they are "frozen". The characteristic temperatures of electron states range about 10^5 deg. K, those of nuclear states about 10^{11} deg. K.

Molecule. The physical state of a molecule is determined by the physical state of its electron cloud, the vibrating motion of the nuclei, the rotation of the molecule as a whole, and the internal state of its nuclei. In so far as these forms of quantized motion can be considered mutually independent, we can write

$$Z^{\text{int}} = Z^{\text{el}} Z^{\text{vib}} Z^{\text{rot}} Z^{\text{nuc}}. \qquad (1.1.21)$$

Similarly as we did for the atom, we assume

$$Z^{\text{nuc}} = \prod_{\alpha} g_{0,\alpha}^{\text{nuc}} \tag{1.1.22}$$

where g_0^{nuc} is the statistical weight of the ground state of the αth nucleus of the molecule. We shall limit our further discussion to diatomic molecules. For such molecules we have to a first approximation (rigid rotor)

$$\varepsilon_J^{\text{rot}} = \frac{h^2}{8\pi^2 I} J(J+1); \qquad g_J^{\text{rot}} = 2J+1;$$

$$T_1^{\text{rot}} \sim 10°\text{K to } 10^2\,°\text{K} \tag{1.1.23}$$

where I is the moment of inertia of the molecule, J the rotational quantum number $J = 1, 2, \ldots,$ T_1^{rot} the characteristic temperature of the first excited rotational state. For the rotational part of the partition function we have [Ref. 1.1.4]

$$Z^{\text{rot}} = \frac{1}{2} \sum_{J=1} (2J+1)\, e^{-\frac{h^2 J(J+1)}{8\pi^2 IkT}} \sim \frac{8\pi^2 IkT}{2h^2}. \tag{1.1.24}$$

The coefficient of symmetry $\frac{1}{2}$ is introduced in molecules with identical nuclei. For the vibrational motion (harmonic oscillator) we have

$$\varepsilon_v^{\text{vib}} = h\nu\left(v + \frac{1}{2}\right); \qquad g_v^{\text{vib}} = 1; \qquad T_1^{\text{vib}} \sim 10^3\,°K \tag{1.1.25}$$

where ν is the fundamental frequency of the oscillator, and v the vibrational quantum number $v = 0, 1, 2, \ldots$. The vibrational energy levels are not degenerated $g_v^{\text{vib}} = 1$. The vibration part of the partition function is [Ref. 1.1.4]:

$$Z^{\text{vib}} = \sum_{v=0} e^{-\frac{h\nu(v+1/2)}{kT}} = \frac{e^{-\frac{h\nu}{kT}}}{1 - e^{-\frac{h\nu}{kT}}} \tag{1.1.26}$$

Finally, as in the atomic case, for the electronic part of the molecular partition function we have

$$Z^{\text{el}} = \sum_{l} g_l^{\text{el}}\, e^{-\frac{\varepsilon_l^{\text{el}}}{kT}}; \qquad g_l^{\text{el}} = 2J+1;$$

$$T_1^{\text{el}} \sim 10^4\,°\text{K to } 10^5\,°\text{K} \tag{1.1.27}$$

where J is the principal quantum number belonging to the excited state $\varepsilon_l^{\text{el}}$.

In the second approximation, weight has to be given to the mutual influence of the different kinds of internal motion in the molecule, and the effects of inharmonic vibrations and deviations from the rigidity of the rotor have to be taken into account (altered moment of inertia of the molecule at large rotation and oscillation quantum numbers). Moreover, the range of indices through which the summation for the partition function is carried out calls for more precise definition.

In Ref. [1.1.2], for instance, the internal partition function of a diatomic molecule is quoted in the following form:

$$Z^{\text{int}} = g_0^{\text{nuc}} \sum_l g_l^{\text{el}} \, e^{-\frac{\varepsilon_l^{\text{el}}}{kT}} \sum_{v=0}^{v=v_{\max}} e^{-\frac{\varepsilon_v^{\text{vib}}(l)}{kT}} \times$$

$$\times \sum_{J=J_{\min}}^{J=J_{\max}} (2J+1)\, e^{-\frac{\varepsilon_J^{\text{rot}}(l,\,v)}{kT}} \qquad (1.1.28)$$

The notation $\varepsilon_J^{\text{rot}}(l, v)$ stands for a rotational level belonging to oscillation level $\varepsilon_v^{\text{vib}}(l)$, which in turn belongs to electron level $\varepsilon_l^{\text{el}}$. The values $J_{\max}$, $v_{\max}$ indicate rotational and vibration states, respectively immediately below the dissociation energy of the molecule.

The problem of the convergence of electronic partition functions is by no means elementary. Since with increasing electronic quantum number the energy levels accumulate infinitely densely round the finite value of the ionization energy, these sums should in fact diverge. This is what would occur if the atom or molecule were alone in an unlimited space. Actually, however, the interaction in the aggregation of particles causes the highest energy levels of the individual particles to be split up and merge, hence the number of clean-cut discrete states of the particles is limited. The problem of the convergence of partition functions is treated in greater detail in Ref. [1.1.20].

The energy values $\varepsilon_l^{\text{el}}$; $\varepsilon_v^{\text{vib}}(l)$; $\varepsilon_J^{\text{rot}}(l, v)$ and the quantities g_l^{el} in Eq. (1.1.28) are obtained from spectroscopic data, which often (e.g. for molecular ions) is very incomplete. In such cases we compute such data from appropriately chosen quantum-mechanical models of the molecules. For instance, Ref. [1.1.2] presents such a calculation for the ionized molecule NO^+; it also illustrates the intricacy of the matter.

1.1.2.3 Thermodynamic functions of an ideal gas

Let us relate all the thermodynamic quantities discussed in Sec. 1.1.1.2 to one mole of gas ("molar functions") without altering their designations. As standard state for the potentials Eqs. (1.1.1) to (1.1.4) we shall use the fictitious state of one mole of ideal gas at 0°K [according to tables Ref. 1.1.3]. We denote its internal energy by $E_0 = N_0 \varepsilon_0$, where ε_0 is the internal energy of the molecule in the ground state, and $N_0 = 6.0228 \times 10^{23}$ molecules per mole is Avogadro's number. Energy ε_0 is the internal-bond energy acting between the constituent particles of the molecule, which equals the sum of the dissociation energies of the interatomic bonds plus the sum of the ionization energies; it also includes the zero-point vibrational energy of the molecule. If we introduce the energy ε_0 into the partition function (1.1.13), this function splits into two parts:

$$Z = Z_T \,.\, Z_0 = \mathrm{e}^{-\frac{\varepsilon_0}{kT}} \sum_i \mathrm{e}^{-\frac{\varepsilon_i}{kT}} \,\Omega(\varepsilon_i), \tag{1.1.29}$$

where ε_i is the total energy of the molecule reduced by its energy in the ground state. The magnitude of ε_i can be changed by inelastic collisions, while the energy ε_0 can only be changed by chemical reactions (ionization, dissociation).

For the statistical sum of the entire system we can write:

$$\mathscr{Z} = \mathscr{Z}_T \,.\, \mathscr{Z}_0 \tag{1.1.30}$$

where $\mathscr{Z}_0$, the chemical part of the partition function, reads

$$\mathscr{Z}_0 = \mathrm{e}^{-\frac{N_0 \varepsilon_0}{kT}}, \tag{1.1.31}$$

N_0 being the total number of molecules in one mole.

Formula (1.1.12) applies to $\mathscr{Z}_T$, the thermal part of the function, where $\mathscr{E}_i$ is the total energy of the gas reduced by the energy of the standard state $E_0 = N_0 \varepsilon_0$.

For the **molar free energy** F of a gas, statistical physics offers a simple relation:

$$F = -kT \ln \mathscr{Z}, \tag{1.1.32}$$

Using formulae (1.1.30) and (1.1.31) we can write

$$F = -kT \ln \mathscr{Z}_T - kT \ln \mathscr{Z}_0 = -kT \ln \mathscr{Z}_T + E_0. \quad (1.1.33)$$

By substituting $\mathscr{Z} = \frac{1}{N_0!} Z^{N_0}$ from expression (1.1.14) in Eq. (1.1.33), using Stirling's formula for $N!$ [Ref. 1.1.4] an dseparating the translational partition function (1.1.18), we obtain a well-known formula, the computation of which is, for instance, found in Ref. [1.1.4]:

$$F - E_0 = -N_0 kT \ln \left\{ \frac{eV}{N_0} \left(\frac{2\pi mkT}{h^2} \right)^{3/2} Z_T^{\text{int}} \right\} \quad (1.1.34)$$

where e is the base of the natural logarithms.

We can derive the **molar free enthalpy** (the Gibbs potential) from the free energy F by using the defining relation $G = F + pV = F + N_0 kT$ (cf. Eqs. (1.1.4) and (1.1.9)):

$$G - E_0 = -N_0 kT \left[\ln \left\{ \frac{eV}{N_0} \left(\frac{2\pi mkT}{h^2} \right)^{3/2} Z_T^{\text{int}} \right\} - 1 \right]. \quad (1.1.35)$$

The **chemical potential** μ is defined as Gibbs' potential for one molecule:

$$\mu - \varepsilon_0 = \frac{G - E_0}{N_0} = -kT \left[\ln \left\{ \frac{eV}{N_0} \left(\frac{2\pi mkT}{h^2} \right)^{3/2} Z_T^{\text{int}} \right\} - 1 \right]. \quad (1.1.36)$$

The **molar internal energy $\mathbf{E}$** may be computed directly from Eq. (1.1.11) as the mean value of energy over all possible states:

$$E = \sum_i \mathscr{E}_i w(\mathscr{E}_i) = \frac{\sum_i \mathscr{E}_i e^{-\frac{\mathscr{E}_i}{kT}} \Omega(\mathscr{E}_i)}{\sum_i e^{-\frac{\mathscr{E}_i}{kT}} \Omega(\mathscr{E}_i)}. \quad (1.1.37)$$

The detailed computation [cf. for instance Ref. 1.1.4] shows that E is related to the partition function $\mathscr{Z}$

$$E = N_0 kT^2 \frac{d(\ln \mathscr{Z})}{dT}; \quad (1.1.38)$$

using Eqs. (1.1.30) and (1.1.31) we obtain

$$E - E_0 = N_0 k T^2 \frac{\mathrm{d}(\ln \mathscr{Z}_T)}{\mathrm{d}T}. \tag{1.1.39}$$

We would arrive at the same result by using the well-known thermodynamic relation

$$E = F - T\left(\frac{\partial F}{\partial T}\right)_V \tag{1.1.40}$$

which is obtained by the combination of Eqs. (1.1.3) and (1.1.5).

The **molar enthalpy** H is obtained from the defining relation $H = E + pV = E + N_0 kT$:

$$H - E_0 = N_0 kT\left[T\frac{\mathrm{d}(\ln \mathscr{Z}_T)}{\mathrm{d}T} + 1\right]. \tag{1.1.41}$$

The **molar entropy** S is calculated from the free energy F according to Eqs. (1.1.5) and (1.1.34). Since in the product, Z_T^{int} appears under a logarithm sign, the contribution of the internal states to the entropy can be separated (which, in fact, also applies to the quantities $F - E_0$, $G - G_0$, $E - E_0$, $H - E_0$); this gives:

$$S = -\left(\frac{\partial F}{\partial T}\right)_V = S^{\text{tr}} + S^{\text{int}}. \tag{1.1.42}$$

The entropy of the standard state selected by us is zero, in accordance with the Third Law of Thermodynamics.

The entropy connected with translational motion obeys the well-known formula

$$S^{\text{tr}} = \frac{5}{2} N_0 k \ln T - N_0 k \ln p + N_0 \ln\left(\frac{2\pi m}{h^2}\right)^{3/2} k^{5/2} + \frac{5N_0 k}{2} \tag{1.1.43}$$

which follows from Eqs. (1.1.34), (1.1.42) and the equation of state $pV = N_0 kT$.

Finally, **the molar heat capacities C_V and C_p** at constant volume and pressure, respectively are given by the following formulae:

$$C_V = \left(\frac{\partial E_T}{\partial T}\right)_V = C_V^{tr} + C_V^{int} = \frac{3}{2} N_0 k + \left(\frac{\partial E_T^{int}}{\partial T}\right)_V \quad (1.1.44)$$

$$C_p = \left(\frac{\partial H_T}{\partial T}\right)_p = C_p^{tr} + C_p^{int} = \frac{5}{2} N_0 k + \left(\frac{\partial H_T^{int}}{\partial T}\right)_p . \quad (1.1.45)$$

Thus the calculation of all the quantities listed can be reduced to the computation of the partition function Z_T^{int} of the molecule or its first and second temperature derivatives. For the most common gases and at temperatures within limits (up to several thousand degrees) the quantities mentioned can be found in tables. In the discussions of this Section we had in mind tables Ref. [1.1.3], although in some details we depart from the symbols used there.

1.1.3 Dissociation Equilibrium

1.1.3.1 Chemical equilibrium and the Law of Mass Action

If a reversible chemical reaction occurs between the molecules of the individual constituents of a mixture of ideal gases in perfect thermodynamic equilibrium, then chemical equilibrium is also established in the mixture. For a given number $\mathcal{N}$ of moles of the mixture, given total pressure p and temperature T—a unique composition is characteristic of this equilibrium.

Let us write the chemical reaction in the form

$$\sum_i \nu_i K_i = 0 \quad (1.1.46)$$

where K_i are the chemical symbols of the reacting substances (the components of the mixture), and ν_i the stoichiometric coefficients indicating the number (integer, as a rule) of moles of each component entering into the reaction (cf. e.g. Eq. 1.1.62). The coefficients ν_i are negative for substances that are dissociated in the reaction, and positive for substances that are formed.

If the system is in equilibrium at constant temperature and pressure, the chemical potentials μ_i of the components satisfy the following relation:

$$\sum_i \nu_i \mu_i = 0. \tag{1.1.47}$$

In a mixture of ideal gases, each gas behaves as if it were occupying the whole volume of the vessel and had the partial pressure

$$p_i = \frac{N_i}{N} p \tag{1.1.48}$$

where N_i is the total number of molecules of type i, and N the total number of molecules of the mixture.

The chemical potential (1.1.36) can be written as follows:

$$\mu_i = kT \ln p_i + \chi_i(T) \tag{1.1.49}$$

where $\chi_i(T)$ is only a function of temperature, and not of partial pressure:

$$\chi_i(T) = -\frac{5}{2} kT \ln T - kT \ln\left[\left(\frac{2\pi m}{h^2}\right)^{3/2} k^{5/2}\right] - kT \ln Z_T^{\text{int}} + \varepsilon_0 \tag{1.1.50}$$

By substituting Eq. (1.1.49) into the condition for equilibrium (1.1.47) we obtain

$$kT \sum_i \nu_i \ln p_i + \sum_i \nu_i \chi_i(T) = 0 \tag{1.1.51}$$

and a simple calculation yields Guldberg and Waage's Law of Mass Action:

$$\prod_i (p_i)^{\nu_i} = K(T) \tag{1.1.52}$$

where the chemical-equilibrium constant $K(T)$* is given by the relation

$$\ln K(T) = -\frac{\sum_i \nu_i \chi_i(T)}{kT}. \tag{1.1.53}$$

---

* By the symbol $K(T)$ we designate in this chapter the equilibrium constant which in textbooks of chemical thermodynamics is usually denoted $K_p(T)$.

For practical calculations it is sometimes found convenient to express Eq. (1.1.52) in molar fractions (molar concentrations)

$$x_i = \frac{p_i}{p} = \frac{N_i}{N} = \frac{\mathcal{N}_i}{\mathcal{N}} \qquad (1.1.54)$$

where $\mathcal{N}_i$ is the number of moles of component i in the reaction, and $\mathcal{N}$ the total number of moles of the mixture:

$$\prod_i (x_i)^{\nu_i} = p^{-\sum_i \nu_i} K(T) = K(p, T). \qquad (1.1.55)$$

Equation (1.1.55) together with Dalton's Law and the equations of mass balance permits the unique computation of the molar fractions (cf. Chap. 1.1.5).

The reader will find more detailed information about chemical equilibrium in the textbooks Refs. [1.1.9] and [1.1.11].

1.1.3.2 Determining the equilibrium constant

Constant $K(p, T)$ in Eq. (1.1.55) is usually expressed by the change $\Delta G(p, T)$ of the Gibbs potential (free enthalpy) in the reaction. By logarithmic calculation we obtain from Eq. (1.1.55)

$$\ln K(p, T) = -\sum_i \nu_i \ln p + \ln K(T). \qquad (1.1.56)$$

By substituting from Eq. (1.1.53) we arrive at

$$\ln K(p, T) = -\frac{\sum_i \nu_i[kT \ln p + \chi_i(T)]}{kT} = -\frac{\sum_i \nu_i \mu_i(T, p)}{kT} \qquad (1.1.57)$$

where $\mu_i(T, p)$ is the chemical potential of component i at temperature T and pressure p. After multiplying the numerator and denominator of the right-hand side in Eq. (1.1.57) with Avogadro's number, we have

$$\ln K(p, T) = -\frac{\sum_i \nu_i[N_0 \mu_i(T, p)]}{N_0 kT} \qquad (1.1.58)$$

where the expressions in square brackets indicate the molar free enthalpy $G_i(p, T)$ of component i, the chemical potential being the free enthalpy per particle. Hence the summation in the numerator (Eq. 1.1.58) indicates the total change $\Delta G(p, T)$ in free enthalpy due to the reaction. We can thus write

$$\ln K(p, T) = -\frac{\Delta G(p, T)}{RT} \tag{1.1.59}$$

where $R = N_0 k$ is the universal gas constant; $\Delta G(p, T) = \sum_i \nu_i G_i(p, T)$ is the difference between the free enthalpies of substances dissociated and substances formed in the reaction. If we separate the "chemical" and the "thermal" portion of this difference, we can write:

$$\Delta G(p, T) = \sum_i \nu_i (E_0)_i + \sum_i \nu_i [G_i(p, T) - (E_0)_i] \tag{1.1.60}$$

where $(E_0)_i$ is the symbol for the standard state of component i. The formula

$$\Delta E_0 = \sum_i \nu_i (E_0)_i \tag{1.1.61}$$

indicates the total change due to the chemical reaction in the free enthalpy of the standard states of the individual components, which appears as heat of reaction. As a rule, this change is far greater than the change in the thermal portion of the free enthalpy as given by the second term on the right-hand side of Eq. (1.1.60).

For a number of reactions, the values of E_0 are tabled, or they can by calculated from the heat of formation or the bond energies of the molecules (e.g. the dissociation or ionization energies). Moreover, for some gases, tables also indicate the values of what is called the G function, defined as $\frac{G(p, T) - E_0}{T}$ [see e.g. Ref. 1.1.3]. The values of the equilibrium constants at a pressure of $p = 1$ atm. (consequently $K(p, T) = K(T)$) are also tabled for some reactions.

Equation (1.1.59) enables us to calculate the equilibrium constants $K(p, T)$ by means of the reaction heat ΔE_0 and the corresponding change in the G function. For very high temperatures tables of T, G-functions are mostly not available. In such cases we either use Eq. (1.1.35) to calculate the tables from the known internal partition functions Z_T^{int} of the molecule, or we compute directly $K(T)$ values according to Eqs. (1.1.50) and (1.1.53). The first way has been taken for instance in Refs. [1.1.6] and [1.1.7] in the calculation of the equilibrium composition of mixtures resulting from the dissociation of some hydrocarbons at temperatures of 1000°K to 50 000°K, whereas the second way has been used for calculating the equilibrium composition of the air at temperatures from 1000°K to 12 000°K, as reported in Ref. [1.1.2].

1.1.3.3. Dissociation equilibrium of diatomic molecules

Let us consider an illustrative example of simple chemical equilibrium such as the dissociation of nitrogen which follows the pattern

$$N_2 \rightleftarrows 2N. \qquad (1.162)$$

The stoichiometric coefficients of this reaction are $\nu_{(N_2)} = -1$, $\nu_{(N)} = 2$. Their summation, $\sum_i \nu_i = -1 + 2 = +1$, indicates the change in the number of moles due to the reaction. In this case, the Law of Mass Action will have the following form:

$$K(T) = \frac{p^2_{(N)}}{p_{(N_2)}} \qquad (1.1.63)$$

or if Eq. (1.1.55) is used:

$$\frac{x_{(N)}}{x_{(N_2)}} = \frac{K(T)}{p}. \qquad (1.1.64)$$

It follows from Eq. (1.1.64) that with rising pressure p the relative amount of N atoms decreases and the relative amount of undissociated N_2 molecules increases. This fact is more accurately described by the degree of dissociation α, indicating which proportion of the total original number $N^{\circ}_{(N_2)}$ of molecules (or moles) has already been dissociated. For the composition of the reacting mixture we have then

$$N_{(N_2)} = N^{\circ}_{(N_2)}(1 - \alpha), \qquad N_{(N)} = N^{\circ}_{(N_2)}\, 2\alpha. \qquad (1.1.65)$$

The total number N of the molecules is given by the relation

$$N = N_{(N)} + N_{(N_2)} = N^{\circ}_{(N_2)}(1 + \alpha). \qquad (1.1.66)$$

Expressed in terms of the degree of dissociation, the molar fractions $x_{(N)}$ and $x_{(N_2)}$, respectively take the form

$$x_{(N)} = \frac{2\alpha}{1 + \alpha}; \qquad x_{(N_2)} = \frac{1 - \alpha}{1 + \alpha}. \qquad (1.1.67)$$

Equation (1.1.64) can thus be re-written in the form

$$p \frac{4\alpha^2}{1-\alpha^2} = K(T). \qquad (1.1.68)$$

If the numerical value of $K(T)$ is known, Eq. (1.1.68) can be solved with respect to α:

$$\alpha = \sqrt{\frac{K(T)}{K(T)+4p}}. \qquad (1.1.69)$$

Hence it is obvious that with rising pressure the degree of dissociation decreases, i.e. the number of molecules in the mixture increases and the number of atoms drops.

The chemical-equilibrium constant for the dissociation of diatomic gases with identical nuclei is usually written in the following form (here for nitrogen):

$$K(T) = \text{const}\; T^{5/2} \frac{(Z^{\text{int}}_{(N)})^2}{Z^{\text{int}}_{(N_2)}} e^{-\frac{\varepsilon^{\text{diss}}_{(N)}}{kT}} \qquad (1.1.70)$$

where $Z^{\text{int}}_{(N)}$, $Z^{\text{int}}_{(N_2)}$ are the internal partition functions of nitrogen atoms and molecules, respectively (for numerical values cf.Ref. [1.1.28]), $\varepsilon^{\text{diss}}_{(N)}$ the dissociation energy of the nitrogen molecule (= 9.764 eV; see e.g. Ref. [1.1.8]). The value of the constant in Eq. (1.1.70) is 0.474 542 for nitrogen [Ref. 1.1.2]. By substituting the expression for $K(T)$ from

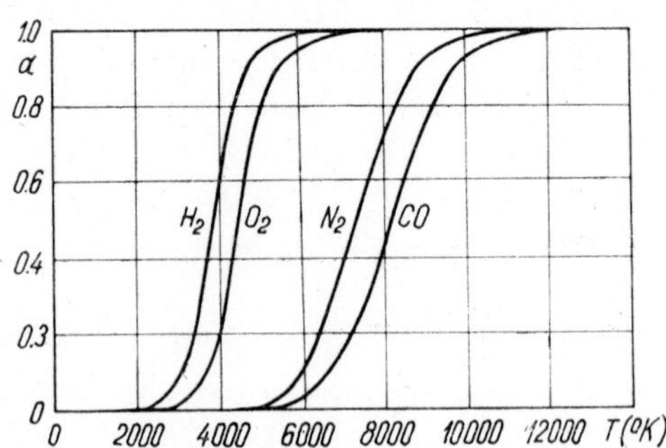

Fig. 1 Degree of dissociation (α) of some gases vs. temperature

Eq. (1.1.70) into Eq. (1.1.69), we obtain the dependence of the degree of dissociation upon temperature and pressure. For $p = 1$ atm. this dependence is found in Chart 1 (drawn from [Ref. 1.1.10]), together with similar dependences for molecular hydrogen, oxygen and carbon monoxide. It is worth noting that perceptible dissociation of nitrogen and carbon monoxide only sets in at temperatures higher than 6000°K, whereas hydrogen and oxygen already dissociate at 4000°K.

1.1.4 Ionization Equilibrium

1.1.4.1 Ionization as a chemical reaction

The concentration of charged particles – ions and electrons – is perhaps the principal quantity characteristic of plasma as a fourth state of aggregation. In a plasma, charged particles come into existence as products from a variety of elementary processes (collision ionization, ionization by stages, photoionization, etc.). All these processes can be conceived as elementary reactions of the type

$$\mathrm{A} \rightleftarrows \mathrm{A}^+ + \mathrm{e} \tag{1.1.71}$$

where A is the chemical symbol of the atom (for simplicity's sake we limit the discussion of this Section to monatomic gases), A^+ of a singly ionized atom, and e of an electron. The stoichiometric coefficients are $\nu_{(\mathrm{A})} = -1$; $\nu_{(\mathrm{A}+)} = 1$; $\nu_{(\mathrm{e})} = 1$; $\sum_i \nu_i = 1$.

Together with thermodynamic equilibrium, ionization equilibrium is established in a plasma. Let x_A, x_1, x_e be the molar concentrations of neutral atoms, singly ionized atoms and electrons, respectively. The Law of Mass Action takes the following form in the case of ionization (1.1.71):

$$\frac{x_1 x_\mathrm{e}}{x_\mathrm{A}} = \frac{1}{p} K^{(1)}(T) \tag{1.1.72}$$

where p is the total pressure and $K^{(1)}(T)$ the chemical-equilibrium constant for the (first) ionization of the atom. By a procedure similar to the one by which we derived Eq. (1.1.69) from Eq. (1.1.64), we can obtain the dependence of the degree of ionization $\alpha = \alpha(T, p)$ on temperature and pressure (the degree of ionization α indicates what proportion N_1 of the total original number of atoms N_A° is ionized; $N_1 = \alpha N_\mathrm{A}^\circ$). Then we can according to Eq. (1.1.59) compute the constant

$K^{(1)}(T)$ from the change in the Gibbs potential $\Delta G(p, T)$ due to ionization. Although the ionization equilibrium was actually first solved by a similar thermodynamic procedure [Ref. 1.1.14, 1.1.15], we prefer the simpler and more elegant procedure Ref. [1.1.12] which is directly based on the general law of statistical physics as expressed in Eq. (1.1.11).

1.1.4.2 Saha's equation

The procedure mentioned above consists in the consideration of the natural and ionized atoms as two distinct states of the same subsystem {positive ion plus electron}. Eq. (1.1.11) applied to this two-particle subsystem gives the following expression for the probability $w_A(\mathscr{E}_A)$ that the subsystem will be a neutral atom with total energy $\mathscr{E}_A = \mathscr{E}_A^{tr} + \mathscr{E}_A^{int}$:

$$w_A(\mathscr{E}_A) = \frac{1}{\mathscr{Z}} e^{-\frac{\mathscr{E}_A}{kT}} \Omega(\mathscr{E}_A) \tag{1.1.73}$$

where $\Omega(\mathscr{E}_A)$ is the number of different ways in which the state of a neutral atom of energy $\mathscr{E}_A$ can be realized; $\mathscr{Z}$ is the partition function summarized over all possible states of the subsystem {ion + electron}, including ionized states.

On condition that the physical states of the ion and the free electron are mutually independent, the same formula (1.1.11) gives Eq. (1.1.74) for the probability $w_{i+e}(\mathscr{E}_{i+e})$ of an ionized state of the said subsystem having the total energy $\mathscr{E}_{i+e} = \varepsilon_i + \varepsilon_e + \varepsilon^{ion}$ (where ε_i, ε_e, ε^{ion} are the total energy of the ion, the energy of the electron and the ionization energy, respectively).

$$w_{i+e}(\mathscr{E}_{i+e}) = \frac{1}{\mathscr{Z}} e^{-\frac{\varepsilon_i}{kT}} \Omega(\varepsilon_i) \, . \, e^{-\frac{\varepsilon_e}{kT}} \Omega(\varepsilon_e) \, . \, e^{-\frac{\varepsilon^{ion}}{kT}} ; \tag{1.1.74}$$

$\Omega(\varepsilon_i)$, $\Omega(\varepsilon_e)$ are expressions analogous to $\Omega(\mathscr{E}_A)$.

The probability w_A that the subsystem will constitute a neutral atom (irrespective of its energy) is obtained by the summation of Eq. (1.1.74) over all energies $\mathscr{E}_A$:

$$w_A = \frac{Z_A}{\mathscr{Z}} \tag{1.1.75}$$

where Z_A is the partition function of the neutral atom.

Similarly, the probability that the system will be in the first ionized state irrespective of the values for ε_i, ε_e is obtained from Eq. (1.1.74) by summation through all values of ε_i, ε_e:

$$w_1 = \frac{Z_1 Z_e}{\mathscr{Z}} e^{-\frac{\varepsilon^{ion}}{kT}} \qquad (1.1.76)$$

where Z_1 and Z_e are the partition functions of a single ion and electron, respectively. Since the concentrations n_A, n_1 of neutral and ionized atoms are proportional to the probabilities w_A, w_1, we obtain the following ratio of concentrations:

$$\frac{n_1}{n_A} = \frac{Z_1 Z_e}{Z_A} e^{-\frac{\varepsilon^{ion}}{kT}}. \qquad (1.1.77)$$

The partition function $Z_e = Z_e^{tr} \times Z_e^{int}$, where $Z_e^{int} = 2$ (2 possible directions of spin) is given by:

$$Z_e^{tr} = \frac{V}{h^3} (2\pi m_e kT)^{3/2} \qquad (1.1.78)$$

m_e being the mass of the electron, and V the average volume of plasma per one electron. This volume is calculated from the equation $n_e V = 1$. By substituting Eq. (1.1.78) into Eq. (1.1.77) we obtain Saha's equation for the equilibrium composition of singly ionized gas:

$$\frac{n_1 n_e}{n_A} = \frac{2Z_1}{Z_A} \frac{(2\pi m_e kT)^{3/2}}{h^3} e^{-\frac{\varepsilon^{ion}}{kT}}. \qquad (1.1.79)$$

The whole procedure can be repeated for the computation of equilibrium concentrations in the ionization of an r times ionized atom. Thus we obtain:

$$\frac{n_{r+1} \cdot n_e}{n_r} = \frac{2Z_{r+1}}{Z_r} \frac{(2\pi m_e)^{3/2}}{h^3} (kT)^{3/2} \cdot e^{-\frac{\varepsilon^{ion}}{kT}} \qquad (1.1.80)$$

where the indices r and $r + 1$ denote quantities corresponding to the r and $r + 1$ times ionized atom, and n_e is the concentration of all electrons (i.e. not only those arising from the $r + 1^{st}$ ionization).

1.1.4.3 **Different forms of Saha's equation**

Equations (1.1.79), (1.1.80) can be multiplied by kT. Since the electron partial pressure is described by

$$p_e = n_e kT \tag{1.1.81}$$

we obtain Saha's equation in the forms:

$$\frac{n_1}{n_A} p_e = \frac{2Z_1}{Z_A} \frac{(2\pi m_e)^{3/2}}{h^3} (kT)^{5/2} e^{-\frac{\varepsilon_1^{ion}}{kT}} \tag{1.1.79'}$$

$$\frac{n_{r+1}}{n_r} p_e = \frac{2Z_{r+1}}{Z_r} \frac{(2\pi m_e)^{3/2}}{h^3} (kT)^{5/2} e^{-\frac{\varepsilon_r^{ion}}{kT}}. \tag{1.1.80'}$$

For numerical computation we often use the form

$$\log \frac{n_{r+1}}{n_r} p_e = -\varepsilon_r^{ion} . \frac{5040}{T} + \frac{5}{2} \log T - 0{,}48 + \log \frac{Z_{r+1}}{Z_r} . 2 \tag{1.1.82}$$

[Ref. 1.1.12]; we obtain it from Eq. (1.1.80) byt aking the logarithm and substituting numerical values for k, h, m_e. The pressure p_e is measured in bars (1 bar $= 10^5$ Nm^{-2}), and ε_r^{ion} is substituted in eV. The computation is facilitated by the use of nomographic charts found in Ref. [1.1.12].

Let us now introduce the degree of ionization α by equations

$$n_1 = \alpha n_A^\circ; \qquad n_A = (1 - \alpha) n_A^\circ; \qquad n_e = \alpha n_A^\circ \tag{1.1.83}$$

where n_A° is the original density of neutral atoms.

By substituting these expressions in equation (1.1.79), we obtain

$$\frac{\alpha}{1 - \alpha} p_e = \frac{2Z_1}{Z_A} \frac{(2\pi m_e)^{3/2}}{h^3} (kT)^{5/2} e^{-\frac{\varepsilon_1^{ion}}{kT}}. \tag{1.1.84}$$

Let us divide both sides of this equation by the total pressure p

$$p = p_A + p_e + p_1 = nkT \tag{1.1.85}$$

where n is the total number of all particles in the unit volume ($n = n_1 + n_A + n_e = (1 + \alpha) n_A^\circ$). Then we obtain $\frac{p_e}{p} = \frac{n_e}{n} =$

$$\frac{\alpha}{1 + \alpha} : \frac{\alpha^2}{1 - \alpha^2} = \frac{1}{p} . \frac{2Z_1}{Z_A} \frac{(2\pi m_e)^{3/2}}{h^3} (kT)^{5/2} e^{-\frac{\varepsilon_1^{ion}}{kT}}. \tag{1.1.86}$$

The function $\alpha(T, p)$ for the ionization of silver, copper, iron, mercury, oxygen, hydrogen, nitrogen is plotted in Chart 2 [Ref. 1.1.10].

Considering that the molar concentrations x_A, x_1, x_e in Sec. 1.1.4.1 depend on the degree of ionization α according to the relations

$$x_A = \frac{n_A}{n} = \frac{1-\alpha}{1+\alpha};$$

$$x_1 = \frac{n_1}{n} = \frac{\alpha}{1+\alpha}; \qquad x_e = \frac{\alpha}{1+\alpha}$$

we can re-write Eq. (1.1.72) in the form

$$\frac{\alpha^2}{1-\alpha^2} = \frac{1}{p} K_1(T). \tag{1.1.87}$$

By comparing Eqs. (1.1.86) and (1.1.87) we obtain for the equilibrium constant of ionization

$$K_1(T) = \frac{2Z_1}{Z_A} \frac{(2\pi m_e)^{3/2}}{h^3} (kT)^{5/2} e^{-\frac{\varepsilon_{1\,\text{ion}}}{kT}}; \tag{1.1.88}$$

this similarly applies to the rth ionization.

Eq. (1.1.88) is analogous to Eq. (1.1.70) for the chemical-equilibrium constant in the dissociation of diatomic molecules. This

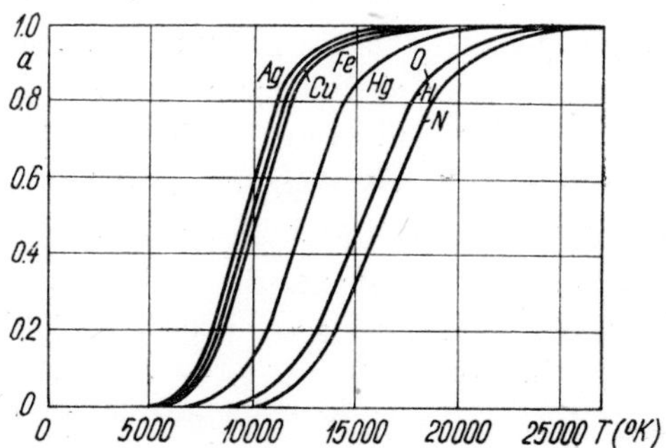

Fig. 2 Degree of ionization (α) of some gases vs. temperature

similarity is not fortuitous; for one way of obtaining Eq. (1.1.70) is the procedure by which we derived Saha's equation in Sec. 1.1.4.2. In this case normal and dissociated molecules are considered different states of the same diatomic subsystem.

If multiple ionizations occur in the gas, this raises the problem of the simultaneous chemical equilibrium (cf. Chap. 1.1.15 – several interdependent reactions take effect simultaneously) which is a rather complex problem in any case. However, the detailed numerical computation of the constants of ionization equilibrium shows that, as a rule,

$$K_{r+1}(T) \gg K_r(T) \tag{1.1.89}$$

where index r is again the order of the ionization. According to Eq. (1.1.89) ionization takes its course by stages as temperature increases. First the gas is completely singly ionized, then completely doubly ionized, etc. To a rough approximation, an ionized monatomic gas is always a mixture of only three kinds of particles: r-fold ions, $r+1$-fold ions and electrons, the order r of the ionization depending on temperature. This approximation is more thoroughly discussed in Ref. [1.1.2]; formulae presented there for the degrees of ionization $\alpha_1, \alpha_2, \ldots, \alpha_r$ describe the proportion of once, twice, ... r times ionized atoms.

1.1.4.4 Remarks concerning Saha's equation

We based the derivation of Saha's equation on the tacit assumption that – except for collisions – there is no permanent interaction between the subsystem {ion + electron} and the rest of the particles. Actually, – owing to the Coulombic nature of the action of charged particles, which takes effect in spite of their considerable separation – every such subsystem in a plasma of high electron density is in interaction with numerous particles, or theoretically with all of them ("collective interaction"). Attempts (made e.g. in Refs. [1.1.16, 1.1.17, 1.1.18, 1.1.19, 1.1.22]) at the construction of more or less exact and general statistical thermodynamics of such a system go far beyond the scope of this book.

The Coulombic interaction is mostly described (cf. e.g. Refs. [1.1.20, 1.1.21, 1.1.30]) in Debye-Hückel's approximation known from the theory of electrolytes. This approximation takes regard of the screening-off of the Coulombic field of positive ions by the cloud of free electrons surrounding the ion (polarization of plasma). The resultant field of the positive ion plus the electron cloud has then the following form:

$$\varphi(r) = \frac{q}{r} \mathrm{e}^{-\frac{r}{D}} + \text{const} \tag{1.1.90}$$

where q is the charge of the ion; D is the Debye (screening) radius dependent on the temperature T and density n_e of the screening electrons, according to the formula:

$$D \sim \left(\frac{\varepsilon_0 kT}{4\pi n_e (q+1) e^2} \right)^{\frac{1}{2}}. \tag{1.1.91}$$

The constant in Eq. (1.1.90) is usually selected in such a way that for non-polarizable plasma ($D \to \infty$) the potential $\varphi(r)$ should have the Coulombic form $\varphi(r) = \frac{q}{r}$.

In Debye-Hückel's approximation, the subsystem {ion + + electron} is in interaction with various centres of force having potentials of the type described by Eq. (1.1.90). These centres are statistically distributed (in time and space) around the subsystem. The interaction of the subsystem and the centres of forces (Eq. 1.1.90) will have several consequences:

- The number of possible discrete energy levels of the atom is reduced, as can easily be proved by the solution of the pertinent Schrödinger equation [Ref. 1.1.20]. This results in a finite number of energy levels in the internal partition function (1.1.29) of the atom, which automatically makes for the convergence of this function.
- At large quantum numbers n, the discrete levels of the subsystem are blurred and split into separate elementary states (removal of degeneracy), or else the discrete levels of adjoining ions overlap in zones, which causes a specific "zonal conductivity" of the plasma.
- The ionization energy ε^{ion} of the atom is reduced, because the transition of the electron into the zone of closely packed discrete levels is sufficient to liberate it from the atom.

The reduction of the atomic ionization energy in the plasma has been theoretically discussed for instance in Refs. [1.1.20, 1.1.21].

Brunner [1.1.21] arrives at the following value $\Delta\varepsilon^{ion}$ of this reduction:

$$\Delta\varepsilon^{ion} = \left\{0.121 \sqrt[3]{\frac{n_e}{10^{15}}} + 0.025 \sqrt{10^{-12} \frac{n_e}{T}}\right\} \text{eV} \tag{1.1.92}$$

where n_e is the number of electrons in 1 cm^3, and T the temperature [°K].

For the ionization energy we substitute then in Saha's equation the value $\varepsilon^{ion} - \Delta\varepsilon^{ion}$, which in fact is a function of the required solution n_e of this equation. The practical procedure is to compute first the concentration n_e by solving Saha's equation for non-reduced ε^{ion}, to calculate from this n_e the correction $\Delta\varepsilon^{ion}$ (Eq. 1.1.92), and once more to solve Saha's equation for the corrected value of ionization potential $\varepsilon^{ion} - \Delta\varepsilon^{ion}$. As a rule, this iteration cycle rapidly converges.*

1.1.5 Dissociation and Ionization Equilibrium in Gas Mixtures

1.1.5.1 Calculating the composion of the mixture

In plasma torches, MHD generators, arc furnaces for cracking natural gas into acetylene, and many other installations operating at extremely high temperatures the plasma is a mixture of gases of different particles: of molecules, compound particles originating from their dissociation, atoms, electrons, positive and negative ions. Various chemical reactions occur between these gases. Thermodynamic equilibrium is accompanied by chemical equilibrium, of which the unique equilibrium composition of the mixture is characteristic. Since the products of one reaction may react with the starting substances or the products of some other reaction we refer to this state as "simultaneous chemical equilibrium". The calculation of such an equilibrium is complicated and tiresome but in principle possible and with electronic computers practically feasible. We limit ourselves to a general formulation of the problem in this

---

* The systemization of various kinds of plasma in the temperature range $T(0.025 \text{ eV} \leqq T \leqq 100 \text{ keV})$ and the specific-volume range $(0.1 \text{ cm}^3 \text{ g}^{-1} \leqq v \leqq \leqq 12{,}000 \text{ cm}^3 \text{ g}^{-1})$ has been attempted for hydrogen plasma in Ref. [1.1.13]. The diagrams in this paper may be taken as a rough indication in deciding whether for a specific kind of plasma the approximation of an ideal gas as used in this book will be applicable.

Section, and recommend for more details for instance Refs. [1.1.2, 1.1.5, 1.1.6, 1.1.7].

Let us first consider a non-ionized gas mixture with a total of k components. We procede in principle by the way taken in paper [Ref. 1.1.2].

We denote the molecule of component i by the symbol K_i and (also considering the atom a monatomic molecule) we write

$$K_i = \sum_m A_{n_{im}}^{(m)} \qquad (1.1.93)$$

where $A^{(m)}$ stands for atoms of the mth type of which the molecule consists, and n_{im} for the number of atoms $A^{(m)}$ in the ith molecule. (For instance: $K_i = H_2O$; $A^{(1)} = H$; $A^{(2)} = O$; $n_{i1} = 2$; $n_{i2} = 1$.) Index i assumes the values $i = 1, 2 \ldots k$; index $m = 1, 2, \ldots a$, where a is the number of types of atoms occurring in the mixture.

By r let us denote the number of chemical reactions occurring in the mixture; this number $r = k - a$. We denote the stoichiometric coefficient for molecule K_i in the sth reaction by v_{si} (s being $s = 1, 2, \ldots r$); the rth reaction can then be written in the following form:

$$\sum_{i=1}^{k} v_{si} K_i = 0. \qquad (1.1.94)$$

The complete set of equations of the chemical equilibrium consists of

1. 1 equation expressing **Dalton's Law:**

$$\sum_{i=1}^{k} x_i = 1 \qquad (1.1.95)$$

where x_i is the molar concentration of component i.

2. a — 1 equations of the **mass balance** of mth-type atoms

$$\frac{\sum\limits_{i=1}^{k} n_{im} x_i}{\sum\limits_{i=1}^{k} n_{i1} x_1} = q_{m1}; \qquad m = 2, 3 \ldots a \qquad (1.1.96)$$

where the constant q_{m_1} is found from the total number of atoms of the mth type at the beginning of the reaction. In Eq. (1.1.96) the balance of m-type atoms is related to the balance of atoms of the preselected 1st kind. There is a total of only $a - 1$ equations, because the concentrations x_i are mutually linked by Eq. (1.1.95).

3. $k - a$ equations of the **Law of Mass Action** (cf. Eq. 1.1.55)

$$\prod_{i=1}^{k} (x_i)^{\nu_{si}} = K_{a+s}(T)_p^{-\nu_s}; \qquad s = 1, 2 \dots r = k - a. \qquad (1.1.97)$$

In Eq. (1.1.97) $\nu_s = \sum_{i=1}^{k} \nu_{si}$ is the change in the number of moles in the sth reaction, and $K_{a+s}(T)$ is the constant of chemical equilibrium for the sth reaction. Index a in the symbol for this constant signifies that from the total number of unknown quantities x_i, $k - a$ quantities can be found from Eq. (1.1.97) with the aid of the remaining a unknown quantities obtained from Eqs. (1.1.95), (1.1.96) or the symmetrized Eqs. (1.1.99):

$$\prod_{i=a+1}^{i=k} (x_i)^{\nu_{si}} = \frac{K_{a+s}(T)\, p^{-\nu_s}}{\prod_{i=1}^{i=a} (x_i)^{\nu_{si}}}. \qquad (1.1.98)$$

The set of Eqs. (1.1.95), (1.1.96), (1.1.98) is complete and allows the unique expression of the concentrations $x_1, x_2, \dots x_k$. Their numerical solution is hampered by the assymmetry of Eqs. (1.1.95) and (1.1.96), which can be eliminated for instance by selecting some a-components as reference components and expressing the concentration x_m as a function of concentrations $x_1, x_2, \dots x_m$:

$$x_m = x_m(x_1, x_2 \dots x_k), \qquad m = 1, 2 \dots a. \qquad (1.1.99)$$

The form of the functions (they are linear) and how to select the reference components is described in Ref. [1.1.2].

For computing the composition of ionized mixtures, the set of Eqs. (1.1.95) to (1.1.97) or the symmetrized system (1.1.98), (1.1.99) has to be supplemented by a set of Saha's equations for the ionization of all components at all levels. Let $x_{k+1}, x_{2k+1} \dots x_{nk+1}$ be the concentration of the once, twice, up to n times ionized first component, $x_{k+2}, x_{2k+2} \dots x_{nk+2}$ the same concentration for the second component, etc. Similarly, we write $K_{k+1}(T)$, $K_{2k+1}(T) \dots K_{nk+1}(T)$ for the ionization of the first component, $K_{k+k}(T)$, $K_{2k+k}(T) \ldots K_{nk+k}(T)$ for the ionization of the kth component; x_e is the electron concentration. Then we can write the set of Saha's equations in the following form:

$$\begin{aligned}
\frac{x_{k+1}x_e}{x_1} &= \frac{1}{p}\tilde{K}_{k+1}(T), \quad \ldots \quad \frac{x_{nk+1}x_e^n}{x_1} = \frac{1}{p^n}\tilde{K}_{nk+1}(T) \\
&\vdots \qquad\qquad \ldots \\
\frac{x_{k+k}x_e}{x_k} &= \frac{1}{p}\tilde{K}_{k+k}(T), \quad \ldots \quad \frac{x_{nk+k}x_e^n}{x_k} = \frac{1}{p^n}\tilde{K}_{nk+k}(T).
\end{aligned} \tag{1.1.100}$$

The denominators on the left-hand side are the concentrations of the neutral molecules of the corresponding component instead of the usual concentration of components on the preceding ionization level [cf. Eq. 1.1.80′). The ionization constants $\tilde{K}_n(T)$ are connected with the constants $K_n(T)$ for the nth ionization by the relation

$$\tilde{K}_n(T) = \prod_{\omega=1}^{\omega=n-1} K_\omega(T). \tag{1.1.101}$$

The Law of Conservation of Electric Charge gives

$$x_e = \sum_{\omega=1}^{n} \omega \sum_{i=1}^{k} x_{\omega k+i} \tag{1.1.102}$$

where ω is the summation index.

If we multiply this equation by the electron concentration x_e and use Eqs. (1.1.100) for expressing the products $x_{\omega k+i} \cdot x_e$, then we obtain the electron concentration x_e as an implicit function of the concentration of the neutral particles, the pressure p and the constants of chemical equilibrium:

$$x_e^2 = \sum_{\omega=1}^{n} \frac{1}{x_e^{\omega-1}} \frac{\omega}{p^\omega} \sum_{i=1}^{k} K_{\omega k+i}(T)\, x_i\,. \tag{1.1.103}$$

The schematic procedure for solving the general system of chemical equilibrium is then as follows:

1) we select m suitable reference components and assign to them the roughly estimated concentrations $x_1, x_2 \ldots x_m$ [Ref. 1.1.2];

2) we compute $x_{m+1}, x_{m+2} \dots x_k$ from Eqs. (1.1.98);

3) we find x_e from Eq. (1.1.103);

4) we calculate $x_{k+1}, x_{k+2} \dots x_{nk+k}$ from Eqs. (1.1.100)

5) the values thus obtained are substituted in Eq. (1.1.99) which is solved approximately by Newton's method. Thus we obtain more exact concentration values $x_1, x_2 \dots x_m$ for the reference components. The whole procedure is then repeated until the requisite accuracy is obtained.

1.1.5.2 Thermodynamic functions of dissociated and ionized gas mixtures

Apart from temperature and pressure, almost every thermodynamic quantity of the equilibrium mixture (e.g. E, F, G, H, S, C_p, C_v) is additive; this means that its value for the whole mixture is given by the addition of the quantities for the individual components. The molar values of these quantities have been computed according to the formulae of statistical physics in Chap. 1.1.2, the molar concentrations $x_i = \frac{\mathcal{N}_i}{\mathcal{N}} = \frac{p_i}{p}$ are found by the procedure discussed in the preceding Section. The total number of moles of the mixture $\mathcal{N} = \sum_{i=1}^{i=k} \mathcal{N}_i$ is obtained from the equation of state of the ideal gas:

$$p \, . \, V = RT \sum_{i=1}^{k} \mathcal{N}_i \tag{1.1.104}$$

The number of moles $\mathcal{N}_i$ of the i component is

$$\mathcal{N}_i = x_i \mathcal{N} = x_i \frac{pV}{RT} \, . \tag{1.1.105}$$

The molar enthalpy (i.e. the enthalpy of one mole of mixture) is obtained from Eq. (1.1.41):

$$\begin{aligned} H_{\text{mix}} &= \frac{1}{\mathcal{N}} \sum_{i=1}^{k} \mathcal{N}_i \left\{ N_0 kT \left[T \frac{\mathrm{d}(\ln \mathscr{Z}_T)}{\mathrm{d}T} + 1 \right] + E_0 \right\}_i \\ &= \sum_{i=1}^{k} x_i \left\{ N_0 kT \left[T \frac{\mathrm{d}(\ln \mathscr{Z}_T)}{\mathrm{d}T} + 1 \right] \right\}_i + \sum_{i=1}^{k} x_i \{E_0\}_i \, . \end{aligned} \tag{1.1.106}$$

The last summation in Eq. (1.1.106) can be resolved into the energy $[E_0]_{mix}$ of the standard state of the mixture ($T = 0$ °K, same values for x_i), and the latent enthalpy of dissociation and ionization $[E_0]_{lat}$ corresponding to bonds between various particles which have already been reversibly broken. Let us denote the bond energy of the lth bond corresponding to one particle of the ith component by $\frac{1}{2}\{\varepsilon_0\}_{il}$*, the number of lth bonds of particles of the ith component by n_{il}. Then we can write:

$$\sum_{i=1}^{k} x_i\{E_0\}_i = [E_0]_{mix} + [E_0]_{lat} =$$

$$= \sum_{i=1}^{k} x_i \frac{1}{2} \sum_{l=1}^{l=a_i} N_{0\,il}\{\varepsilon_0\}_{il} + \sum_{i=1}^{k} x_i \frac{1}{2} \sum_{l=a_i+1}^{l=b_i} N_0 n_{il}\{\varepsilon_0\}_{il} \quad (1.1.107)$$

where index $l \leqq a_i$ denotes solved bonds, index $a_i < l \leqq b_i$ existing bonds, $a_i + b_i$ is the number of all possible bonds of an ith particle, and N_0 Avogadro's number. The enthalpy of the mixture related to the standard state will then be the following quantity:

$$H_{mix} - [E_0]_{mix} = \sum_{i=1}^{k} x_i \left\{N_0 kT\left[T\frac{\mathrm{d}(\ln \mathscr{Z}_T)}{\mathrm{d}T}\right] + 1\right\}_i$$

$$+ \frac{N_0}{2} \sum_{i=1}^{k} x_i \sum_{i=1}^{l=a_i} n_{il}\{\varepsilon_0\}_{il}. \quad (1.1.108)$$

As an example, let us take the dissociation of nitrogen. At temperatures between 6,000 °K and 12,000 °K and a pressure of $p =$ $= 1$ atm., nitrogen is a mixture of N_2-molecules, N-atoms, N^+-ions, N_2^+-ions and electrons. The latent enthalpy of the mixture will consist of:

(a) the dissociation enthalpy which we attribute to atomic nitrogen and N^+ ions:

$$\frac{N_0}{2} x_{(N)}\{\varepsilon_0^{dis}\}_{(N_2)} + \frac{N_0}{2} x_{(N^+)}\{\varepsilon_0^{dis}\}_{(N_2)} \quad (1.1.109)$$

where $\{\varepsilon_0^{dis}\}_{(N_2)}$ is the dissociation energy of the N_2 molecules, and $x_{(N)}, x_{(N^+)}$ are the concentrations of N and N^+, respectively;

* By bond energy we mean the mean value of the dissociation energy of the bond in various diss ociationevents [Ref. 1.1.11]. The coefficient $\frac{1}{2}$ before $\{\varepsilon_0\}_{il}$ prevents the individual bond from being counted twice,

(b) the ionization enthalpy which, as a rule, is fully attributed to ion gases (electrons are only attributed the enthalpy of their translational motion):

$$N_0 x_{(N^+)}\{\varepsilon^{ion} - \Delta\varepsilon^{ion}\}_{(N)} + N_0 x_{(N^+{}_2)}\{\varepsilon^{ion} - \Delta\varepsilon^{ion}\}_{(N_2)} \quad (1.1.110)$$

$\varepsilon^{ion} - \Delta\varepsilon^{ion}$ being the ionization energy corrected according to Eq. (1.1.92).

In order to compute the total enthalpy Eq. (1.1.108), the enthalpy corresponding to the internal degrees of freedom of the individual particles and the enthalpy of their translational motion would have to be added to the dissociation and ionization enthalpies. The enthalpy of the excited electron states is usually neglected [cf. e.g. Refs. 1.1.23, 1.1.24].

Subsequently we proceed—starting from Eqs. (1.1.39), (1.1.34), (1.1.35)—to compute internal energy, free energy and free enthalpy (Gibbs' potential). It is often convenient to express the values of the thermodynamic functions of the mixture using the degree α of the reaction (dissociation or ionization) [cf. e.g. Ref. 1.1.1, 1.1.3].

The molar entropy of a mixture of ideal gases is also obtained by the summation of the entropies of the various components in accordance with Eq. (1.1.42). Since S^{int} Eq. (1.1.42) is only dependent on temperature and not on pressure, we can (using Eq. 1.1.43) write for the molar entropy of the ith component

$$(S)_i = -N_0 k \ln p_i + S_i(T) \quad (1.1.111)$$

where p_i is the partial pressure of the ith component, and $S_i(T)$ its entropy for $p_i = 1$ atm. For the entropy of the mixture we have

$$[S]_{mix} = \sum_{i=1}^{k} x_i (S)_i. \quad (1.1.112)$$

Substituting Eq. (1.1.111) into Eq. (1.1.112) and considering that $\frac{p_i}{p} = x_i$, we have

$$[S]_{mix} = \sum_{i=1}^{k} x_i S_i(T) - N_0 k \sum_{i=1}^{k} x_i \ln p - N_0 k \sum_{i=1}^{k} x_i \ln x_i =$$

$$= \sum_{i=1}^{k} x_i[-N_0 k \ln p + S_i(T)] - N_0 k \sum_{i=1}^{k} x_i \ln x_i. \quad (1.1.113)$$

The last member is called the mixing entropy, which indicates that it describes the variation which (reversible) mixing produces in the entropy of the system of separate ideal gases.

In calculating derived quantities – such as the molar heat capacity of a mixture at constant volume and pressure $[C_V]_{mix}$ and $[C_p]_{mix}$ – we have to keep in mind the presence of molar concentrations x_i which are functions of temperature and pressure in Eq. (1.1.108) for the enthalpy of the mixture and in the analogous formula for the internal energy of the mixture.

Thus for instance, if we want the molar heat capacity at constant pressure, the expression

$$[C_p]_{mix} = \left(\frac{\partial H_{mix}}{\partial T}\right)_p = \left(\frac{\partial \sum_{i=1}^{k} x_i H_i}{\partial T}\right)_p \tag{1.1.114}$$

has to be derived as the summation:

$$[C_p]_{mix} = \sum_{i=1}^{k} x_i \left\{\frac{\partial (H)_i}{\partial T}\right\}_p + \sum_{i=1}^{k} \left(\frac{\partial x_i}{\partial T}\right)_p \cdot \{H\}_i . \tag{1.1.115}$$

The first member $\sum_{i=1}^{k} x_i \{C_p\}_i$ is the summation of the molar-heat capacities of the individual components, and the second member expresses the effect of the temperature-dependent composition of the mixture. Similarly, for molar heat capacity at constant volume we obtain:

$$[C_V]_{mix} = \left(\frac{\partial E_{mix}}{\partial T}\right)_V = \sum_{i=1}^{k} x_i \left\{\frac{\partial (E_i)}{\partial T}\right\}_V = \sum_{i=1}^{k} \left(\frac{\partial x_i}{\partial T}\right)_V \cdot \{E\}_i . \tag{1.1.116}$$

In the case of simple ionization or dissociation reactions it is usually convenient to use the degree α of ionization or dissociation instead of the concentration x_i. Eqs. (1.1.108), (1.1.113), (1.1.115), (1.1.116) and similar formulae for other thermodynamic functions are then often expressed by relatively complex functions of α and its derivatives $\left(\frac{\partial \alpha}{\partial T}\right)_p$, $\left(\frac{\partial \alpha}{\partial T}\right)_V$. Many expressions of this type are found in Ref. [1.1.3].

The thermodynamic functions of more complex dissociated or ionized mixtures are usually calculated by electronic digital computers simultaneously with the composition of the mixtures. The programming procedure (for an IBM-704 computer) is outlined in Ref. [1.1.5].

1.2 Non-Equilibrium Processes in Plasma

1.2.1 Deviations from Thermodynamic Equilibrium

1.2.1.1 Macroscopic aspects

Let us imagine—as in Sec. 1.1.1.2—an elementary volume filled with flowing plasma and isolated by walls; these, however, can be pervious to some kind of particles, to heat, and (partly or totally) to electromagnetic radiation. Moreover, let these walls be perfectly plastic, that is deformable under the action of flow in accordance with the field of macroscopic velocities ("flowing volume of plasma"). Various physical quantities—mass, energy, momentum, electric charge, etc.—can be exchanged between such a volume and the adjoining ones, and these phenomena, such as diffusion, heat conduction, internal friction, conduction of electric current, etc. are called transport phenomena in the plasma. Various chemical reactions, including ionization, dissociation, recombination and similar processes liberating or binding reaction energies can occur within the element.

The thermodynamic equilibrium discussed in Division 1.1 has the essential property that both transport phenomena and chemical reactions take reversible courses resulting in zero flows of physical quantities (cf. Sec. 1.2.3.3). In principle, all processes tending to change the energy of the plasma irreversibly—i.e. by one-sided supply or removal—can be said to affect the equilibrium unfavourably. Such are radiation, heat conduction, Joule heating and similar processes. As long as they are negligible in comparison with processes involving the mutual exchange between different forms of energy within the element, the deviations from thermodynamic equilibrium are also negligible.

Sometimes partial equilibria can be established in the plasma, in the sense that two different components of the plasma may each be in a state very near to equilibrium without, however, being in mutual equilibrium. A typical example are the electron and ion gases in the plasma of low-pressure discharges, where the temperature of the electrons is by several orders of magnitude higher than that of the ions or neutral particles, while each of these gases separately conforms with Maxwell's (i.e. equilibrium) distribution.

There are many kinds of irreversible processes and several of them (e.g. diffusion, heat conduction, chemical reactions) may occur simultaneously in a plasma. Common to all of them is the fact that entropy (cf. Sec. 1.2.3.3) is produced in the plasma. Unless transferred to neighbouring elements, the entropy increases owing to irreversible processes until an equilibrium (at maximum entropy) is established. The production of entropy is on one hand characteristic of the rate at which equilibrium is set up, on the other hand it is the starting point for the classification of various types of irreversible processes occurring in the plasma. This will be discussed in greater detail in Sec. 1.2.3.3.

1.2.1.2 **Disturbances affecting statistical equilibrium**

In Sec. 1.1.1.3 we have pointed out that perfect thermodynamic equilibrium is always accompanied by statistical equilibrium, whose characteristic feature is the unique distribution of particles according to types, and within the given type according to coordinates, momenta, and internal states of the individual particles. Such distributions are exemplified by Eqs. (1.1.6), (1.1.7), (1.1.8), (1.1.79), (1.1.80) or by the general formula (1.1.11) from which the former equations can be derived. If irreversible processes unbalance the thermodynamic equilibrium, such a disturbance shows also in the equilibrium distributions mentioned above. If the disturbance is slight, these distributions often retain their form (i.e. the kind of their functional dependence), and only their T parameter (i.e. temperature) changes.

In non-equilibrium plasma, temperatures T belonging to the individual distributions can mutually differ. Such a plasma is called nonisothermal and we speak of kinetic temperature (of electrons T_e, ions T_i, neutral particles T_n), excitation temperature T_{exc}, radiation temperature T_{rad}, etc. Where the partial equilibrium of some component is concerned, the concept of the temperature of the component makes sense. Often, however, temperature is defined as a conventional quantity, which is based on the consciously incorrect extrapolation of some relations valid in the state of statistical equilibrium. Thus we define, for instance, the temperature T_ε of a monoenergetic beam of accelerated particles having energy ε by the relation $\varepsilon = kT_\varepsilon$, although this is a state of distinct non-equilibrium.

Let us now discuss some kinds of temperature which are most frequently used to describe a plasma [Refs. 1.2.1, 1.2.2, 1.2.3].

The **kinetic temperature** $-T_k$ of the kth component is in equilibrium defined as a parameter of Maxwell's velocity distribution:

$$f_k^{(0)}(\mathbf{v}_k) = n_k \left(\frac{m_k}{2\pi k T_k}\right)^{3/2} \mathrm{e}^{-\frac{m_k v_k^2}{2kT_k}}; \tag{1.2.1}$$

it indicates the number of kth particles (having mass m_k) whose velocity in the unit elementary velocity interval ranges about the value of $\mathbf{v}_k$ (cf. Sec. 1.2.1.1); n_k is the concentration of kth particles per cm^3. The temperature T_k is connected with the mean kinetic energy of random (thermal) motion of the particles by the well-known relation

$$\frac{1}{2} m_k \overline{v_k^2} = \frac{3}{2} kT_k, \tag{1.2.2}$$

where the mean square velocity $\overline{v_k^2}$ is given by the following equation, a special instance of Eq. (1.2.10):

$$\overline{v_k^2} = \frac{1}{n_k} \int_0^\infty v_k^2 f_k^{(0)}(\mathbf{v}_k)\, \mathrm{d}\mathbf{v}_k. \tag{1.2.3}$$

In Eq. (1.2.3) we integrate over the entire velocity space of the kth particle. The disturbance of the equilibrium can show in two ways:

1) By a change in the form of the distribution function $f_k^{(0)}(\mathbf{v}_k)$. The isotropic distribution (Eq. 1.2.1) (in which no velocity direction is preferred) is disturbed owing to the directed flow of particles, and an anisotropic distribution, $f_k(\mathbf{v}_k)$ having a maximum in the direction of the

macroscopic velocity $\bar{\mathbf{v}}_k$ of these flows is established. The macroscopic velocities $\bar{\mathbf{v}}_k$ are defined as mean velocity values $\mathbf{v}_k$:

$$\bar{\mathbf{v}}_k = \frac{1}{n_k} \int_0^\infty \mathbf{v}_k f_k(\mathbf{v}_k)\, \mathrm{d}\mathbf{v}_k . \tag{1.2.4}$$

The kinetic temperature of k-type particles is then defined by Eq. (1.2.5) in terms of the mean kinetic energy corresponding to the thermal velocities $\mathbf{v}_k - \bar{\mathbf{v}}_k$ of the particles:

$$\frac{3}{2} kT_k = \frac{1}{2} m_k \cdot \frac{1}{n_k} \int_0^\infty (\mathbf{v}_k - \bar{\mathbf{v}}_k)^2 f_k(\mathbf{v}_k)\, \mathrm{d}\mathbf{v}_k . \tag{1.2.5}$$

It thus depends largely on the form of the distribution function $f_k(\mathbf{v}_k)$.

2) The distribution form (1.2.1) does not change, but the value of parameter T_k does in such a way that it differs for different components. Every component then constitutes a gas which – taken by itself – is almost in equilibrium, whereas no equilibrium exists between the components. Such a state is set up if the exchange of energies and momenta within the component is more effective than the exchange between the components.

The electron gas in low-pressure discharges is an example of such a state. Owing to the small mass m_e of the electrons, it extracts energy from the electric field and is soon randomized in velocity by the effective Coulombic interaction of the electrons [Ref. 1.2.4]. In view of the small mass ratio of the electrons and the heavy particles $\frac{m_e}{M}$, the electrons are not efficient in transferring the gained energy to the heavy particles. Thus the temperature of the electrons stabilizes at such a level that the energy gained in unit time is, within the same time interval, transferred in collisions with heavy particles. The author of Ref. [1.2.1] suggests the following equation for the relative difference between the kinetic temperatures:

$$\frac{T_e - T_n}{T_e} = \frac{M}{4m_e} \frac{(\bar{l}eE)^2}{\left(\frac{3}{2} kT_e\right)^2}, \tag{1.2.6}$$

where T_e, T_n, m_e, M are the temperature [°K] and mass [kg] of electrons and heavy particles, respectively; the product $\bar{l}eE$ is the energy gained by the electron in a field of intensity E (V m^{-1}) on the mean free path $\bar{l}$ [m]; $\frac{3}{2}kT_e$ is the thermal energy per electron, and e its charge. Under the conditions prevailing in plasma torches, the relative temperature difference (Eq. 1.2.6) usually amounts to several per cent.

The **excitation temperature** of the mth level is defined by the Boltzmann distribution (Eq. 1.1.7)

$$\frac{n_m}{n_0} = \frac{g_m}{g_0} e^{-\frac{\varepsilon_m}{kT}}. \tag{1.2.7}$$

In accordance with the principle of detailed equilibrium (cf. Sec. 1.1.1.3), the number of excitations by absorption of photons in the plasma equals the number of returns to ground state by the emission of photons (i.e. the sum of spontaneous and induced emission). Similarly, the number of excitations due to collisions of the first kind equals the number of returns to ground state due to collisions of the second kind. If the radiation can freely escape from the plasma (because the walls of the plasma element are pervious to radiation), the density of radiation in the plasma is greatly reduced. Consequently, the number of excitations by absorption of photons decreases at a higher rate than the number of induced emissions, and the number of excited particles (at virtually unchanged kinetic temperature of the particles) therefore decreases.

Thus the detailed equilibrium (Sec. 1.1.1.3) is upset: the resultant density of excited particles n_m settles at such a level that the **sum** of excitations (per second) due to collisions and photon absorptions equals the **sum** of returns (per second) due to collisions and photon emissions. A thorough analysis of this problem together with a rough quantitative estimation for the equilibrium between translational motion and excitation is found in Ref. [1.2.1]. Here we limit ourselves to the statement that owing to the phenomena mentioned, the temperature T derived from the given density n_m by using Eq. (1.2.7)—the so-called **excitation temperature** T_{exc}—is usually smaller than the kinetic temperature of the particles.

The **ionization temperature** is defined by Saha's equation (Eq. 1.1.79). Detailed equilibrium between direct processes (collision ionization, photoionization) and indirect ones (recombination by triple collisions and photorecombination) is established if a plasma is in thermodynamic equilibrium. As in the preceding example, the number of photoionizations will be reduced and the equilibrium upset by irreversible escape of radiation if the walls of the plasma element become transparent. The density of charged particles (or degree of ionization) is therefore usually lower than would correspond to the kinetic temperature of the particles. In other words, the temperature derived (according to Eq. 1.1.79) from the given density n_e – the so-called ionization temperature – is usually lower than the kinetic temperature of the particles. A thorough anylysis is again found in Ref. [1.2.1], and the general application of Saha's equation to non-equilibrium states due to the escape of radiation from the plasma is discussed in Ref. [1.2.5].

Chemical reactions in the plasma can also upset the equilibrium if they occur at a higher rate than the exchange of energy between the translational and internal degrees of freedom of the individual molecules [Ref. 1.2.3]. In this case, too, Maxwell's distribution (Eq. 1.2.1) is upset (with its isotropy preserved), and so is Boltzmann's distribution (Eq. 1.2.7) which, moreover, depends on whether electronic, vibrational or rotational levels are concerned. Accordingly, we refer to the **vibrational, rotational** and **electron temperatures** of the plasma.

The concept of the "partial" temperature T_s of some plasma component – and even more so the concept of the temperature of the plasma as a whole – loses its physical meaning if the temperature gradient grad T_s is too great. Or putting it more precisely: if some component of a plasma is to be considered a thermodynamic system (and attributed a temperature T_s), the temperature difference along the mean free path $\bar{l}_s$ in the process causing the energy exchange must be small in comparison with the total temperature value [Refs. 1.2.6, 1.2.1]:

$$\frac{\bar{l}_s \operatorname{grad} T_s}{T_s} \ll 1. \tag{1.2.8}$$

The mean free path $\bar{l}_s$ in Eq. (1.2.8) can refer to elastic collisions of particles (kinetic temperature), excitation collisions between particles (excitation temperature), ionization collisions (ionization temperature), various chemical reactions, etc. Here we also include the mean free path of photons, related to photoexcitation and photoionization. The magnitudes of these paths (or of their inverse values $Q_s = \frac{1}{\bar{l}_s}$, the effective cross-sections) are decisive for the corresponding partial equilibria and are closely connected with the relaxation times for the establishment of these equilibria (cf. Sec. 1.2.8.1).

1.2.2 Fundamentals of the Kinetic Theory of Transport Phenomena

1.2.2.1 Basic concepts of kinetic theory

If we consider plasma a mixture of ideal gases, we may describe its physical state by the methods of the ordinary kinetic theory of gases. By now, this theory is very thoroughly elaborated and affords a consistent and integrated idea of the relationships between the statistical frequencies of various elementary processes in the plasma (collisions of particles) on one hand, and the macroscopic quantities and processes of plasma as a whole (pressure, temperature, electrical conductivity, etc.) on the other hand.

The basic concept of kinetic theory is the **distribution function** of particles of the ith type: $f_i(\mathbf{r}, \mathbf{v}_i, t)$. This function indicates the number of ith type particles which are present in the unit-volume element about point $\mathbf{r}$ at time t, and have a velocity in the unit interval of velocity about point $\mathbf{v}_i$ in the velocity space corresponding to i^{th} particles. At equilibrium, the distribution functions af all types of particles are Maxwellian (cf. Eq. 1.2.1):

$$f_i^{(0)}(\mathbf{r}, \mathbf{v}_i, t) = n_i \left(\frac{m_i}{2\pi kT} \right)^{3/2} \mathrm{e}^{-\frac{m_i v_i^2}{2kT}} . \tag{1.2.9}$$

In the general case, functions $f_i(\mathbf{r}, \mathbf{v}_i, t)$ can depend on time t as well as on direction $\mathbf{v}_i$ in the velocity space (hence not only on $|\mathbf{v}_i| = V_i$ as in Eq. 1.2.9); these functions may be obtained by solving Boltzmann's kinetic equations (cf. Sec. 1.2.2.2).

Given the distribution functions $f_i(\mathbf{r}, \mathbf{v}_i, t)$, we can calculate the mean values of the macroscopic quantities which describe the state of the gas as a whole. Our starting point is the statistical definition of the mean value, which for any quantity $\alpha(\mathbf{r}, \mathbf{v}_i, t)$ reads:

$$\overline{\alpha(\mathbf{r}, t)} = \frac{1}{n_i(\mathbf{r}, t)} \int \alpha(\mathbf{r}, \mathbf{v}_i, t) f_i(\mathbf{r}, \mathbf{v}_i, t) \, d\mathbf{v}_i, \tag{1.2.10}$$

where we integrate over the entire velocity space ($d\mathbf{v}_i = dv_x \, dv_y \, dv_z$ is a volume element of the velocity space; $n_i(\mathbf{r}, t)$ is the density of ith particles in point $\mathbf{r}$ at time t).

The **macroscopic velocity** $\bar{\mathbf{v}}_i$ of ith particles is given by Eq. (1.2.4):

$$\bar{\mathbf{v}}_i(\mathbf{r}, t) = \frac{1}{n_i} \int \mathbf{v}_i f_i(\mathbf{r}, \mathbf{v}_i, t) \, d\mathbf{v}_i. \tag{1.2.11}$$

The **mass-average velocity** $\mathbf{v}_0(\mathbf{r}, t)$ of a gas mixture, determined only by the resultant of external forces (hence the name) and the pressure gradient acting upon the mixture, is defined by the equation:

$$\mathbf{v}_0(\mathbf{r}, t) = \frac{1}{\sum_i n_i m_i} \sum_i n_i m_i \bar{\mathbf{v}}_i = \frac{1}{\varrho} \sum_i \varrho_i \bar{\mathbf{v}}_i, \tag{1.2.12}$$

where ϱ_i and ϱ are the partial and total densities, respectively.

The diffusion **velocity** of the ith component $\bar{\mathbf{V}}_i(\mathbf{r}, t)$ is defined as the flow velocity of that component relative to a coordinate system moving with mass-average velocity $\mathbf{v}_0(\mathbf{r}, t)$:

$$\bar{\mathbf{V}}_i(\mathbf{r}, t) = \bar{\mathbf{v}}_i(\mathbf{r}, t) - \mathbf{v}_0(\mathbf{r}, t). \tag{1.2.13}$$

The **(kinetic) temperature** of a gas mixture is defined by what is called the thermal velocities $\mathbf{V}_i(\mathbf{V}_i(\mathbf{r}, t) = \mathbf{v}_i(\mathbf{r}, t) - \mathbf{v}_0(\mathbf{r}, t))$ of the individual molecules (i.e. the particle velocities in a coordinate system moving at velocity $\mathbf{v}_0$):

$$\frac{3}{2} kT = \frac{1}{n} \sum_i n_i \left(\frac{1}{2} m_i \overline{\mathbf{V}_i^2} \right) = \frac{1}{n} \sum_i n_i \left(\frac{1}{2} m_i \overline{|\mathbf{v}_i - \mathbf{v}_0|^2} \right). \tag{1.2.14}$$

Eq. (1.2.14) can be considered a generalisation of Eq. (1.2.5); $n = \sum n_i$ is the total density of particles.

To every plasma particle belongs a number of physical quantities: its mass m_i, momentum due to macroscopic motion $m_i\overline{\mathbf{v}}_i$, momentum due to thermal motion $m_i(\mathbf{v}_i - \mathbf{v}_0)$, the kinetic energy of this motion $\frac{1}{2} m_i(\mathbf{v}_i - \mathbf{v}_0)(\mathbf{v}_i - \mathbf{v}_0)$ etc. As they move in the plasma, the particles transfer these quantities from one place to another. In non-homogeneous plasma (e.g. grad $T \neq 0$, grad $n_i \neq 0$, etc.) this transfer has the character of macroscopic processes (heat conduction, diffusion, etc.). The transfer of any molecular quantity ψ_i is mathematically described by the vector of flow density $\boldsymbol{\Psi}_i$ which indicates what amount of quantity ψ_i flows in unit time through a perpendicular unit area moving together with the plasma at the mass-average velocity $\mathbf{v}_0$. The flow density $\boldsymbol{\Psi}_i$ can be expressed by the distribution function f_i

$$\boldsymbol{\Psi}_i(\mathbf{r}, t) = \int (\mathbf{v}_i - \mathbf{v}_0)\, \psi_i f_i \, d\mathbf{v}_i, \tag{1.2.15}$$

[Refs. 1.2.4, 1.2.8] where we integrate over the whole velocity space of the ith particle.

Let us explicitly express the flow densities for the transfer of mass, momentum and kinetic energy of particles.

Mass transfer (diffusion): $\psi_i = m_i$

$$\boldsymbol{\Psi}_i = m_i \int f_i(\mathbf{v}_i - \mathbf{v}_0)\, d\mathbf{v}_i, \tag{1.2.16}$$

and with the aid of Eqs. (1.2.13, 1.2.11) we obtain

$$\boldsymbol{\Psi}_i = m_i n_i(\overline{\mathbf{v}}_i - \mathbf{v}_0) = \varrho_i \overline{\mathbf{V}}_i = \mathbf{J}_i(\mathbf{r}, t); \tag{1.2.17}$$

$\mathbf{J}_i(\mathbf{r}, t)$ is the density of the diffusion flow of ith particles.

Momentum transfer:

$$\psi_i = m_i(\mathbf{v}_i - \mathbf{v}_0)$$

$$\boldsymbol{\Psi}_i = m_i \int f_i(\mathbf{v}_i - \mathbf{v}_0)(\mathbf{v}_i - \mathbf{v}_0)\, d\mathbf{v}_i, \tag{1.2.18}$$

which—averaging the quantity $(\mathbf{v}_i - \mathbf{v}_0)(\mathbf{v}_i - \mathbf{v}_0)$ according to Eq. (1.2.10)—can be written

$$\boldsymbol{\Psi}_i = n_i m_i \overline{(\mathbf{v}_i - \mathbf{v}_0)(\mathbf{v}_i - \mathbf{v}_0)} = \mathsf{P}_i(\mathbf{r}, t). \tag{1.2.19}$$

The quantity P_i (called the tensor of partial pressure) has nine components of the following type:

$$(\mathsf{P}_i)_{xy} = n_i m_i \overline{(\mathbf{v}_i - \mathbf{v}_0)_x (\mathbf{v}_i - \mathbf{v}_0)_y}, \tag{1.2.20}$$

which in this case describe the transfer of the xth component of momentum in the direction of the y axis. When the gas is in equilibrium, the mixed components $(\mathsf{P}_i)_{xy} = (\mathsf{P}_i)_{xz} = \ldots = (\mathsf{P}_i)_y = 0$ vanish and there remain only members of the type $(\mathsf{P}_i)_{xx} = (\mathsf{P}_i)_{yy} = (\mathsf{P}_i)_{zz} = p$ which describe the partial hydrostatic pressure of the ith component (i.e. the transfer of momentum on a unit area of a plate immersed in the gas).

The tensor P of the mixture equals the sum of the partial-pressure tensors:

$$\mathsf{P}(\mathbf{r}, t) = \sum_i \mathsf{P}_i = \sum_i n_i m_i \overline{(\mathbf{v}_i - \mathbf{v}_0)(\mathbf{v}_i - \mathbf{v}_0)}. \tag{1.2.21}$$

Transfer of the kinetic energy of thermal motion of particles

$$\psi_i = \frac{1}{2} m_i \, |\mathbf{v}_i - \mathbf{v}_0|^2$$

$$\Psi_i = \frac{1}{2} \int f_i m_i \, |\mathbf{v}_i - \mathbf{v}_0|^2 (\mathbf{v}_i - \mathbf{v}_0) \, d\mathbf{v}_i = \mathbf{Q}_i(\mathbf{r}, t), \tag{1.2.22}$$

where the vector of heat-flux density can be expressed by using Eq. (1.2.10):

$$\mathbf{Q}_i(\mathbf{r}, t) = \frac{1}{2} n_i m_i \overline{|\mathbf{v}_i - \mathbf{v}_0|^2 (\mathbf{v}_i - \mathbf{v}_0)}. \tag{1.2.23}$$

The total heat-flux density is given by vector

$$\mathbf{Q}(\mathbf{r}, t) = \sum_i \mathbf{Q}_i(\mathbf{r}, t) = \sum_i \frac{1}{2} n_i m_i \overline{|\mathbf{v}_i - \mathbf{v}_0|^2 (\mathbf{v}_i - \mathbf{v}_0)}. \tag{1.2.24}$$

Thus employing the distribution function $f_i(\mathbf{r}, \mathbf{v}_i, t)$, we can by the averaging operation involved in Eq. (1.2.10) express various useful macroscopic quantities. The determination of the function f_i will be discussed in later paragraphs.

1.2.2.2 Boltzmann's kinetic equation

Let us assume a gas mixture sufficiently dilute to confer the character of binary collisions (only two particles participating in any interaction) on all interactions. This assumption also involves concentrations of charged particles low enough to minimize the effect of Coulombic collisions (which in view of a theoretically infinite effective cross-section for collisions can never be binary collisions) and make it negligible compared with collisions of the types electron-molecule, molecule-molecule, ion-molecule. Moreover, let the gas mixture be non-homogeneous (grad $f_i \neq 0$) and let an outside force $\mathbf{F}_i$ (either independent of velocity or of Lorentz type) act on the ith particles.

The distribution function $f_i(\mathbf{r}, \mathbf{v}_i, t)$ indicates, roughly speaking, the number of ith particles in the elementary space $\mathrm{d}\mathbf{r}$ about point $\mathbf{r}$ which have a velocity in the elementary range $\mathrm{d}\mathbf{v}_i$ about $\mathbf{v}_i$ in the velocity space. The total time derivative of this function is

$$\frac{\mathrm{D}f_i}{\mathrm{D}t} = \frac{\partial f_i}{\partial t} + \frac{\partial f_i}{\partial x}\frac{\mathrm{d}x}{\mathrm{d}t} + \frac{\partial f_i}{\partial y}\frac{\mathrm{d}y}{\mathrm{d}t} + \frac{\partial f_i}{\partial z}\frac{\mathrm{d}z}{\mathrm{d}t} + \\ + \frac{\partial f_i}{\partial v_x}\frac{\mathrm{d}v_x}{\mathrm{d}t} + \frac{\partial f_i}{\partial v_y}\frac{\mathrm{d}v_y}{\mathrm{d}t} + \frac{\partial f_i}{\partial v_z}\frac{\mathrm{d}v_z}{\mathrm{d}t}. \qquad (1.2.25)$$

Written in a more concentrated form (using $\frac{\mathrm{d}x}{\mathrm{d}t} = v_x$; $\frac{\mathrm{d}v_x}{\mathrm{d}t} = \frac{F_{ix}}{m_i}$, where m_i is the mass of ith-type particles; $\frac{\partial f_i}{\partial x} = \mathrm{grad}_x f_i$; $\frac{\partial f_i}{\partial v_x} = \mathrm{grad}_{v_x} f_i$) this gives

$$\frac{\mathrm{D}f_i}{\mathrm{D}t} = \frac{\partial f_i}{\partial t} + \mathbf{v}_i \,.\, \mathrm{grad}_r f_i + \frac{\mathbf{F}_i}{m_i} \,.\, \mathrm{grad}_{\mathbf{v}_i} f_i. \qquad (1.2.26)$$

The total derivative $\frac{\mathrm{D}f_i}{\mathrm{D}t}$ indicates how in the range mentioned ($\mathrm{d}\mathbf{r}\,\mathrm{d}\mathbf{v}_i$) the number of particles varies due to the explicit time dependence $\left(\text{member } \frac{\partial f_i}{\partial t}\right)$ due to coordinates varying as the particles move with velocity $\mathbf{v}_i$ (member $\mathbf{v}_i \,.\, \mathrm{grad}_{\mathbf{r}} f_i$), and due to velocity changing by the action of force $\mathbf{F}_i$ $\left(\text{member } \frac{\mathbf{F}_i}{n_i} \,.\, \mathrm{grad}_{\mathbf{v}_i} f_i\right)$. This total derivative must equal the difference between the number Γ_{ij}^{+} of ith particles which in unit time have entered the element $d\mathbf{r}$, $d\mathbf{v}_i$—owing to a change in velocity caused by collisions with type j particles—and the number Γ_{ij}^{-} of ith particles which owing to similar collisions have leaked from the element:

$$\frac{Df_i}{Dt} = \frac{\partial f_i}{\partial t} + \mathbf{v}_i \cdot \text{grad}_{\mathbf{r}} f_i + \frac{\mathbf{F}_i}{m_i} \cdot \text{grad}_{\mathbf{v}_i} f_i = \sum_j (\Gamma_{ij}^{+} - \Gamma_{ij}^{-}). \quad (1.2.27)$$

Eq. (1.2.27) indicates the time change of the distribution function and is called Boltzmann's kinetic equation. The expression

$$\left[\frac{\delta f_i}{\delta t}\right]_{\text{coll}} = \sum_j (\Gamma_{ij}^{+} - \Gamma_{ij}^{-}) \quad (1.2.28)$$

is called the collision term. In a later paragraph we shall find that it contains the unknown function f_i under the integral. The equation obtained is thus integro-differential and the gas mixture is described by a system of such equations ($i = 1, 2, \ldots$), mutually connected by collision terms of the type written in Eq. (1.2.28), each of them including all unknown functions f_j ($j = 1, 2, \ldots i \ldots$) under the integration sign.

In calculating Γ_{ij}^{-} we start from Fig. 3. We consider the i-type particle to be fixed and exposed to a flow of incident j-type particles of flow

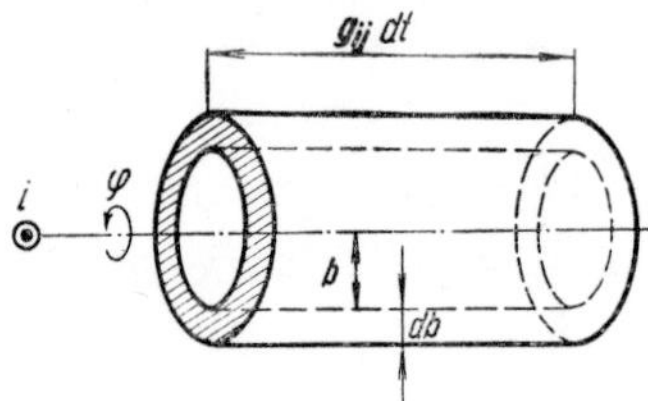

Fig. 3 Computing the collision term of Boltzmann's equation

density $f_j\mathbf{g}_{ij}$, where $\mathbf{g}_{ij}$ is the relative velocity of particles i and j. The i particle will in unit time obviously be in collision with all j particles contained in a cylinder of length $g_{ij} = |\mathbf{g}_{ij}|$ and of a radius given by the reach of the interparticle forces. The number of particles in collision with the ith particle within a time interval dt in an elementary coaxial shell of radius b (impact parameter) and thickness db equals

$$\int_{\varphi=0}^{\varphi=2\pi} f_j(\mathbf{r}, \mathbf{v}_j, t)\, g_{ij} b \, db \, d\varphi \, dt \, d\mathbf{v}_j .$$

The total number of collisions of i-type particles in an element $d\mathbf{r}\, d\mathbf{v}_i$ in a time interval dt is proportional to the density of ith particles in this element $f_i(\mathbf{r}, \mathbf{v}_i, t)$ and is given by the integral over all velocity values $\mathbf{v}_j$, parameters b and angles φ

$$\Gamma_{ij}^{-}\, dt = \int_{\varphi=0}^{\varphi=2\pi} \int_{b=0}^{b_{\max}} \int_{-\infty}^{\infty} f_j(\mathbf{r}, \mathbf{v}_j, t) f_i(\mathbf{r}, \mathbf{v}_i, t)\, g_{ij} b\, d\varphi\, db\, d\mathbf{v}_j\, dt, \tag{1.2.29}$$

$b_{\max}$ is the maximum value of the parameter b at which interaction between particles occurs. This value must be small in comparison with the mean free path (assumption of binary collisions).

The value Γ_{ij}^{-} (Eq. 1.2.29) can be written in the following lucid form:

$$\Gamma_{ij}^{-} = \iiint f_i f_j g_{ij}\, b\, db\, d\varphi\, d\mathbf{v}_j. \tag{1.2.30}$$

In a similar way we can evaluate the term Γ_{ij}^{+}. Using the laws of symmetry peculiar to equations that describe the collision of particles, we can write the following formula for the collision term (given a spherically symmetric potential of the interaction of the particles):

$$\left[\frac{\delta f_i}{\delta t}\right]_{coll} = \sum_j \iiint (f_i' f_j' - f_i f_j)\, g_{ij} b\, db\, d\varphi\, d\mathbf{v}_j \tag{1.2.31}$$

[Ref. 1.2.8], where the velocities $\mathbf{v}_i'$, $\mathbf{v}_j'$ in the arguments of the functions $f_i'(\mathbf{r}, \mathbf{v}_i', t) f_j'(\mathbf{r}, \mathbf{v}_j', t)$ are functions of the velocities $\mathbf{v}_i$, $\mathbf{v}_{jn}$; the form of these functions $\mathbf{v}_i' = \mathbf{v}_i'(\mathbf{v}_i, \mathbf{v}_j)$, $\mathbf{v}_j'(\mathbf{v}_i, \mathbf{v}_j)$ is given by the equations of conservation of momentum and energy in the collision. The summation in Eq. (1.2.31) is made over all kinds of j particles including the case $j = i$.

1.2.2.3 Solving the Boltzmann equations

It is evident at first sight that the Boltzmann equation

$$\frac{\partial f_i}{\partial t} + \mathbf{v}_i . \operatorname{grad}_{\mathbf{r}} f_i + \frac{\mathbf{F}_i}{m_i} . \operatorname{grad}_{\mathbf{v}_i} f_i =$$

$$= \sum_j \iiint (f_i' f_j' - f_i f_j)\, g_{ij} b\, db\, d\varphi\, d\mathbf{v}_j \tag{1.2.32}$$

is by no means easy to solve. The chief difficulties are due to the fact that the integral operation carried out on the right-hand side of Eq. (1.2.32) is not linear (products $f_i f_j$, $f'_i f'_j$), hence the principle of superposition is not applicable. The general solution of Eq. (1.2.32) – usually carried out by the Chapman-Enskog method of successive approximations (for instance in [Refs. 1.2.7, 1.2.8]) or Grad's method of moments [Ref. 1.2.10] – is very complex and goes beyond the scope of this book. Here we shall content ourselves with a few remarks concerning the Chapman-Enskog procedure.

At thermodynamic equilibrium $\dfrac{\partial f_i}{\partial t} = 0$; $\mathrm{grad}_{\mathbf{r}} f_i = 0$, and with the forces $\mathbf{F}_i$ equalling zero, the state of the gas is only determined by the collision of particles. Under these conditions, the Boltzmann equation has the following form:

$$\sum_j \iiint (f'_i f'_j - f_i f_j)\, g_{ij} b \, \mathrm{d}b \, \mathrm{d}\varphi \, \mathrm{d}\mathbf{v}_j = 0. \tag{1.2.33}$$

In this case Maxwell's distribution function $f_i^{(0)}$ (Eq. 1.2.34) is the general solution of the equation, as can be proved using for instance Boltzmann's H-function,

$$f_i^{(0)} = n_i \left(\frac{m_i}{2\pi k T} \right)^{3/2} \mathrm{e}^{-m_i V_i^2 / 2kT}, \tag{1.2.34}$$

where $V_i = | \mathbf{v}_i - \mathbf{v}_0 |$ is the absolute value of the thermal velocity of particle i.

Let us now assume local thermodynamic equilibrium, that is a state in which the parameters n, $\mathbf{v}_0$ and T as encountered in the distribution Eq. (1.2.34) are functions of position and time. This is the instance when the distribution function will be "nearly Maxwellian", meaning that the effect of non-homogeneities and external forces will only be evident as slight perturbation $\Phi_i(\mathbf{r}, \mathbf{v}_i, t)$ of the function $f_i^{(0)}$. Then we can write for the distribution function f_i (as a first approximation)

$$f_i(\mathbf{r}, \mathbf{v}_i, t) = f_i^{(0)}(\mathbf{r}, \mathbf{v}_i, t) + \Phi_i f_i^{(0)}(\mathbf{r}, \mathbf{v}_i, t). \tag{1.2.35}$$

In the same first approximation we then assume that the left side of the Boltzmann equation is determined by the Maxwell function $f_i^{(0)}(\mathbf{r}, \mathbf{v}_i, t)$

only. On this assumption we derive the perturbation function $\Phi_i(\mathbf{r}, \mathbf{v}_i, t)$ from the following equation [cf. Ref. 1.2.8]:

$$\frac{\partial f_i^{(0)}}{\partial t} + \mathbf{v}_i \,.\, \text{grad}_\mathbf{r}\, f_i^{(0)} + \frac{\mathbf{F}_i}{m_i}\, \text{grad}_{\mathbf{v}_i}\, f_i^{(0)} =$$

$$= \sum_j \iiint f_i^{(0)} f_j^{(0)} [\Phi_i' + \Phi_j' - \Phi_i - \Phi_j]\, g_{ij} b \,\mathrm{d}b\, \mathrm{d}\varphi\, \mathrm{d}\mathbf{v}_j \qquad (1.2.36)$$

This integral equation, in which $f_i^{(0)}$ is a known function of the coordinates, is linear with respect to the unknown perturbation functions Φ and (unlike Eq. 1.2.32) does not contain any members with partial derivatives such as $\frac{\partial \Phi}{\partial t}$; $\text{grad}_\mathbf{r}\, \Phi$, $\text{grad}_{\mathbf{v}_i}\, \Phi$. If we substitute from Eq. (1.2.34) into the left side of Eq. (1.2.36), we express it by the macroscopic functions $n_i(\mathbf{r}, t)$, $T(\mathbf{r}, t)$ and their derivatives. As we are primarily interested in the effect of the gradients $\text{grad}_\mathbf{r}\, n_i$, $\text{grad}_\mathbf{r}\, T$, $\text{grad}_\mathbf{r}\, p$, $\text{grad}_\mathbf{r}\, \mathbf{v}_0$ and the external forces $\mathbf{F}_i$ upon the disturbance Φ_i of the distribution function, we express the left-hand side of Eq. (1.2.36) by using the equations of conservation of mass, momentum and energy (cf. Sec. 1.2.3.2). This left-hand side is linear in the gradients and we find the solution $\Phi_i(\mathbf{r}, \mathbf{v}_i, t)$ in the form of a similar linear combination:*

$$\varphi_i = \mathbf{A}_i \,.\, \text{grad}_\mathbf{r}\, T + \mathbf{B}_i : \text{grad}_\mathbf{r}\, \mathbf{v}_0 + \mathbf{C}_i \,.\, \text{grad}_\mathbf{r}\, p + \mathbf{E}_i \,.\, \text{grad}_\mathbf{r}\, n_i + \mathbf{D}_i \,.\, \mathbf{F}_i, \qquad (1.2.37)$$

where the first, third, fourth and fifth members are scalar products of vectors, and the second member is a binary tensor product ($\text{grad}_\mathbf{r}\, \mathbf{v}_0$ is a 2nd order tensor). By substituting Eq. (1.2.37) into Eq. (1.2.36), with the left side also expressed as a linear combination of the same type as Eq. (1.2.37), and comparing the coefficients of corresponding gradients, we arrive at an integral equation for the functions **A**, **B**, **C**, **D**, **E**.

These equations are then solved by expansion, using spherical functions (Sonin's polynomials) which are particularly sui-

---

* The dependence of Φ_i on the gradients of the said quantities is, in fact, more complicated. Eq. (1.2.37) is no more than a symbolic notation indicating the causes for the disturbance Φ_i of the distribution function $f_i^{(0)}$.

table for describing anisotropies in the velocity space. Limiting the number of expansion members, we obtain more or less exact values of the functions $\mathbf{A}_i$, $\mathbf{B}_i$, $\mathbf{C}_i$, $\mathbf{D}_i$, $\mathbf{E}_i$ and thereby of the transport coefficients (i.e. coefficients of thermal conductivity, internal friction, diffusion, thermal diffusion, etc.). Since the integral on the right-hand side of Eq. (1.2.36) depends (via integration with respect to $b\,db$) on the detailed course of the interparticle collisions, we may express the values of the transport coefficients by the parameters describing the course of the collisions. This expression, though, is very involved and the clear physical concept is lost in it. In view of the very good agreement with experiments at normal and elevated temperatures, these exact kinetic methods may be applicable (with the aid of electronic computers) in high-temperature regions inaccessible to direct measurement.

1.2.2.4 Effects of charged particles

In Sec. 1.1.4.4 we gave due consideration to the equilibrium composition of a plasma by introducing into the Saha equations (1.1.79, 1.1.80) the ionization potential $\varepsilon_i^{\text{ion}}$ corrected by the value $\Delta\varepsilon_i^{\text{ion}}$ ensuing from Eq. (1.1.92). Now we have to express the (Coulombic) effect of charged particles on transport and relaxation phenomena in the plasma. In view of the slow decrease of interaction forces with increasing distance $\left(F \sim -\dfrac{1}{r^2}\right)$, we can no longer consider the collisions as binary; for a charged test particle selected at random is at every instant in interaction with a large number of other particles (or theoretically with all of them). In Debye's approximation (Sec. 1.1.4.4) this particle experiences the effects of all other charged particles surrounding it in a sphere of the Debye radius D.

Let us limit our discussion to the case of a fully ionized plasma consisting of a single type of positive ions and electrons of equal concentration (quasi-neutrality: $n_i \sim n_e$). The electron and ion gases will then be described by the distribution functions $f_i(\mathbf{r}, \mathbf{v}_i \,.\, t)$ and $f_e(\mathbf{r}, \mathbf{v}_e, t)$ the indices e and i standing for electrons and positive ions. In the Boltzmann equation for the partition function of electrons

$$\frac{\partial f_e}{\partial t} + \mathbf{v}_e \,.\, \text{grad}_{\mathbf{r}} f_e + \frac{\mathbf{F}_e}{m_e} \,.\, \text{grad}_{\mathbf{v}_e} f_e = \left[\frac{\delta f_e}{\delta t}\right]_{\text{coll}}, \qquad (1.2.38)$$

the collision term will (even at comparatively low densities n_e) no longer have the form given by Eq. (1.2.31) for binary collisions. The particles found in the element d$\mathbf{v}$ of the velocity space enter or leave this element owing to weak very frequent "impacts" of particles present in a sphere of radius D (the radius referring of course to the coordinate space). These impacts cause the representative points of the particles in the velocity space to execute a specific Brownian motion; this motion leads to the diffusion of the particles in the velocity space and makes a considerable contribution to the total entry–vs.–exit balance of particles in the elementary range d$\mathbf{r}$ d$\mathbf{v}$. Applying the stochastic theory of the Brownian motion to the diffusion of the representative points in the velocity space, we arrive at the Focker–Planck equation for the distribution function f_e [cf. e.g. 1.2.4, 1.2.11]

$$\frac{\partial f_e}{\partial t} + \mathbf{v}_e \cdot \mathrm{grad}_{\mathbf{r}} f_e + \frac{\mathbf{F}_e}{m_e} \cdot \mathrm{grad}_{\mathbf{v}_e} f_e = -(\mathrm{grad}_{\mathbf{v}_e} \cdot f_e \langle \Delta \mathbf{v}_e \rangle) + + \frac{1}{2}(\mathrm{grad}_{\mathbf{v}_e}\, \mathrm{grad}_{\mathbf{v}_e} : f_e \langle \Delta \mathbf{v}_e\, \Delta \mathbf{v}_e \rangle), \tag{1.2.39}$$

where statistical dispersions of components of the vector $\mathbf{v}_e$ are the components of vector $\langle \Delta \mathbf{v}_e \rangle$, and statistical dispersions of the tensor $\mathbf{v}_e \mathbf{v}_e$ are components of the second-order tensor $\langle \Delta \mathbf{v}_e\, \Delta \mathbf{v}_e \rangle$.

The symbol: stands for the binary product of the tensors.

The right-hand side of Eq. (1.2.39) is also non-linear with respect to the function f_e (the expressions $\langle \Delta \mathbf{v}_e \rangle$ $\langle \Delta \mathbf{v}_e\, \Delta \mathbf{v}_e \rangle$ contain the unknown function f_e under the integration sign). Finding a general solution for a system of independent equations of type (1.2.39) is almost hopeless; in certain simple cases, however, Eq. (1.2.39) can be solved and brings valuable results (e.g. the temperature dependence of the electrical conductivity of a fully ionized gas).

In conclusion, let us note a few parameters used for the classification of different types of plasma:

1. The mean distance h between particles is obtained from condition $nh^3 = 1$:

$$h = n^{-1/3}. \tag{1.2.40}$$

2. Debye's length D indicates the separation over which the Coulombic field of a positive ion is screened off by a polarized plasma.

$$D = \sqrt{\frac{\varepsilon_0 kT}{4\pi n_e e^2}} . \tag{1.2.41}$$

Unlike in Eq. (1.1.91) we have neglected the screening effect of the ions, which – because of their large mass – cannot become effective during the short period of the collision.

3. The critical impact parameter $\bar{b}_0$ indicates such a parameter in Fig. 1.2.1 which causes an electron approaching with thermal velocity to deviate by 90°; the target particle is assumed to be at rest. Collisions with parameter $b \leqq \bar{b}_0$ are called short-range, collisions with $b > \bar{b}_0$ – long-range. Short-range collisions involve a collision term of Boltzmann type, and where they are the only ones to be considered, we often can assume the collisions to be binary. Long-range collisions involve collision members of Focker-Planck type. The theory of Coulombic collisions gives the following value for the critical impact parameter:

$$b_0 = \frac{e^2}{3\varepsilon_0 kT} . \tag{1.2.42}$$

4. To the critical parameter $\bar{b}_0$ corresponds the mean free path $\bar{l}_{90°}$, after traversing which the electron is deflected by more than 90°.

$$\bar{l}_{90°} = \frac{1}{\sqrt{2}n\pi\bar{b}_0^2} . \tag{1.2.43}$$

1.2.3 Fundamental Concepts of the Phenomenological Thermodynamics of Irreversible Processes

1.2.3.1 A macroscopic plasma model

One of the greatest difficulties encountered in the physical description of non-equilibrium ionized and dissociated gas mixtures is the great number of processes contributory to this state (such as diffusion, heat conduction, macroscopic motion of the plasma as a whole, chemical reactions). These processes affect each other to such an extent that it is not clear, at first sight, which of them are the causes and which the effects. Hence the urgent need for a uniform classification of the macroscopic processes occurring in a plasma and of their mutual relationships. This is the subject of the branch of physics called phenomenological thermodynamics of irreversible processes. The term "phenomenological" stresses the fact that this branch is concerned with the formal description of phenomena without considering their microscopic mechanism.

Thermodynamics of irreversible processes is based on broad general assumptions; its methods are applicable to all processes in which the volume elements of the medium to be examined can be considered to be thermodynamic systems capable of exchanging some physical quantities (mass, heat, etc.) with neighbouring elements. Under this assumption, the thermodynamic state of every element is fully described for instance by the temperature T, pressure p and chemical potentials μ_i of all i-type components; the remaining thermodynamic quantities of the element can be calculated from the independent variables T, p, μ_i with the aid of known thermodynamic relationships. The thermodynamic state of the plasma as a whole is then described by quoting for every element of the medium the values of the macroscopic quantities as functions of the position $\mathbf{r}$ and the time t. We thus obtain what is called fields of macroscopic quantities.

Computations in the thermodynamics of irreversible processes follow then the formal pattern of the procedure outlined below:

1. We describe the physical quantities in space and time by expressing them in fields. These are of two types:
   a) fields of quantities of state: $T(\mathbf{r}, t)$, $p(\mathbf{r}, t)$ $\mu(\mathbf{r}, t)$, etc.
   b) fields of densities, for instance of density of mass $\varrho(\mathbf{r}, t)$, density of energy $\varrho u(\mathbf{r}, t)$, density of mass of the ith component $\varrho_i(\mathbf{r}, t)$, density of entropy $\varrho s(\mathbf{r}, t)$, where $u(\mathbf{r}, t)$ and $s(\mathbf{r}, t)$ are energy and entropy per kg.
2. With the aid of the quantities mentioned, we write the basic laws of conservation (of mass, momentum, total energy, entropy) and the continuity equation of the individual components as differential equations. These equations, however, contain – in addition to the functions mentioned in point 1 – new variables describing the transport

of quantities between adjoining elements; they are called the densities of macroscopic flows of these quantities (diffusion flow density, heat flow density, etc.).

3. We write – in a general form – the relationships between the densities of macroscopic fluxes and the "forces" which generate them. These relationships constitute a second set of equations known as the **phenomenological equations** of the thermodynamics of irreversible processes. Fourier's Law of Heat Conduction, Fick's Law of Diffusion, Newton's Law of Internal Friction are examples of such equations. In the phenomenological equations we encounter coefficients of proportionality, such as the coefficient of heat conduction, diffusion coefficient etc.
4. By comparison with kinetic theory we estimate the values of the phenomenological coefficients as functions of the quantities of state, and hence as functions of $\boldsymbol{r}$ and t (we reduce, for instance, the computation to a determination of effective cross-sections in the collision of two particles).
5. We solve the system of equations thus constructed, with due consideration to initial and boundary conditions. In view of the great number of equations, simplifying assumptions are made in their actual solution. A complete system of equations is used as starting point for establishing the physical similarity, which in turn is employed for a semi-empirical treatment of many actual physical systems.

1.2.3.2 Equations of conservation

Let us denote by ε any quantity (specific mass, energy, entropy, etc.) relating to a unit mass of plasma. If a mass element of plasma moves with mass-average velocity $\boldsymbol{v}_0(\boldsymbol{r}, t)$ (cf. Sec. 1.2.2.1) and is deformed accordingly, there are two possible causes for changes in the value of ε: the first is the transport of quantity ε through the surface confining the element, and the second the generation or decay of quantity ε owing to the processes occurring within the element.

Let the mass of the element of plasma be Δm and its volume ΔV. Then the total amount of quantity ε corresponding to Δm is $\mathscr{E} = \varepsilon\, \Delta m = \varepsilon \varrho\, \Delta V$, where ϱ is the density of the plasma. Since the equation of continuity $\mathrm{d}(\varrho V) = 0$ applies to the plasma volume, we may write as follows for the time variation of quantity $\mathscr{E}$ in flowing ($\mathrm{d}\mathscr{E} = \mathrm{d}\varepsilon\varrho\,\Delta V$):

$$\varrho\frac{\mathrm{d}\varepsilon}{\mathrm{d}t}\Delta V = -S_\varepsilon + \sigma_\varepsilon\,\Delta V, \tag{1.2.44}$$

S_ε is the total flow of the quantity through the limiting surface of the element, and σ_ε is our symbol for the production of quantity ε (the amount of quantity ε generated in 1 m^3 in 1 sec). If we divide the whole equation by ΔV, let $\Delta V \to 0$, and keep in mind that $\lim\limits_{\Delta V\to 0}\frac{S_\varepsilon}{\Delta V}$ defines the divergence of the flow density $\boldsymbol{J}_\varepsilon$ of quantity ε, then we obtain Eq. (1.2.44) in the differential form which is suitable for calculating the space- and time-dependence of $\varepsilon = \varepsilon(\boldsymbol{r}, t)$

$$\varrho\frac{\mathrm{d}\varepsilon}{\mathrm{d}t} = -\operatorname{div}\boldsymbol{J}_\varepsilon + \sigma_\varepsilon. \tag{1.2.45}$$

We can then split the total derivative on the right-hand side of the equation into a local part $\frac{\partial\varepsilon}{\partial t}$ and a convective part $\boldsymbol{v}_0$ grad ε:

$$\frac{\mathrm{d}\varepsilon}{\mathrm{d}t} = \frac{\partial\varepsilon}{\partial t} + \boldsymbol{v}_0\,.\,\operatorname{grad}\varepsilon. \tag{1.2.46}$$

All equations of conservation can be written in this form. Here we quote only some of the most frequent forms obtained by the transformation of Eq. (1.2.45) (for such transformations cf. e.g. Ref. [1.2.6]).

(a) **Partial density of the ith component:** $\varepsilon = \frac{\varrho_i}{\varrho} = \gamma_i$

$$\varrho\frac{\mathrm{d}\gamma_i}{\mathrm{d}t} = -\operatorname{div}\boldsymbol{J}_i + \Gamma_i, \tag{1.2.47}$$

where $\boldsymbol{J}_i$ is the diffusion flow of the ith component according to Eq. (1.2.17), and Γ_i indicates the production of the concentration of the ith component due to chemical reactions. Since the total mass always remains constant (and considering Eqs. 1.2.17, 1.2.12),

$$\sum_{i=1}^{i=n}\Gamma_i = 0, \qquad \sum_{i=1}^{i=n}\boldsymbol{J}_i = 0. \tag{1.2.48}$$

(b) **Momentum:** $\mathbf{v}_0$ (momentum of unit mass)

$$\varrho \frac{d\mathbf{v}_0}{dt} = -\mathrm{Div}\,\mathsf{P} + \sum \varrho_i \mathbf{F}_i \qquad (1.2.49)$$

where P is the pressure tensor as defined by Eq. (1.2.21). Div denotes the divergence of a tensor; in a non-viscous liquid $\mathrm{Div}\,\mathsf{P} = \mathrm{grad}\,p$, p standing for the isotropic hydrostatic pressure. $\mathbf{F}_i$ is an external force affecting one kilogram of the ith component. According to Eq. (1.2.49), the time change of the mass-average velocity is determined by the resultant of external and pressure forces acting on the element of plasma. Thus Eq. (1.2.49) is the equation of motion of the plasma element (Navier-Stokes equation).

(c) **Energy:** $\varepsilon = u + \frac{1}{2} v_0^2$ (where u is the internal energy per unit mass of the plasma):

$$\varrho \frac{d}{dt}\left(\frac{1}{2} v_0^2 + u\right) = -\mathrm{div}\,[\mathsf{P} \cdot \mathbf{v}_0 + \mathbf{Q}] + \left(\sum \varrho_i \mathbf{F}_i\right) \cdot \mathbf{v}_0 + \sum (\mathbf{F}_i \cdot \mathbf{J}_i). \qquad (1.2.50)$$

$\mathbf{Q}$ is the density of heat flow as defined in Eq. (1.2.24) or (1.2.47). The expression $\mathsf{P} \cdot \mathbf{v}_0$ describes the energy transfer due to pressure and tension work. The member $(\Sigma\, \varrho \mathbf{F}_i) \cdot \mathbf{v}_0$ indicates the production of specific kinetic energy $\frac{1}{2} v_0^2$ of the plasma, and the last member – the production of internal energy u, originating from the work of external forces applied to diffusion flows (e.g. ohmic heating).

Similarly we could write the equations of conservation for the electric charge, the internal energy of the plasma, electromagnetic forms of energy [cf. e.g. Ref. 1.2.12], etc. More particulars may be found in Refs. [1.2.1, 1.2.6, 1.2.8]. All balance equations can also take what is called their local form, in which the total derivative is written as in Eq. (1.2.46) and the convection member $\mathbf{v}_0\,\mathrm{grad}\,\varepsilon$ is combined with the other members of the right-hand side. The local form of the density balance, for instance, is the following well-known equation:

$$\frac{\partial \varrho}{\partial t} = -\mathrm{div}\,(\varrho \mathbf{v}_0). \qquad (1.2.51)$$

1.2.3.3 Entropy balance

The entropy S is a convenient yardstick for the quantitative description of irreversible processes such as heat conduction, diffusion, internal friction.

Every physical state of a plasma – i.e. every configuration of positions, momenta and internal energies of its constituent particles – has at a given volume V, given internal energy U and given masses m_i of the individual components, a certain probability of occurrence w, its maximum w_0 corresponding to the equilibrium configuration.

In principle, the probabilities w, w_0 can be computed by methods of statistical physics. If we construct a quantity

$$S = k \ln w, \qquad S_0 = k \ln w_0 \tag{1.2.52}$$

where k is the Boltzmann constant, this statistical quantity proves to have all the properties of a macroscopic quantity – entropy. Thus for the variation dS_0 of the equilibrium entropy as a function of (U, V, m_i), the following well-known (Gibbs) relation is valid:

$$dS_0 = \frac{1}{T} dU + \frac{p}{T} dV - \sum \frac{\mu_i}{T} dm_i, \tag{1.2.53}$$

where T is the absolute temperature, $p = \Sigma p_i$ the total pressure, and n_i the chemical potential of the ith component.

The natural tendency of the system to attain the state of the most probable distribution of positions, momenta and internal energies of particles (i.e. equilibrium distribution) causes the probability w and hence (cf. Eq. 1.2.52) the entropy S to increase. Irreversible processes such as diffusion or heat conduction are the macroscopic expression of this tendency.

The entropy of a plasma element can change on account of two reasons: it can flow in or out through the surface limiting the element, or it can arise as the product of irreversible processes going on in the plasma. By substituting the specific entropy s_0 (entropy per unit mass) for ε in the balance equation (1.2.45), i.e. $\varepsilon = s_0$, and denoting the production of entropy due to irreversible processes by σ_{s_0}, we obtain an entropy balance of the following form:

$$\varrho \frac{\mathrm{d}s_0}{\mathrm{d}t} = -\mathrm{div}\, \boldsymbol{J}_{s_0} + \sigma_{s_0}, \tag{1.2.54}$$

where $\boldsymbol{J}_{s_0}$ stands for the density of the entropy flow.

If we substitute in Eq. (1.2.54) – which is written for the specific quantities u, v, γ_i – from Eq. (1.2.53) for $\mathrm{d}s_0$, and from the balance equations for the specific internal energy u, for the concentration of the ith component γ_i (cf. Eq. 1.2.47), and for the specific volume $v = \frac{1}{\varrho}$, then we obtain [Ref. 1.2.6]:

$$\boldsymbol{J}_{s_0} = \frac{1}{T}\left(\mathbf{Q} - \sum_{i=1}^{n} \mu_i \boldsymbol{J}_i\right), \tag{1.2.55}$$

$$\sigma_{s_0} = \mathbf{Q} \cdot \mathrm{grad}\,\frac{1}{T} + \sum_{i=1}^{n} \boldsymbol{J}_i \cdot \left(\frac{\boldsymbol{F}_i}{T} - \mathrm{grad}\,\frac{\mu_i}{T}\right) - $$
$$- (\mathsf{P} - p\mathsf{I}) : \frac{1}{T}\,\mathrm{Grad}\,\boldsymbol{v}_0 - \sum_{i=1}^{n} \Gamma_i \frac{\mu_i}{T}. \tag{1.2.56}$$

In Eqs. (1.2.55), (1.2.56) $\mathbf{Q}$ stands for the density of heat flow (Eq. 1.2.24), $\boldsymbol{J}_i$ is the diffusion flow of the ith component, $\boldsymbol{F}_i$ the external force acting on the ith component, P the pressure tensor (Eq. 1.2.21), p the pressure, I the unit tensor, Grad $\boldsymbol{v}_0$ the gradient (2nd-order tensor) of the vector of mass-average velocity, Γ_i the production of the ith component due to chemical reactions (according to Eq. 1.2.47), and the sign : stands for the binary tensor product. It can be proved that in accordance with the law of increasing entropy in irreversible processes, the expression for σ_{s_0} has never a negative value. Moreover, Eq. (1.2.56) is the starting point for the classification of irreversible processes as mentioned in the introductory paragraph of Chap. 1.2.3. For if we separate the products of tensorial and vectorial quantities indicated in Eq. (1.2.56) into scalar components, the production of entropy can be written in the following form:

$$\sigma_{s_0} = \sum_{j=1}^{m} Y_j X_j, \tag{1.2.57}$$

where X_j (called the generalized thermodynamic forces) are components of the quantities

$$\operatorname{grad}\frac{1}{T}, \quad \frac{\mathbf{F}_i}{T}, \quad -\operatorname{grad}\frac{\mu_i}{T}, \quad -\frac{1}{T}\operatorname{Grad}\mathbf{v}_0, \quad -\frac{\mu_i}{T},$$

and Y_j (the "scalar fluxes") are components of the quantities

$$\mathbf{Q}, \mathbf{J}_i, \mathsf{P} - p\mathsf{I}, \Gamma_i.$$

We know from experience that a thermal gradient causes heat flux, a gradient of the concentration of the ith component causes diffusion flux of this component, an electric field causes electric current, – generally speaking: every force X_j in Eq. (1.2.57) causes the corresponding scalar flux Y_j. The force X_j may, in fact, also cause fluxes Y_k corresponding to other forces X_k. The thermal gradient, for instance causes diffusion fluxes (thermal diffusion) in addition to thermal flow. The relationships between fluxes and forces are linear in the first approximation (to which the thermodynamics of irreversible processes restricts itself) and can be written in a general form (the "phenomenological equation"):

$$Y_j = \sum_{k=1}^{m} L_{jk} X_k; \tag{1.2.58}$$

in this formula L_{jk} are the coefficients of proportionality which are closely related with the coefficient of thermal conductivity, with electrical conductivity, viscosity and other more common quantities. These coefficients are the general functions of variables – particularly temperature – determining the local thermodynamic equilibrium of the plasma.

Depending on the behaviour of the quantities Y_j, X_j with respect to the transformation of time $t \to -t$ and on their tensorial character (scalar, vector, second-order tensor), some of the coefficients L_{jk} may vanish ($L_{jk} = 0$, whereupon Y_j does not depend on X_k), some may be symmetric ($L_{jk} = L_{kj}$; Onsager coefficients), some antisymmetric ($L_{jk} = -L_{kj}$; Casimir coefficients). Macroscopic thermodynamics of irreversible processes does not offer the numerical values of coefficients L_{jk}, which must be obtained by computations based on the kinetic theory of gases. Some partial results of such computations will be quoted in the subsequent sections of this Chapter.

Eq. (1.2.58) together with balance equations of the type of Eqs. (1.2.47), (1.2.49), (1.2.50), (1.2.51) and with initial and boundary conditions permit the fields of macroscopic quantities in the non-equilibrium mixture of reacting gases to be computed. A complete system of equations is found in Ref. [1.2.6]; Refs. [1.2.1, 1.2.13, 1.2.32] deal with the application of thermodynamics of irreversible processes to high-pressure plasmas.

1.2.3.4 Plasma state and thermodynamics of irreversible processes

Let us now note some of the peculiarities of phenomenological thermodynamics of irreversible processes as applied to ionized gas.

The equation of conservation (Eq. 1.2.45) does not change in form. The equations of conservation written for charged particles occupy an important place among the balance equations. The diffusion fluxes J_{ion}, J_{el} describe the transport of the electric charge (electric current), Γ_{ion}, Γ_{el} are the characteristics of volume ionization and recombination. External forces in Eq. (1.2.49) include – in addition to gravitation (neglected under laboratory conditions) and the centrifugal force – also the electric force acting on the space charge (usually neglected in view of the condition of quasi-neutrality $n_+ \sim n_-$), and especially the magnetic force affecting the plasma carrying electric current in a magnetic field. An important member in the expression for the production of internal energy in the energy balance equation (1.2.50) is the term describing the ohmic heating of the plasma by an electric current. In instances of high conductivity and strong electromagnetic fields, the energy of the electric and magnetic fields [for equation of conservation cf. Ref. 1.2.12] also participates in the energy balance (and to a lesser extent in the momentum balance).

The balance of entropy (Eq. 1.2.54) retains its form, as do Eqs. (1.2.55), (1.2.56). In the case of plasmas moving in electromagnetic fields, a member describing the dissipation of electromagnetic energy may be expected to appear in Eq. (1.2.56) for the production of entropy. Since a plasma in a magnetic field is an anisotropic medium, the coefficients L_{jk} in Eq. (1.2.58) are tensors and their symmetric properties require therefore closer examination. This applies especially to electric conductivity which has recently been studied from this point of view [Ref. 1.2.14].

Let us add that thermodynamics of irreversible processes can not always be employed to describe a highly ionized gas which deviates to a marked extent from the ideal gas. In some cases an element of such a plasma cannot be considered a thermodynamic system, in other cases the relationships of type Eq. (1.2.58) are not linear.

1.2.4 Conduction of Electricity in a Plasma

1.2.4.1 The electric conductivity of plasma

Owing to the electric field and some other causes (such as the gradients of pressure or temperature), the transport of electric charges in the plasma may be directed so as to produce an electric current. By the density $\boldsymbol{j}$ of an electric current we describe the quantity

$$\boldsymbol{j} = \sum_k z_k n_k \overline{\boldsymbol{V}}_k, \tag{1.2.59}$$

where $z_k, n_k, \overline{\boldsymbol{V}}_k$ are the charge, density and diffusion velocity (Eq. 1.2.13) of a k-type ion. The summation in Eq. (1.2.59) is carried out over all types of ions in the plasma. If only neutral atoms and positive and negative ions are present, which carry elementary charges $\pm e$ and have approximately the same densities $n_+ \cong n_- \cong n$ (quasi-neutrality), then Eq. (1.2.59) assumes the form

$$\boldsymbol{j} = ne(\overline{\boldsymbol{V}}_+ - \overline{\boldsymbol{V}}_-). \tag{1.2.60}$$

Various theories of electric conductivity differ in their ways of calculating the velocities $\overline{\boldsymbol{V}}_+$, $\overline{\boldsymbol{V}}_-$. In this section we quote the simplest among them which is based upon the concept of the mean free path. According to this theory [cf. Ref. 1.2.15] the electron velocity $\overline{\boldsymbol{V}}_e$ in an electric field $\boldsymbol{E}$ is established in such a manner that the energy acquired by an electron in the electric field is compensated for by the average energy loss in elastic collisions of electrons with other particles. If we assume that in the collision the electron loses its total momentum $m_e \overline{\boldsymbol{V}}_e$

and if we denote by τ the time corresponding to the mean free path (i.e. τ^{-1} is the number of collisions of the electron per second), then we have

$$e\mathbf{E} = m_e \overline{\mathbf{V}}_e \tau^{-1}. \qquad (1.2.61)$$

Hence follows

$$\overline{\mathbf{V}}_e = \mu_e \mathbf{E} = \frac{e\tau}{m_e} \mathbf{E}, \qquad (1.2.62)$$

where μ_e is the diffusion velocity of the electrons in the unit field $\mathbf{E}$, and is called electron mobility.

If we know the mobilities μ_+ and μ_- of positive and negative particles (μ_+ and μ_- are usually found by experiment), we can rewrite Eq. (1.2.60) for the current density in the following form:

$$\mathbf{j} = ne(\mu_+ - \mu_-)\,\mathbf{E}. \qquad (1.2.63)$$

In order to calculate the mobility μ_+ or μ_- from Eq. (1.2.62) or from a similar formula written for positive or negative ions, we have first to determine the mean time τ between two collisions. As the directed diffusion velocity in weak fields $\mathbf{E}$ is always far less than the velocity of random motion,—and Eq. (1.2.62) is valid for weak fields only—we may write:

$$\tau_j = \frac{\overline{l}_j}{\overline{c}_j} = \frac{1}{\overline{c}_j} \frac{1}{\sum\limits_k n_k q_{jk}}; \qquad (1.2.64)$$

in this equation τ_j, $\overline{l}_j$, $\overline{c}_j$, q_{jk}, n_k are the mean time between collisions, the mean free path, the mean thermal velocity $\overline{c}_j = \sqrt{\frac{8kT_j}{\pi m_j}}$, the total effective cross section for the collision of j-type particles with k-type particles, and the density of k-type particles, respectively.

Since the mobility μ_e of electrons greatly exceeds the mobility of positive or negative ions, Eq. (1.2.63) is, as a rule, (if $n_e \gtrsim n_-$) written in the form

$$\mathbf{j} = \sigma_e \mathbf{E} = ne\mu_e \mathbf{E}, \qquad (1.2.65)$$

where σ_e is electric conductivity of the plasma caused by electrons only.

Using Eqs. (1.2.65), and (1.2.62), we can write for the conductivity σ_e:

$$\sigma_e = \frac{e^2 n_e}{\sqrt{\frac{8mkT_e}{\pi}}} \frac{1}{\sum_k n_k q_{ek}}; \tag{1.2.66}$$

the interactions of electrons both with neutral atoms and with positive and negative ions have to be included in the summation in the denominator. The magnitude of the effective cross-section for the interaction between electrons and neutral particles is found by experiment or by quantum-mechanical computation. For the value of the effective cross-section q_{ei} of the collision between an electron and a positive ion of charge e, we use what is called the Gvozdover cross-section [Refs. 1.2.1, 1.2.16]

$$q_{ei} = \frac{e^4}{(kT)^2} \ln\left(\frac{kT}{e^2 n_i^{1/3}}\right), \tag{1.2.67}$$

where n_i is the density of positive ions. For the mean effective cross-section of the interaction of electrons with a mixture of once to z times ionized ions, the expressions on the right-hand side of Eq. (1.2.67) have to be multiplied by $(\bar{z})^2$, the square of the effective ionic charge defined by the relation

$$\bar{z} = \frac{\sum n_z z^2}{n_e} = \frac{\sum n_z z^2}{\sum n_z z}, \tag{1.2.68}$$

where n_z is the density of z times ionized ions.

The derivation of Eq. (1.2.66) is based on the balance equation (1.2.61). This equation is a good approximation in cases where the field does not much change along the mean free path (analogy with Meixner's condition Eq. 1.2.8). Where the field does not fulfil this condition, we can—instead of using the balance equation (1.2.61)—make the assumption that on the average the electron loses a $\varkappa^{\text{th}}$ part of its mean energy, in one collision:

$$\frac{1}{2} \varkappa m_e (\bar{c}_e)^2 \cdot \frac{\bar{c}_e}{\bar{l}_e} = eE\bar{V}_e . \tag{1.2.69}$$

On this condition conductivity σ_e is no longer a constant independent of the electric field E, as it is in Eq. (1.2.66), but a function of $E^{1/2}$ [Ref. 1.2.15]:

$$\sigma_e \sim n_e \left(\frac{\varkappa}{2}\right)^{1/4} \left(\frac{e}{m_e}\right)^{1/2} \left(\frac{E}{p}\right)^{1/2}, \tag{1.2.70}$$

where p is the pressure in the plasma $\left(\bar{l}_e \sim \frac{1}{p}\right)$.

In practice, the best starting point for calculating electric conductivity is the formula

$$\sigma_e = e n_e \mu_e, \tag{1.2.71}$$

where for μ_e we substitute mobilities measured by experiment, which are mostly quoted as functions of $\frac{E}{p}$. The electron density is usually computed from Saha's equations written for the given local temperature, with due corrections for non-equilibrium ionization and recombination (including the formation and decay of negative ions).

For the precise calculation of the electric conductivity of a plasma, we have to start from Boltzmann's or Focker-Planck's kinetic equations. This applies particularly to plasmas with magnetic or A.C. electric fields or with both at a time.

1.2.4.2 Electric conductivity of plasmas in magnetic and A. C. fields

Let us now assume that a homogeneous weakly ionized plasma at rest is placed in a magnetic field of induction $\mathbf{B}$, and an A.C. electric field $\mathbf{E} = \mathbf{E}_0\, e^{i\omega t}$ is superimposed on it. Boltzmann's equation for the distribution function $f_e(\mathbf{v})$ of the electron gas will then have the following form:

$$\frac{\partial f_e}{\partial t} - \frac{e}{m_e}(\mathbf{E} + \mathbf{v} \times \mathbf{B}) \,.\, \mathrm{grad}_{\mathbf{v}} f_e = \left[\frac{\delta f_e}{\delta t}\right]_{\mathrm{coll.}} \tag{1.2.72}$$

The collision member $\left[\frac{\delta f_e}{\delta t}\right]_{\mathrm{coll.}}$ is described by Eq. (1.2.31) which in the case of electrons in collision with heavy neutral particles $\frac{m_e}{m_a} = 0$

(and neglecting collisions between charged particles) has the form:

$$\left[\frac{\delta f_e}{\delta t}\right]_{\text{coll.}} = n_a v \int_0^{\infty}\int_0^{2\pi} (f'_e - f_e)\, b\, db\, d\varphi; \qquad (1.2.73)$$

in this equation n_a is the density of neutral particles; $\boldsymbol{v}$ the electron velocity in relation to neutral particles at rest; f'_e and f_e are the distribution functions of electrons according to velocities before and after the collision; b is the variable value of the collision parameter; and φ the azimuth angle.

The current density $\boldsymbol{j}$ is defined by the formula:

$$\boldsymbol{j} = n_e e \overline{\boldsymbol{V}}_e, \qquad (1.2.74)$$

where the drift velocity $\overline{\boldsymbol{V}}_e$

$$\overline{\boldsymbol{V}}_e = \frac{1}{n_e}\int \boldsymbol{v}_e f_e(\boldsymbol{v}_e)\, d\boldsymbol{v}_e \qquad (1.2.75)$$

is related to the electron distribution function by Eqs. (1.2.11) and (1.2.13) (for plasma at rest, $\boldsymbol{v}_0 = 0$).

Equation (1.2.72) with the collision term in the form of Eq. (1.2.73) is solved by developing the function f_e into the series using spherical functions:

$$f_e = f_e^{(0)} + \sum_{l,m} \{\alpha_{lm} v_e^l P_{lm}(\Theta) \cos m\varphi + \beta_{lm} v_e^l P_{lm}(\Theta) \sin m\varphi\}; \qquad (1.2.76)$$

$P_{lm}(\Theta)$ are Legendre's polynomials. The coefficients α_{lm}, β_{lm} depend on time via $e^{i\omega t}$.

If in expanding Eq. (1.2.76) we restrict ourselves to the members $l = 0$, $l = 1$, we obtain the isotropic (in the velocity space) part of the distribution function $f_e^{(0)}$ (Maxwell's distribution) with superimposed first-order anisotropies due to the directed electron motion. The spherical functions can be proved [Ref. 1.2.4] to be eigenfunctions of the integral operator (Eq. 1.2.73), hence the integrodifferential equation (1.2.72) can be reduced to differential equations for α_{lm}, β_{lm}.

Moreover, Eq. (1.2.75) for the directed electron velocity can be shown [Ref. 1.2.4] to have the following form:

$$\bar{\mathbf{V}}_e = \frac{4\pi}{3}\frac{1}{n_e}\int_0^{\infty}(\alpha_{11}\mathbf{i} + \beta_{11}\mathbf{j} + \alpha_{10}\mathbf{k})\, v_e^4\, d\mathbf{v}_e \tag{1.2.77}$$

where **i**, **j**, **k** are unit vectors in the direction of the coordinate axes of the velocity space (**k** ∥ **B**). Drift velocity $\bar{\mathbf{V}}_e$ thus computed proves not to be collinear with **E**, and if we want to keep Ohm's Law valid in the form

$$\mathbf{j}_e = n_e e \bar{\mathbf{V}}_e = \sigma_e \mathbf{E}, \tag{1.2.78}$$

conductivity σ_e has to be an antisymmetric 2nd-order tensor [Ref. 1.2.4]

$$\sigma_e = \begin{vmatrix} \sigma_1 & \sigma_2 & 0 \\ -\sigma_2 & \sigma_1 & 0 \\ 0 & 0 & \sigma_0 \end{vmatrix}, \tag{1.2.79}$$

where

$$\sigma_0 = \frac{4\pi e^2}{3m_e}\int_{v=0}^{v=\infty} f_e^{(0)} \frac{d}{dv}\left(\frac{1}{j\omega + \nu_1}\right) dv, \tag{1.2.80}$$

$$\sigma_1 = \frac{4}{3}\frac{\pi e^2}{m_e}\int_{v=0}^{v=\infty} f_e^{(0)} \frac{d}{dv}\left[\frac{j\omega + \nu_1}{(j\omega + \nu_1)^2 + \omega_b^2}\right] dv, \tag{1.2.81}$$

$$\sigma_2 = \frac{4}{3}\frac{\pi e^2}{m_e}\int_{v=0}^{v=\infty} f_e^{(0)} \frac{d}{dv}\left[\frac{\omega_b}{(j\omega + \nu_1)^2 + \omega_b^2}\right] dv. \tag{1.2.82}$$

In Eqs. (1.2.80) to (1.2.82) $f_e^{(0)}$ is the isotropic (Maxwellian) part of the distribution function, which is fully determined by the electron temperature T_e (Eq. 1.2.34); ω_b is the cyclotron frequency $\omega_b = \frac{eB}{m_e}$, ω the frequency of the superimposed electric field, and ν_1 the number of electron collisions per second:

$$\nu_1 = 2\pi n_a v_e \int_0^{\infty} (1 - \cos\chi)\, b\, db, \tag{1.2.83}$$

χ being the angle by which the electron deviates from its original direction owing to a collision of parameter b. Though the magnitude of ν_1 depends on the electron velocity v_e (in Ref. [1.2.17] this dependance is specified for some gases of technical importance), in the first approximation we often consider ν_1 a constant.

The tensorial conductivity assumes a particularly simple form in the case of $\nu_1 = \text{const}$, $\omega = 0$ (D.C. electric field):

$$\sigma_e = \sigma_0 \begin{vmatrix} \frac{\nu_1^2}{\nu_1^2 + \omega_b^2}, & \frac{\omega_b \nu_1}{\nu_1^2 + \omega_b^2}, & 0 \\ -\frac{\omega_b \nu_1}{\nu_1^2 + \omega_b^2}, & \frac{\nu_1^2}{\nu_1^2 + \omega_b^2}, & 0 \\ 0, & 0, & 1 \end{vmatrix}, \tag{1.2.84}$$

where $\sigma_0 = \dfrac{n_e e^2}{m_e \nu_1}$ is the conductivity of a plasma without a magnetic field, computed for $\tau = \dfrac{1}{\nu_1}$ from Eqs. (1.2.62), (1.2.65) in accordance with "mean-free-path kinetic theory".

The component $\sigma_{33} = \sigma_0$ is the conductivity in the direction of the magnetic field. The diagonal components σ_{11}, σ_{22} are the components of conductivity in the direction of the electric field and differ from σ_0 by the factor $\dfrac{1}{1 + \left(\dfrac{\omega_b}{\nu_1}\right)^2}$. The mixed components represent the conductivity in the direction perpendicular to both **B** and **E** and differ from σ_0 by a factor of $\dfrac{\omega_b/\nu_1}{1 + \omega_b/\nu_1}$. The fraction $\dfrac{\omega_b}{\nu_1} = \dfrac{eB}{m_e}\tau$ indicates the distance (measured in radians) traversed by the electron in its spiral motion in the magnetic field, within the mean time τ between two collisions. Conductivity in the direction of the magnetic field is not affected by the rotation of the electron in the magnetic field, as this motion occurs in a plane perpendicular to **B**.

Using Eq. (1.2.84) we may rewrite Ohm's Law (Eq. 1.2.78) in a vectorial form:

$$\mathbf{j} = \sigma_0 \mathbf{E}_{\parallel} + \frac{\sigma_0}{1 + \left(\dfrac{\omega_b}{\nu_1}\right)^2} \mathbf{E}_0 + \sigma_0 \frac{\dfrac{\omega_b}{\nu_1}}{1 + \left(\dfrac{\omega_b}{\nu_1}\right)^2} \frac{\mathbf{B}}{B} \times \mathbf{E}_{\perp}, \quad (1.2.85)$$

where $\mathbf{E}_{\parallel}$, $\mathbf{E}_0$ denote the intensity components of the superimposed electric field in the directions parallel and perpendicular to $\mathbf{B}$. The last member in Eq. (1.2.85) is called the density of the Hall current. For $\dfrac{\omega_b}{\nu_1} \ll 1$, the Hall current is negligible, the first two members on the right-hand side are comparable, and the magnetic field hardly affects the current. With $\dfrac{\omega_b}{\nu_1} \gg 1$, conductivity is greatly reduced in the directions perpendicular to $\mathbf{B}$, particularly in the direction of $\mathbf{E}_{\perp}$.

1.2.4.3 Electric conductivity of fully ionized gas

In deriving Eqs. (1.2.80) (1.2.82) for the electric conductivity of a plasma, we assumed a density of charged particles low enough to permit the effect of mutual collisions of these particles to be neglected, and the collision member in Boltzmann's equation to be taken in the form of Eq. (1.2.73) which is valid for collisions between electrons and heavy neutral particles $\left(\dfrac{m_e}{M} = 0\right)$. In the opposite case, that is with considerable densities of charged particles, the general starting point in computing the electric conductivity of a plasma is the Focker-Planck equation (1.2.39). Many such calculations in various approximations have been carried out. We quote – without deriving it – Spitzer's by now classical formula for the electric conductivity σ of a fully ionized gas composed of electrons and Z times ionized ions [Ref. 1.2.18]:

$$\eta = \sigma^{-1} = 38 \frac{Z \ln \Lambda}{\gamma(Z)\, T^{3/2}} \quad [\text{ohm . m}], \quad (1.2.86)$$

where η, σ are the resistivity and conductivity of the plasma, T the temperature [°K], $\Lambda = \dfrac{D}{b_0}$ the mean value of the ratio Debye's length D

(Eq. 1.2.41) to the critical parameter b_0 (Eq. 1.2.42); $\gamma(Z)$ is given by the following table:

ion charge Z	1	2	4	16	
$\gamma(Z)$	0.582	0.683	0.785	0.923	1.000

In the case of a fully ionized gas consisting of ions of different charges, the effective value $\bar{Z}$ of the ion charge [cf. e.g. Ref. 1.2.4] is introduced by the relationship

$$n_e\bar{Z} = \sum_k n_k Z_k^2, \tag{1.2.87}$$

where n_e is the electron density, and n_k, Z_k are the densities and charges of k-type ions.

For practical calculations of partially ionized plasmas consisting of electrons, ions and neutral particles, the so-called additive conductivity σ_{add} is sometimes introduced by the relation

$$\sigma_{add} = \left[\frac{1}{\sigma_0} + \frac{1}{\sigma_{ei}}\right]^{-1}; \tag{1.2.88}$$

σ_0 is the conductivity determined by the collisions of electrons with neutral particles, and σ_{ei} conductivity due to Coulombic collisions in the plasma.

1.2.5 Diffusion and Ambipolar Diffusion

1.2.5.1 Diffusion in a binary gas mixture

To describe a flowing mixture of dissociated and ionized gases, let us use a coordinate system that moves together with a given volume element of the mixture with mass-average velocity $\mathbf{v}_0(\mathbf{r}, t)$ (cf. Eq. 1.2.12). This velocity is determined by the resultant of the external forces and the pressure gradient acting on the element and can be calculated from the equation of motion of the mixture as a whole. Various gaseous components move with respect to this system at macroscopic velocities $\bar{\mathbf{V}}_i(\mathbf{r}, t)$ ("diffusion velocities"). These macroscopic velocities are derived from irreversible processes taking place in a non-homogeneous plasma and from that part of the external forces and the pressure gradient which does not affect the macroscopic motion of the plasma as a whole (e.g. the action of an electric field on a quasi-neutral plasma).

The velocities $\overline{\mathbf{V}}_i(\mathbf{r}, t)$ can be computed by solving the complete system of the equations of irreversible thermodynamics. Exact kinetic theory is then used to find the phenomenological coefficients. For the sake of simplicity, we limit our discussion to a binary mixture (of components 1 and 2) with no external forces present. In this case the solution of the Boltzmann equation takes the following form [Ref. 1.2.7]:

$$\overline{\mathbf{V}}_1 - \overline{\mathbf{V}}_2 = \frac{n^2}{n_1 n_2} \mathscr{D}_{12} \left[\operatorname{grad} \left(\frac{n_1}{n} \right) + \right.$$

$$\left. + \frac{n_1 n_2 (M_2 - M_1)}{\varrho n} \operatorname{grad} (\ln p) + k_T \operatorname{grad} (\ln T) \right], \qquad (1.2.89)$$

$$\varrho_1 \overline{\mathbf{V}}_1 + \varrho_2 \overline{\mathbf{V}}_2 = 0. \qquad (1.2.90)$$

In Eqs. (1.2.89) (1.2.90) n_1, n_2 denotes the number of moles of components 1 and 2 per m^3, n the total number of moles per m^3, $M_1 M_2$ the molecular weights of the two components, ϱ_1, ϱ_2 their densities and ϱ the density of the mixture, p the pressure, and T—temperature. $\mathscr{D}_{12}$ is the coefficient of binary diffusion, $k_T = \frac{D_T}{\mathscr{D}_{12}}$ the ratio of the coefficien of thermal diffusion D_T to the coefficient of binary diffusion $\mathscr{D}_{12}$. Equation (1.2.90) follows directly from the definition of diffusion velocities (cf. Eq. 1.2.13).

By introducing weight concentrations

$$K_i = \frac{M_i n_i}{Mn} = \frac{\varrho_i}{\varrho}, \qquad (1.2.91)$$

where M is the "effective" molecular weight of the mixture:

$$\frac{1}{M} = \sum \frac{K_i}{M_i}, \qquad (1.2.92)$$

we can rewrite Eq. (1.2.89) in the following form [Ref. 1.2.19]:

$$\bar{\boldsymbol{V}}_1 = -\mathscr{D}_{12}\left\{\operatorname{grad}(\ln K_1) + \frac{M_2 - M_1}{M} K_2 \operatorname{grad}(\ln p) + \right.$$
$$\left. + \frac{M_1 M_2}{K_1 M^2} k_T \operatorname{grad}(\ln T)\right\}, \tag{1.2.93}$$

$$\bar{\boldsymbol{V}}_2 = -\mathscr{D}_{12}\left\{\operatorname{grad}(\ln K_2) - \frac{M_2 - M_1}{M} K_1 \operatorname{grad}(\ln p) - \right.$$
$$\left. - \frac{M_1 M_2}{K_2 M^2} k_T \operatorname{grad}(\ln T)\right\}. \tag{1.2.94}$$

Eqs. (1.2.93) (1.2.94) separate the effects of the individual factors causing mass transport by diffusion, i.e. the gradients of concentration, pressure and temperature. The effect of the other two members on the right-hand side is often negligible; the diffusion is then well described by Fick's Law

$$\varrho K_i \bar{\boldsymbol{V}}_i = -\varrho \mathscr{D}_{12} \operatorname{grad} K_i, \qquad i = 1, 2, \tag{1.2.95}$$

expressing the proportionality between the diffusion flow of mass and the gradient of concentration.

Kinetic theory expresses the coefficients $\mathscr{D}_{12}$ and k_T by so-called collision integrals $\Omega_{1,2}^{(1,s)*}(T_{12}^*)$ [Ref. 1.2.8]. These integrals (for their tabulation cf. Ref. 1.2.8) describe the interaction (collision) of type-1 and type-2 particles and are based on the preselected form of the interaction potential of particles. Thus in the first approximation the diffusion coefficient $\mathscr{D}_{12}$ has the following form [Ref. 1.2.8]:

$$\mathscr{D}_{12} = 2.628 \,.\, 10^{-7} \sqrt{\frac{1}{2} \frac{M_1 + M_2}{M_1 M_2}} \cdot \frac{T^{3/2}}{p \sigma_{12}^2 \Omega_{1,2}^{(1,1)*}(T_{1,2}^*)} \mathrm{m}^2 \mathrm{s}^{-1}; \tag{1.2.96}$$

p stands here for the pressure [atm], T for temperature [°K], $M_1 M_2$ for molecular weights, $T_{1,2}^* = kT/\varepsilon_{12}$; σ_{12} [Å], ε_{12} [°K] are the parameters of the potential energy of interaction of particles, which have been tabled for various types of interaction potentials [Ref. 1.2.8].

Thy physical significance of the collision integrals $\Omega_{1,2}^{(l,s)*}(T_{1,2}^*)$ is in the description they give of the molecular model selected by us (i.e. the selected dependence of the interaction potential) as distinct from the model of rigid elastic spheres of radius $\sigma_{1,2}$, where $\sigma_{12} = = \frac{1}{2}(\sigma_1 + \sigma_2)$; σ_1, σ_2 are the radii of molecules 1 and 2. Thus for a gas model consisting of rigid elastic spheres the following formula is valid:

$$\mathscr{D}_{12} = 2.628 \,.\, 10^{-7} \sqrt{\frac{1}{2} \frac{M_1 + M_2}{M_1 M_2}} \frac{T^{3/2}}{p\sigma_{12}^2} \text{ m}^2\text{s}^{-1}, \tag{1.2.97}$$

which in the case of self-diffusion (in a single-component mixture; $M_1 = M_2 = M$) is reduced to the well-known form of the coefficient of self-diffusion

$$\mathscr{D}_{12} = \mathscr{D} = 2.628 \,.\, 10^{-7} \frac{\sqrt{\frac{T^3}{M}}}{p\sigma^2} \text{ m}^2\text{s}^{-1}. \tag{1.2.98}$$

Far more complex is the functional dependence on the collision integrals $\Omega_{1,2}^{(l,s)*}(T_{1,2}^*)$ for the thermal diffusion ratio k_T. Since in kinetic theory thermal diffusion is a second-order effect which cannot be adequately described by the elementary kinetic theory of gases, there are no simple expressions for k_T, of a type such as Eqs. (1.2.99), (1.2.123) or (1.2.129) for the coefficients of binary diffusion, internal friction or thermal conductivity respectively. Fortunately, thermal diffusion may be neglected in many practical applications of kinetic theory.

"Meanfree-path kinetic theory" offers a simple formula for the coefficient of self-diffusion:

$$\mathscr{D} = \frac{1}{3} \overline{c l} \tag{1.2.99}$$

where $\bar{c}$ is the mean thermal velocity $\bar{c} = \sqrt{\frac{8kT}{\pi m}}$, and $\bar{l}$ is the mean free path of the gas molecules.

1.2.5.2 Diffusion in multicomponent gas mixtures

In a multicomponent mixture, the diffusion flow $\varrho_i \overline{\mathbf{V}}_i$ is determined not merely by the concentration gradient of this component, but by the concentration gradients of the other components too. If there are no gradients of pressure and temperature and no external forces are active, then [Ref. 1.2.8]

$$\varrho_i \overline{\mathbf{V}}_i = \frac{n^2}{\varrho} \sum_{j=1}^{\nu} M_i M_j D_{ij} \operatorname{grad} \frac{n_j}{n}, \tag{1.2.100}$$

the notations being the same as in Eq. (1.2.89). The coefficients $\mathscr{D}_{ij}$ are called multicomponent diffusion coefficients (as distinct from the binary diffusion coefficients $\mathscr{D}_{ij}$), and by using coefficients $\mathscr{D}_{ij}$ they can be expressed by the following formula:

$$\mathscr{D}_{ij} = \frac{1}{M_j} \sum_k x_k M_k \frac{K^{ji} - K^{ii}}{|K|}, \tag{1.2.101}$$

where x_k is the molar fraction of the kth component, and K^{ji}, K^{ii} cofactors of determinant $|K|$ whose elements are

$$K_{ij} = \frac{x_i}{\mathscr{D}_{ij}} + \frac{M_j}{M_i} \sum_{k \neq i} \frac{x_k}{\mathscr{D}_{ik}} \quad \text{for} \quad i \neq j, \tag{1.2.102}$$

$$K_{ii} = 0. \tag{1.2.102$'$}$$

Similarly we may write the equations for diffusion due to a pressure gradient or external forces. Thermal diffusion, caused by temperature gradients, is described by thermal diffusion coefficients (D_i^T, which are highly involved functions of the collision integrals $\Omega_{12}^{(l,s)*}(T_{1,2}^*)$. It is obvious that the values of multicomponent diffusion coefficients D_{ij} as well as thermal diffusion coefficients D_i^T can only be calculated with digital computers. Numerical values for the parameters of molecular interaction are obtained either by extrapolating known (experimental) low-temperature values of such parameters into the high-temperature region (Sutherland's formula, e.g. in Ref. [1.2.21]); or by theoretical computation [Ref. 1.2.22]; or else by experiment (molecular-beam technique, cf. e.g. [Ref. 1.2.23]). In recent years a great amount of work has gone toward the extension of the exact kinetic theory of transport phenomena to mixtures of polar gases and mixtures containing ionized and excited particles [Ref. 1.2.24].

1.2.5.3 Ambipolar diffusion

Let us consider the diffusion of charged particles (ions and electrons) in a gas of neutral (electropositive) particles, assuming that the degree of ionization is small and hence the number of collisions between charged particles inconsiderable. The dependence of the diffusion coefficients of ions and electrons, $\mathscr{D}_i$ and $\mathscr{D}_e$, on the reduced mass $\frac{1}{M} = \frac{1}{M_1} + \frac{1}{M_2}$ of the colliding particles, on temperature and on pressure is the same as in Eqs. (1.2.96) and (1.2.98). However, in determining the effective radius for the collision of charged and uncharged particles, we have to take into consideration the polarization effects of the Coulombic field of ions or molecules. For instance, owing to the effects of polarization, the diffusion coefficient $\mathscr{D}_i$ of the ions in the gas mixture is only one fourth or one fifth of the corresponding coefficient of self-diffusion of the neutral particles in the same gas.

Since in a slightly ionized gas electrical conductivity and diffusion have the same mechanism (i.e. diffusion flow—due to an external force in the first instance and to a concentration gradient in the second one), so-called Einstein relations can be proved to exist between the diffusion coefficients $\mathscr{D}_i$, $\mathscr{D}_e$ and the mobilities μ_i, μ_e (cf. Eqs. 1.2.62 and 1.2.63).

$$\frac{\mathscr{D}_e}{|\mu_e|} = \frac{kT_e}{e}; \qquad \frac{\mathscr{D}_i}{|\mu_i|} = \frac{kT_i}{Ze} \tag{1.2.103}$$

where Ze is the charge of the positive ion; T_e, T_i are the temperatures of electrons and ions; and k is the Boltzmann constant.

From the dependence of $\mathscr{D}_e$ and $\mathscr{D}_i$ on the reduced mass ($\mathscr{D} \sim M^{-\frac{1}{2}}$) follows the approximate relation

$$\left|\frac{\mu_e}{\mu_i}\right| \sim \frac{\mathscr{D}_e}{\mathscr{D}_i} \sim \sqrt{\frac{m_i}{m_e}}. \tag{1.2.104}$$

From Eq. (1.2.104) we note that $\mathscr{D}_e \gg \mathscr{D}_i$ (similarly to $\mu_e \gg \mu_i$), i.e. the lightweight electrons have a higher rate of diffusion than the heavy ions. Owing to the differing rates of diffusion, the electrons are quicker in escaping from a region of higher concentration of charged particles and leave behind an excess of positive ions flowing at a slower rate. The space charge thus set up creates an electric field E which decelerates the electrons and accelerates the ions until their rates of diffusion are equal; the particles of both kinds then move together, as a single gas of charged particles in relation to the gas of neutral particles. The diffusion flow of electrons and ions, respectively is determined by the equations

$$n_e \bar{\mathbf{V}}_e = -\mathscr{D}_e \operatorname{grad} n_e + n_e \mu_e \mathbf{E}, \tag{1.2.105}$$

$$n_i \bar{\mathbf{V}}_i = -\mathscr{D}_i \operatorname{grad} n_i + n_i \mu_i \mathbf{E}, \tag{1.2.106}$$

where $\mu_e < 0$, $\mu_i > 0$.

If ions and electrons flow at the same rate, then

$$n_e \bar{\mathbf{V}}_e = n_i \bar{\mathbf{V}}_i = \mathbf{J}. \tag{1.2.107}$$

By using Eq. (1.2.107) and eliminating $\mathbf{E}$ from Eqs. (1.2.105) and (1.2.106) we obtain

$$(n_i \mu_i - n_e \mu_e) \mathbf{J} = -(n_i \mu_i \mathscr{D}_e \operatorname{grad} n_e - n_e \mu_e \mathscr{D}_i \operatorname{grad} n_i). \tag{1.2.108}$$

Allis' hypothesis of proportionality [Ref. 1.2.4]

$$\frac{\operatorname{grad} n_e}{n_e} = \frac{\operatorname{grad} n_i}{n_i} \tag{1.2.109}$$

enables us to transform Eq. (1.2.108) into

$$\mathbf{J} = -\mathscr{D}_s \operatorname{grad} n_e, \tag{1.2.110}$$

where the coefficient of "simultaneous" diffusion $\mathscr{D}_s$ is defined by the equation

$$\mathscr{D}_s = \frac{\mu_i \mathscr{D}_e - \mu_e \mathscr{D}_i}{\mu_i - \mu_e} \left(1 - \mu_e \frac{\varrho_+}{\sigma}\right), \tag{1.2.111}$$

the density of the space charge ϱ_+ and the conductivity σ being given by the following expressions (for single ions):

$$\varrho_+ = (n_i - n_e)\, e \tag{1.2.112}$$

$$\sigma = (n_i \mu_i - n_e \mu_e)\, e. \tag{1.2.113}$$

Assuming that the bond between ions and electrons is strong enough for ϱ_+ to be practically negligible—i.e. the condition of quasi-neutrality

$$n_i = n_e = n \tag{1.2.114}$$

to be fulfilled – we may neglect the second term in brackets in Eq. (1.2.111) and write:

$$n_e \bar{\mathbf{V}}_e = n_i \bar{\mathbf{V}}_i = \mathbf{J} = -\mathscr{D}_a \operatorname{grad} n, \qquad (1.2.115)$$

where the coefficient of ambipolar diffusion is given by the following equation (note: $\mu_e < 0$):

$$\mathscr{D}_a = \frac{\mu_i \mathscr{D}_e - \mu_e \mathscr{D}_i}{\mu_i - \mu_e}. \qquad (1.2.116)$$

As $|\mu_i| \ll |\mu_e|$, for $T_e \sim T_i$

$$\mathscr{D}_a \sim 2\mathscr{D}_i, \qquad (1.2.117)$$

hence the limiting factor in the rate of ambipolar diffusion is the low diffusion rate of the ions.

The electric field created in ambipolar diffusion has the form:

$$\operatorname{div} \mathbf{E} = \frac{1}{\varepsilon_0} \varrho_+ = \frac{1}{\varepsilon_0} (n_+ - n_-) e. \qquad (1.2.118)$$

Substitution from Eq. (1.2.110) into Eq. (1.2.105) yields

$$\mathbf{E} = \frac{\mathscr{D}_e - \mathscr{D}_s}{n_e \mu_e} \operatorname{grad} n_e \qquad (1.2.119)$$

and for the density of the space charge

$$\varrho_+ = \varepsilon_0 \operatorname{div} \mathbf{E} = \varepsilon_0 \frac{\mathscr{D}_e - \mathscr{D}_s}{\mu_e} \operatorname{div} \left(\frac{\operatorname{grad} n_e}{n_e} \right). \qquad (1.2.120)$$

By substituting for ϱ_+ in Eq. (1.2.111) and solving for $\mathscr{D}_s$, we obtain:

$$\mathscr{D}_s = \mathscr{D}_a \frac{\dfrac{\sigma}{\varepsilon_0} - \mathscr{D}_e \operatorname{div} \left(\dfrac{\operatorname{grad} n_e}{n_e} \right)}{\dfrac{\sigma}{\varepsilon_0} - \mathscr{D}_a \operatorname{div} \left(\dfrac{\operatorname{grad} n_e}{n_e} \right)}. \qquad (1.2.121)$$

Thus the dependence of the diffusion coefficient $\mathscr{D}_s$ on the unknown n_e is very complicated; the coefficient $\mathscr{D}_s$ causes the equations to be non-linear and aggravates the difficulty of their solution.

1.2.6 Viscosity

1.2.6.1 Fundamental concepts

As a plasma flows in a nozzle, and especially in the boundary layer next to a solid, the flow field displays marked gradients of mass-average velocity. The inherent tendency of the plasma to equalize these velocities appears externally as internal friction. Thus the surface bounding an elementary volume of moving plasma is subject to surface forces due to the difference in the mean momenta of the molecules moving perpendicularly to that surface. The action of such forces is fully described by the pressure tensor P (Eq. 1.2.21) which in the simple case of flow in parallel layers has the well-known form

$$\mathsf{P} = \begin{vmatrix} p, & 0, & -\eta \dfrac{\partial v_{0x}}{\partial z} \\ 0, & p, & 0 \\ \eta \dfrac{\partial v_{0x}}{\partial z}, & 0, & p \end{vmatrix}, \tag{1.2.122}$$

p being the pressure (force per 1 m^2 acting in the direction perpendicular to the reference surface), v_{0x} the only non-zero component of the velocity vector, $\dfrac{\partial v_{0x}}{\partial z}$ the gradient of this component in the direction perpendicular to it, and η the coefficient of dynamic viscosity.

If velocity fluctuations arise in a flowing plasma, then—depending on the magnitude of the coefficient of dynamic viscosity η—either internal-friction forces suppress these fluctuations (large η), or hydrodynamic instabilities develop (e.g. turbulent flow). The coefficient of viscosity is essential for the description of a plasma flow accompanied by heat transfer, for it appears as a factor in both Reynolds' and Prandtl's numbers $\left(\mathrm{Re} = \dfrac{\varrho V d}{\eta},\ \mathrm{Pr} = \dfrac{C_p \eta}{\varkappa}\right)$.

Elementary kinetic theory offers the following formula [Ref. 1.2.20] for the viscosity of a gas mixture:

$$\eta = \sum_i \frac{1}{3} n_i m_i \bar{c}_i \bar{l}_i; \tag{1.2.123}$$

in this equation n_i is the particle density, m_i the mass, $\bar{c}_i$ the mean thermal velocity, and $\bar{l}_i$ the mean free path of i-type particles. This mean free path is given by the formula below for a gas-mixture model consisting of perfectly elastic rigid spheres:

$$\frac{1}{\bar{l}_i} = \sum_k \left(1 - \frac{1}{2} p_{ki}\right) n_i q_{ki} \sqrt{1 + \frac{m_i}{m_k}}, \tag{1.2.124}$$

where q_{ki} is the effective cross-section for the interaction between particles i and k of masses m_i and m_k, and p_{ki} is the so-called persistence of the velocities:

$$p_{ki} = \frac{m_k - 0.2m_i}{m_k + m_i}. \tag{1.2.125}$$

The effective cross-sections q_{ki} are either measured or calculated according to a suitably chosen model for the collision of particles. If the gas contains electrons, we usually do not include in Eq. (1.2.124) collisions of the types atom → electron or ion → electron, since the heavy particle loses virtually no momentum in such collisions. We do, however, include collisions of the types electron → ion and electron → atom.

1.2.6.2 Computing viscosity by exact kinetic theory

Exact kinetic theory, based on the solution of the Boltzmann equation, converts the computation of dynamic viscosity η into the computations of two types of collision integrals $\Omega_{ij}^{(1,1)*}(T_{ij}^*)$ and $\Omega_{ij}^{(2,2)*}(T_{ij}^*)$ and of the diffusion coefficients $\mathscr{D}_{ij}$ of a binary mixture. We do not quote the explicit expression for the viscosity of a multicomponent mixture, because it is very complex [Ref. 1.2.8].

For the viscosity of a pure gas, exact kinetic theory offers the following formula:

$$\eta = 266.93 \,.\, 10^{-6} \frac{\sqrt{MT}}{\sigma^2 \Omega^{(2,2)*}(T^*)} \text{ kg m}^{-1}\text{ s}^{-1}, \tag{1.2.126}$$

where M is the molecular weight, T temperature, σ the effective cross-section for the collision of the molecule, $\Omega^{(2.2)*}(T^*)$ a collision integral which is also found tabled in Ref. [1.2.8], and T^* reduced temperature. (These designations and units are the same as in Eq. 1.2.96).

For calculating the viscosity of a multicomponent mixture, an approximate formula [Ref. 1.2.26] is used far more frequently than the involved exact expression. It reads

$$(\eta)_{\text{mixt.}} = \sum_{i=1}^{\nu} \frac{\eta_i}{1 + \sum\limits_{\substack{j=1 \\ i \neq j}} \Phi_{ij} \dfrac{x_j}{x_i}}, \tag{1.2.127}$$

where η_i are the viscosities of the individual components (Eq. 1.2.126), and x_i, x_j the molar fractions of components i and j. The coefficients Φ_{ij} are functions of the viscosities η_i, η_j of the components and of the molecular masses M_i, M_j:

$$\Phi_{ij} = \frac{\left[1 + \left(\dfrac{\eta_i}{\eta_j}\right)^{1/2} \left(\dfrac{M_j}{M_i}\right)^{1/4}\right]^2}{2\sqrt{2}\left(1 + \dfrac{M_i}{M_j}\right)^{1/2}}. \tag{1.2.128}$$

Brokaw [Ref. 1.2.27] asserts that the probable difference between the viscosity values of a binary mixture calculated according to Eq. (1.2.127) and according to the formula of exact theory is less than 1·9 %, supposing that the gas is not highly polar and does not contain excited or ionized particles. Ref. [1.2.26] presents various nomograms for calculating coefficients Φ_{ij} of non-polar gas mixtures.

1.2.7 Heat Conduction

1.2.7.1 Basic concepts

The following mechanisms of heat transfer exist in the plasma:

1 – Heat conduction, which is affected solely by the existence of thermal gradients and does not depend on diffusion flows in the plasma; chemical reactions are considered to be frozen.

2 – Heat transfer by diffusion, which is connected with the transfer of the reaction energy of particles in their diffusion flows, caused by external forces and gradients of pressure, concentration and

temperature. To every particle we can attribute a certain amount of enthalpy (the enthalpy of one mole divided by the number of particles in one mole); moving from places of higher temperature a particle transports its enthalpy and transfers it to particles in cooler places or to the wall of the vessel.

3 – Heat transfer by convection, connected with the transfer of the enthalpy of entire plasma elements in the flow of plasma as a whole (i.e. with mass-average velocity). This mechanism is very intensive and largely prevails, for instance in turbulent plasmas.

4 – Heat transfer by radiation.

1.2.7.2 Heat conduction

For the coefficient of thermal conductivity $\varkappa_{J=0}$ in a gas mixture, elementary kinetic theory offers the following equation [Ref. 1.2.1]:

$$\varkappa_{J=0} = \sum_{k=1} \frac{1}{3} n_k \bar{l}_k \bar{c}_k \left(\frac{5}{2} c^k_{\text{transl}} + c^k_{\text{int}} \right); \tag{1.2.129}$$

n_k is the particle density [m^{-3}], $\bar{l}_k$ the mean free path, $\bar{c}_k$ the mean thermal velocity, and c^k_{transl}, c^k_{int} are heat capacities (at constant volume) of the translational and internal degrees of freedom per one k-type particle.

The mean free path $\bar{l}_k$ is again defined by Eq. (1.2.124); we therefore can write

$$\varkappa_{J=0} = \sum_{k=1}^{\nu} \frac{\eta_k}{m_k} \left(\frac{5}{2} c^k_{\text{transl}} + c^k_{\text{int}} \right), \tag{1.2.130}$$

where η_k is the viscosity of the kth component.

If electrons are present in the mixture, the thermal conductivity due to their thermal diffusion (usually neglected in instances of heavy particles) is sometimes taken into consideration when calculating the thermal conductivity $\varkappa^{\text{e}}_{J=0}$ of the electron gas. Maecker – in Ref. [1.2.1] – presents the following formula for the case of slightly ionized gas:

$$\varkappa^{\text{e}}_{J=0} = \frac{1}{3} n_{\text{e}} \bar{l}_{\text{e}} \bar{c}_{\text{e}} 2k(1 + x), \tag{1.2.131}$$

where k is Boltzmann's constant and x the degree of ionization.

The exact kinetic theory of heat conduction in nonreacting mixtures starts from the general solution of the Boltzmann equation and converts the computation of $\varkappa_{J=0}$ into the determination of four kinds of collision integrals and of the coefficients of diffusion in binary mixtures.

The following formula [Ref. 1.2.8] is valid for a simple gas with no internal degrees of freedom:

$$\varkappa_{J=0} = 8.33 \,.\, 10^{-2} \frac{\sqrt{T/M}}{\sigma^2 \Omega^{(2,2)*}(T^*)} \text{ W m}^{-1}\text{s}^{-1}\text{deg}^{-1}. \quad (1.2.132)$$

The designations are the same as in Eq. (1.2.126). By comparing Eqs. (1.2.126) and (1.2.132) we obtain

$$\varkappa_{J=0} = \frac{15}{4} \frac{R}{M} \eta, \quad (1.2.133)$$

where R is the gas constant, and η viscosity as given by Eq. (1.2.126).

For polyatomic molecules, in which energy of translational motion can be transferred to internal degrees of freedom, we use the Eucken formula [Ref. 1.2.8]. We write

$$[\varkappa_{J=0}]^{\text{Eucken}} = \frac{15}{4} \frac{R}{M} \eta \left(\frac{4}{15} \frac{C_v}{R} + \frac{3}{5} \right), \quad (1.2.134)$$

where C_v is the molar heat at constant volume (corresponding to both translational and rotational degrees of freedom), and η viscosity as defined in Eq. (1.2.126). The exact expression for the coefficient of thermal conductivity $\varkappa_{J=0}$ in a multicomponent mixture is even more involved than the analogous expression for viscosity. Approximate equations, similar to Eq. (1.2.127) for the viscosity of such a mixture, are therefore of high practical importance. Brokaw [Ref. 1.2.27] offers such formulae for mixtures of non-reactive gases with internal degrees of freedom. We may consider the thermal conductivity of a gas mixture to be composed of two parts:

$$(\varkappa_{J=0})_{\text{mix}} = (\varkappa_{J=0})'_{\text{mix}} + (\varkappa_{J=0})''_{\text{mix}}, \quad (1.2.135)$$

the first and second terms on the right-hand side describing the transport of the translational and internal energies, respectively of the molecule.

The approximate formula for $(\varkappa_{J=0})'_{\text{mix}}$ has a form analogous to that of Eq. (1.2.127):

$$(\varkappa_{J=0})'_{\text{mix}} = \sum_{i=1}^{\nu} \frac{(\varkappa_{J=0})'_i}{1 + \sum\limits_{\substack{j=1 \\ j \neq i}}^{\nu} \psi_{ij} \frac{x_j}{x_i}}, \tag{1.2.136}$$

where $(\varkappa_{J=0})'_i$ are the conductivities of the components as described by Eq. (1.2.132), x_i and x_j are molar fractions, and coefficients ψ_{ij} are connected with coefficients Φ_{ij} (Eq. 1.2.128) by the approximate relation

$$\psi_{ij} = \Phi_{ij} \left\{ 1 + 2.41 \frac{(M_i - M_j)(M_i - 0.142M_j)}{(M_i + M_j)^2} \right\}; \tag{1.2.137}$$

M_i and M_j are molecular masses.

Thermal conductivity $(\varkappa_{J=0})''_{\text{mix}}$, describing the transport of internal molecular energy, has the following form, in the approximation mentioned:

$$(\varkappa_{J=0})''_{\text{mix}} = \sum_{i=1}^{\nu} \frac{(\varkappa_{J=0})''_i}{1 + \sum\limits_{\substack{j=1 \\ j \neq i}}^{\nu} \frac{\mathscr{D}_i}{\mathscr{D}_{ij}} \frac{x_j}{x_i}}, \tag{1.2.138}$$

where $\mathscr{D}_i$ and $\mathscr{D}_{ji}$ are the coefficients of diffusion and self-diffusion in the binary mixture, x_i and x_j molar fractions, and $(\varkappa_{J=0})''_i$ coefficients of contact conductivity describing the transport of internal energy for the individual components. With the coefficient of self-diffusion these coefficients are connected by a simple relationship:

$$(\varkappa_{J=0})''_i = \varrho \mathscr{D}_i (C_v)_i^{\text{int}}, \tag{1.2.139}$$

where ϱ is the density and $(C_v)_i^{\text{int}}$ the specific heat of the internal degrees of freedom. A more exact formula for $(\varkappa_{J=0})''_i$ may be found in Ref. [1.2.28].

1.2.7.3 Heat transfer by diffusion in dissociating and ionized gas

Typical examples of heat transfer by diffusion are the diffusions of dissociation and ionization energies by the mechanism described below.

In the high-temperature regions of a plasma, where the degree of dissociation is higher, there are more atoms (i.e. dissociated molecules) than in cooler regions where undissociated molecules prevail. The atoms therefore diffuse from warmer to cooler regions where they recombine and thus release recombination heat. Molecules, on the contrary, diffuse into the warmer parts where they absorb heat in their dissociation. In the steady state, the diffusion flows of atoms and molecules in opposite directions are equal and result in the irreversible transport of dissociation energy from warmer to cooler regions (or to the walls) where in the recombination this energy is converted into energy of radiation or of molecular thermal motion.

Let us consider a binary mixture formed by the dissociation of a gas of diatomic molecules of the type N_2, O_2, etc. If we denote by h_m, h_a the enthalpies of one kilogram of molecules and atoms respectively, and by $\mathbf{J}_m$, $\mathbf{J}_a$ the diffusion flows of molecules and atoms, then the energy transported by these flows in the steady state ($\mathbf{J}_m = \mathbf{J}_a$) is

$$\mathbf{Q} = (h_a - h_m)\,\mathbf{J}_a. \tag{1.2.140}$$

The difference of the enthalpies can be written in the form

$$\begin{aligned} h_a - h_m &= \left(\frac{5}{2}\frac{kT}{m_a} + u_a + \frac{1}{2}\frac{\varepsilon_{diss}}{m_a}\right) - \left(\frac{5}{2}\frac{kT}{2m_a} + u_m\right) = \\ &= \frac{1}{2m_a}\left(\frac{5}{2}kT + \varepsilon_{diss}\right) + u_a - u_m, \end{aligned} \tag{1.2.141}$$

where m_a is the mass of one atom; u_a, u_m are the specific energies accumulated in the internal degrees of freedom of the atoms or molecules; k the Boltzmann constant, T temperature [°K]; and ε_{diss} the dissociation energy of the gas molecules.

Disregarding all thermodynamic forces except the concentration gradient K_a, we may write the diffusion flows $\mathbf{J}_a$, $\mathbf{J}_m$ in the form of Eq. (1.2.95):

$$\mathbf{J}_a = -\mathbf{J}_m = \varrho K_a \overline{\mathbf{V}}_a = -\varrho \mathscr{D}_{am} \operatorname{grad} K_a, \tag{1.2.142}$$

where ϱ is the total density of the mixture, K_a the weight concentration (Eq. 1.2.91), and $\mathscr{D}_{am}$ the coefficient of diffusion in the binary mixture of molecules with the atoms produced by their dissociation.

Under conditions of local thermodynamic equilibrium, the concentration K_a is uniquely a function of temperature, and we can therefore write

$$\mathbf{J}_a = -\varrho \mathcal{D}_{am} \frac{\partial K_a}{\partial T} \operatorname{grad} T. \tag{1.2.143}$$

For the diffusion flow of dissociation energy we can then write

$$\mathbf{Q} = \varkappa_{diss} \operatorname{grad} T =$$

$$= -\left\{\left[\frac{1}{2m_a}\left(\frac{5}{2} kT + \varepsilon_{diss}\right) + (u_a - u_m)\right] \varrho \mathcal{D}_{am} \frac{\partial K_a}{\partial T}\right\} \operatorname{grad} T, \tag{1.2.144}$$

where $\varkappa_{diss}$ is the corresponding coefficient of thermal conductivity.

In a similar way we may proceed in the case of ionization energy, with the appropriate changes, of course, in Eq. (1.2.144): $\varepsilon_{diss} \to \varepsilon_{ioniz}$, $\mathcal{D}_{am} \to \mathcal{D}_{ea}$ (coefficient of ambipolar diffusion) $m_a \to \frac{1}{2} m_a$, and setting $u_e = 0$

1.2.7.4 Heat transfer in a multicomponent mixture of reacting gases

Butler and Brokaw [Refs. 1.2.27, 1.2.29] have derived a comparatively simple expression for thermal conductivity by the diffusion of reaction energies in cases when ν independent chemical reactions take place in a gas mixture in chemical equilibrium, with a total of μ components participating. Let us denote the chemical component by X^k (we have called it K_i in Eq. 1.1.93) and write the reaction in the form of Eq. (1.1.94):

$$\sum_{k=1}^{\mu} \nu_{ik} X^k = 0, \qquad i = 1, 2 \ldots j \ldots \nu, \tag{1.2.145}$$

where ν_{ik} are the stoichiometric coefficients of the kth component in the ith reaction. If we denote by symbol ΔH_i the total change in the enthalpy of all components participating in the ith reaction, we can write—in analogy to Eq. (1.1.60) for the change ΔG_i—

$$\Delta H_i = \sum_{k=1}^{\mu} \nu_{ik} H_k; \qquad i = 1, 2 \ldots j \ldots \nu, \tag{1.2.146}$$

where H_k are the enthalpies (Eq. 1.1.41) of the individual components. Such a change in enthalpy, described in Eq. (1.2.141), is an example of a dissociation reaction, although specific enthalpies were discussed there as distinct from the molar enthalpies we are considering now.

The thermal conductivity of such a mixture may be regarded as consisting of two parts:

$$\varkappa = \varkappa_{J=0} + \varkappa_J, \tag{1.2.147}$$

where $\varkappa_{J=0}$ stands for conductivity with all reactions "frozen" (i.e. the mixture having the same composition, but the reactions not in progress), and $\varkappa_J$ thermal conductivity due to diffusion, which may be written as a ratio of determinants:

$$\varkappa_J = -\frac{1}{RT^2}\frac{\begin{vmatrix} A_{11}, \ldots A_{1\nu}, \Delta H_1 \\ \vdots \quad \vdots \quad \vdots \\ A_{1\nu}, \ldots A_{\nu\nu}, \Delta H_\nu \\ \Delta H_1 \ldots \Delta H_\nu, \ 0 \end{vmatrix}}{\begin{vmatrix} A_{11}, \ldots A_{1\nu} \\ \vdots \quad \vdots \\ A_{1\nu}, \ldots A_{\nu\nu} \end{vmatrix}} \tag{1.2.148}$$

In this equation, the terms of determinants A_{kl} are functions of the molar fractions x_k, x_l of the components k and l, of the diffusion coefficients $\mathscr{D}_{kl}$ in binary mixtures, of pressure p and temperature T:

$$A_{ij} = \sum_{k=1}^{\mu-l}\sum_{l=k+1}^{\mu}\frac{RT}{\mathscr{D}_{kl}p}x_k x_l\left(\frac{\nu_{ik}}{x_k}-\frac{\nu_{il}}{x_l}\right)\left(\frac{\nu_{jk}}{x_k}-\frac{\nu_{jl}}{x_l}\right),$$

$$i, j = 1, 2 \ldots, \tag{1.2.149}$$

R being the gas constant.

1.2.8 Relaxation and Recombination

1.2.8.1 Relaxation

Unique particle distribution by kinds, and in every kind unique distribution by coordinates, momenta and internal states is characteristic of plasma in equilibrium. Given the definite energy and external parameters of a system, these distributions are stable, that is spontaneous fluctuations arising in the plasma are automatically counterbalanced by collision mechanisms. If we change very (or theoretically: infinitely) rapidly either the internal energy of the plasma – for instance by heating or cooling – or one or more external parameters (pressure, volume, magnetic and electric fields, etc.), the plasma develops – after a certain period – a new equilibrium corresponding to the altered energy or the altered external parameters. The period required for this transition is called the relaxation time (in the broad sense of this term); it depends on the intensity of the energy exchange between the different components and (within the individual component) between the various degrees of freedom (i.e. on the collision efficiency). Since the exchange between different degrees of freedom or components may take various courses, various relaxation times are characteristic of the transition of the plasma to a new equilibrium. As a rule, the translational degrees of freedom adapt themselves most rapidly to the new conditions*; hence a well-defined "translational" temperature, which soon attains its new equilibrium value, can be attributed to the individual components at every moment. We may then relate the energies of the other degrees of freedom and the plasma composition to the kinetic temperatures. Let us denote the instantaneous (non-equilibrium) value of some "relaxing" quantity by U, and the initial (non-equilibrium) and final (equilibrium) values of this quantity by U_0 and U_∞, respectively. Let us assume that in every instant the rate of change $\frac{\mathrm{d}U}{\mathrm{d}t}$ of quantity U is proportional to the deviation $U - U_\infty$. Then the relaxation of quantity U is in the first approximation described by the equation

$$-\frac{\mathrm{d}U}{\mathrm{d}t} = \frac{1}{\tau}(U - U_\infty), \tag{1.2.150}$$

where τ is the relaxation time. The physical meaning of this quantity becomes obvious if we integrate Eq. (1.2.150):

$$U = (U_0 - U_\infty)\,\mathrm{e}^{-\frac{t}{\tau}} + U_\infty. \tag{1.2.151}$$

---

* This does not apply to very intensive chemical reactions occurring in the plasma.

During the relaxation time, the deviation of quantity U from its equilibrium value decreases by a factor of e^{-1}.

The transfer of energy between various degrees of freedom as well as the formation of new and the decay of old particles (ionization, dissociation, recombination) takes place in collisions. We may, therefore, express the relaxation time τ in terms of the number of collisions z which the particle undergoes in one second and of the probability P that in one of these collisions an exchange of energy or momentum (or else ionization or recombination) will occur:

$$\tau = \frac{1}{zP}. \tag{1.2.152}$$

In this equation P signifies the probability of the recombination of ions. Eq. (1.2.152) indicates the mean lifetime of the ion, i.e. the time between its origin and decay. From the great variety of relaxation processes which may occur in the plasma, we shall discuss only recombination in this Chapter. For a well-arranged survey of relaxation phenomena cf. Calcote's review Ref. [1.2.30].

1.2.8.2 Recombination

In a high-presure plasma, the establishment of ionization equilibrium is due to various elementary processes which result in the formation or decay of charged particles. Let us denote the ground states of atoms A, B and molecule AB by A, B, AB; once or twice excited particles by AB^*, A^*, B^*, AB^{**}, A^{**}, B^{**}; once or twice ionized particles by A^+, B^+, AB^+, A^{++}, B^{++}, AB^{++}; negative ions by A^-, B^-, AB^-; an electron by e^-, and a photon by $h\nu$. We can then schematically write the chief processes resulting in the formation or the decay of charged particles in the form of chemical reactions.

Ionization Processes

collision with electron:

$$A + e^- \rightarrow A^+ + e^- + e^-; \tag{1.2.153}$$

collision between neutral particles:

$$A + A \rightarrow A^+ + A + e^-; \tag{1.2.154}$$

photoionization:

$$A + h\nu \rightarrow A^{+} + e^{-}; \quad (1.2.155)$$

ionization by stages:

$$A^{*} + e^{-} \rightarrow A^{+} + e^{-} + e^{-}; \quad (1.2.156)$$

photoionization by stages:

$$A^{*} + h\nu \rightarrow A^{+} + e^{-}; \quad (1.2.157)$$

formation of negative ions:

$$A + e^{-} \rightarrow A^{-}. \quad (1.2.158)$$

(no charged particle is produced in this last case, but a particle is converted into a qualitatively different one; similar instances are charge transfer and decay of negative ions, unaccompanied by recombination.)

Recombination Processes

radiative recombination:

$$A^{+} + B^{-} \rightarrow AB + h\nu; \quad (1.2.159)$$

mutual recombination

$$A^{+} + B^{-} \rightarrow A + B; \quad (1.2.160)$$

recombination by stages:

$$A^{+} + e^{-} \rightarrow A^{**} \rightarrow A + h\nu; \quad (1.2.161)$$

recombination in ternary collisions (M is the third particle)

$$A^{+} + B^{-} + M \rightarrow AB + M; \quad (1.2.162)$$

dissociative recombination

$$AB^{+} + e^{-} \rightarrow A + B. \quad (1.2.163)$$

Each of these reactions takes place at a specific rate determined by the effective cross section of the collision. In the simples cases – recombination of positive and negative ions (Eqs. 1.2.159, 1.2.160) or of positive ions and electrons (Eqs. 1.2.161, 1.2.163) – we can write for the rate of change in the densities n_{+}, n_{-} of charged particles ($n_{+} \sim n_{-}$):

$$\frac{dn_{+}}{dt} = \frac{dn_{-}}{dt} = -\alpha n_{+} n_{-} \sim -\alpha n_{+}^{2}, \quad (1.2.164)$$

where α is the recombination coefficient [$m^{-3}\ s^{-1}$].

By integration of Eq. (1.2.164) we obtain

$$n_+ = (n_+)_0 \frac{1}{1 + (n_+)_0 \, \alpha t}, \tag{1.2.165}$$

where $(n_+)_0$ denotes the initial density.

By comparing Eq. (1.2.164) with Eq. (1.2.150) ($U \to n_+$, $U_0 \to \to (n_+)_0$, $U_\infty \to 0$), we obtain the following relationship between the relaxation time for the recombination τ_{rec} and the recombination coefficient α:

$$\tau_{ree} = \frac{1}{\alpha n_+} \tag{1.2.166}$$

Recombination in ternary collisions and dissociative recombination are the most important recombination mechanisms in high-pressure plasmas. At present, the recombination coefficients α are usually measured from the decrease in electron density after the interruption of the current in various types of electric discharges; a microwave technique is used for measuring this change in electron density [Ref. 1.2.31]. The form of the functional dependence given by Eq. (1.2.165) indicates whether the decrease in electron density is due to recombination or to other mechanisms, such as diffusion or formation of negative ions.

Calcote [Ref. 1.2.30] tabulates some values of α which we reproduce without analyzing the results of the measurements:

	ion—ion	ion—electron
	α [m^3 s^{-1}]	
radiative recombination	10^{-8}	10^{-6}
mutual recombination	10^{-2}—10^{-3}	10^{-3}—10^{-6}
dielectronic recombination	10^{-2}—10^{-3}	10^{-3}—10^{-6}
recombination in ternary collisions	10^{-1}—10^{-3}	10^{-1}—10^{-4}
dissociative recombination	unknown	1—10^{-3}

For more detailed data on recombination coefficients cf. e.g. Refs. [1.2.9, 1.2.30, 1.2.31].

1.3 Fundamentals of Plasma Dynamics

1.3.1 The Subject of Plasma Dynamics

In the operation of many kinds of apparatus working with high pressure plasma (plasma torches, MHD generators, nozzles of air-blast circuit breakers, etc.), it is the motion of the plasma as a whole, that is the motion with mass-average velocity $\mathbf{v}_0(\mathbf{r}, t)$ that counts (Eq. 1.2.12). According to Eq. (1.2.49), the magnitude and direction of this velocity are determined by the resultant of the external and pressure forces affecting the plasma. P, ϱ, ϱ_i, $\mathbf{F}_i$ are further quantities appearing in Eq. (1.2.49), and their determination (in the general case) requires the solution of the complete system of equations of irreversible thermodynamics, as discussed in Sec. 1.2.3.1. This system is very complex and in the general case unsolvable. Therefore we always aim at reducing the number of equations and terms, including only those relationships which are essential for the given problem. In this Chapter we shall treat the factors which affect the motion of a plasma as a whole and the equations describing this motion; the influence of the remaining factors will be expressed in simplifying assumptions.

The flow of plasma, that is of mixture of dissociated and ionized gases, differs from the flow of ordinary gases mainly in the following points:

- In addition to the normal body forces of gravity, pressure drop and centrifugal force, electric and magnetic forces can also affect a plasma.
- Plasma in motion is capable of producing electric and magnetic fields, and thereby exchanging energy with the surrounding medium.

- In reactions of ionization-, recombination-, dissociation-type, the number of moles moving in the mixture changes; this change can also become evident in the altered mass-average velocity of the plasma.
- At high temperatures, much of the energy accumulated in the internal degrees of freedom of the plasma particles can change into kinetic energy of the plasma as a whole and vice versa.
- The boundary conditions at the plasma-to-solid interface are very complex. In addition to the usual dynamic and thermal boundary layers, several diffusion layers form on the interface, intensive heterogeneous chemical reactions may occur there (surface recombination of charged particles, combustion, etc.). The boundary conditions for electric and magnetic fields must also be taken into consideration.

The interaction of flow and magnetic field greatly complicates the theoretical description of plasmas. Hydrodynamic and magnetic phenomena have to be conceived comprehensively, which means that the systems of equations describing the electromagnetic field (cf. Chap. 1.3.2) and the flow (cf. Chap. 1.3.3) have to be solved simultaneously. The flow of conducting fluids in magnetic fields is the subject of magnetohydrodynamics.

Some typical magnetohydrodynamic phenomena are described in Chap. 1.3.4. They are of fundamental importance for the function of future thermonuclear reactors, whereas in plasma torches and MHD generators most of them are of secondary importance and thermodynamic phenomena are decisive. The term magneto-thermo-hydrodynamics is sometimes used in this connection.

The system of magnetohydrodynamical equations has to include the boundary conditions for both quantities describing electric and magnetic fields and quantities describing flow.

The physical problems of the plasma-to-solid interface are outlined in Chap. 1.3.6, and Chap. 1.3.5 reviews some fundamentals of similarity theory (dimensional analysis) and discusses the applicability of this theory to flowing plasma. Unfortunately, this theory, which should be the starting point for the semi-empirical analysis of the actual technical forms of plasma has not yet been sufficiently developed.

1.3.2 The Equations of Electromagnetism

1.3.2.1 Maxwell's equations

In the International System of Units the complete set of equations describing the electromagnetic field in a medium at rest has the following form [Ref. 1.3.1]:

$$\operatorname{curl} \mathbf{H} - \frac{\partial \mathbf{D}}{\partial t} = \mathbf{j}, \tag{1.3.1}$$

$$\operatorname{div} \mathbf{D} = \varrho_{el}, \tag{1.3.2}$$

$$\operatorname{curl} \mathbf{E} + \frac{\partial \mathbf{B}}{\partial t} = 0, \tag{1.3.3}$$

$$\operatorname{div} \mathbf{B} = 0. \tag{1.3.4}$$

In Eqs. (1.3.1) to (1.3.4) $\mathbf{E}$, $\mathbf{H}$ are the intensities of the electric and magnetic fields, $\mathbf{D}$, $\mathbf{B}$ the vectors of electric displacement and magnetic induction, $\mathbf{j}$ electric-current density, and ϱ_{el} space-charge density.

The vectors $\mathbf{D} - \mathbf{E}$ and $\mathbf{B} - \mathbf{H}$ are related as follows:

$$\mathbf{D} = \varepsilon_{rel}\varepsilon_0 \mathbf{E}, \tag{1.3.5}$$

$$\mathbf{B} = \mu_{rel}\mu_0 \mathbf{H}, \tag{1.3.6}$$

ε_{rel}, μ_{rel} are the relative permittivity and the relative magnetic permeability of the medium, $\varepsilon_0 = 8{\cdot}854 \,.\, 10^{-12}\ \mathrm{C^2\ s^2\ kg^{-1}\ m^{-3}}$ and $\mu_0 = 4\pi \,.\, 10^{-7}\ \mathrm{kg\ m\ C^{-2}}$ are the permittivity and magnetic permeability of vacuum.

The current density $\mathbf{j}$ is determined by Ohm's Law

$$\mathbf{j} = \sigma(\mathbf{E} + \in), \tag{1.3.7}$$

where σ is the electric conductivity of the medium and $\in$ the imposed emf of non-electrical origin.

Equations (1.3.1) to (1.3.4) are universally valid, that is for moving, anisotropic and nonhomogeneous media. Eq. (1.3.1) expresses the fact that the electric current – both conduction and displacement current – generates a magnetic field in its surroundings. Similarly Eq. (1.3.2) demonstrates that the free electric charge ϱ_{el} determines the magnitude of the displacement vector $\mathbf{D}$. Eq. (1.3.3) expresses Faraday's

Law of Induction, and Eq. (1.3.4) indicates the vortex (solenoidal) nature of magnetic induction $\mathbf{B}$.

Equations (1.3.5) and (1.3.6) are only valid for dielectrically and magnetically soft, isotropic, finite-conductivity media at rest. In an ionized gas mixture we may neglect the effect of the magnetic and polarization properties of the individual plasma particles and can in most instances assume $\varepsilon_{rel} = \mu_{rel} = 1$. The vectors $\mathbf{E}$, $\mathbf{D}$, $\mathbf{H}$ and $\mathbf{B}$ are then given by the relations

$$\mathbf{D} = \varepsilon_0 \mathbf{E}, \qquad \mathbf{B} = \mu_0 \mathbf{H}. \tag{1.3.8}$$

The condition $\varepsilon_{rel} = \mu_{rel} = 1$ does not mean that the plasma is devoid of any magnetic or polarization properties. If, for instance, we superimpose an electric field on a plasma, electric surface charges form on the boundaries of the plasma, and their field compensates the external field inside the plasma. Similarly, a plasma in a magnetic field exhibits diamagnetic properties which originate from the uncompensated Larmor currents on the boundary of the plasma [Ref. 1.3.2]. Both these phenomena, however, are not due to the internal structure of the particles, but to the motion of free charges in electric and magnetic fields. This also applies to the anisotropy of plasma in a magnetic field.

In view of the great importance of Ohm's Law, its form for plasma in motion will be discussed in a separate Section (Sec. 1.3.2.2). Here, however, let us note some consequences of Maxwell's equations, which owing to Eq. (1.3.8) assume the following form:

$$\text{curl}\,\mathbf{H} - \varepsilon_0 \frac{\partial \mathbf{E}}{\partial t} = \mathbf{j}, \tag{1.3.1$'$}$$

$$\varepsilon_0 \,\text{div}\,\mathbf{E} = \varrho_{el}. \tag{1.3.2$'$}$$

$$\text{curl}\,\mathbf{E} + \mu_0 \frac{\partial \mathbf{H}}{\partial t} = 0, \tag{1.3.3$'$}$$

$$\mu_0 \,\text{div}\,\mathbf{H} = 0. \tag{1.3.4.$'$}$$

If we apply to Eq. (1.3.1)′ operation div and to Eq. (1.3.2)′ operation $\frac{\partial}{\partial t}$, then by adding the resultant equations we obtain the law of conservation of electric charge (equation of continuity):

$$\frac{\partial \varrho_{el}}{\partial t} + \operatorname{div} \boldsymbol{j} = 0. \tag{1.3.9}$$

Let us assume that a non-zero electric space charge ϱ_{el} which is not maintained by some imposed force $\in$ (hence $\in = 0$) occurs in a certain region of the plasma. This charge produces an electric field $\mathbf{E}$, which results in a balancing current $\boldsymbol{j} = \sigma \mathbf{E}$. We obtain the equation for the time change of the space charge ϱ_{el} by substituting in Eq. (1.3.9) for $\boldsymbol{j} = \sigma \mathbf{E}$ and using Eq. (1.3.2)':

$$\frac{\partial \varrho_{el}}{\partial t} = -\frac{\sigma}{\varepsilon_0} \varrho_{el} \tag{1.3.10}$$

The integral of this equation is the function

$$\varrho_{el}(\mathbf{r}, t) = \varrho_{el}^{0}(\mathbf{r}, 0)\, e^{-\frac{\sigma}{\varepsilon_0} t} \tag{1.3.11}$$

which is typical of a relaxation process (cf. Eq. 1.2.151). The time $\tau_\sigma = \dfrac{1}{\sigma/\varepsilon_0}$ is called the relaxation time for the decay of a space charge. In good conductors, such as metals or plasma, this time is very short (10^{-10} to $10^{-17}\,\mathrm{s}^{-1}$) which is the cause of the rapid establishment of quasi-neutrality in metals and plasma. The mechanism which establishes quasi-neutrality is frequently complicated by the occurence of oscillations; moreover, τ_σ can only be introduced for regions which are large in comparison with the Debye radius (Eq. 1.2.41).

In magnetohydrodynamics, the term $\varepsilon_0 \dfrac{\partial \mathbf{E}}{\partial t}$ in Eq. (1.3.1)' can mostly be neglected in comparison with the term $\boldsymbol{j}$ ("quasistationary approximation"). The ratio of these two terms is

$$\frac{\varepsilon_0 \dfrac{\partial E}{\partial t}}{j} \sim \frac{\dfrac{\varepsilon_0 E}{t_1}}{\sigma E} = \left(\frac{\sigma}{\varepsilon_0} t_1\right)^{-1} \ll 1. \tag{1.3.12}$$

This last inequality is valid unless the characteristic time t_1 for a magnetohydrodynamic process is too short or the conductivity σ too small.

1.3.2.2 Ohm's Law

For use in plasma physics, Ohm's Law (Eq. 1.3.7) has to be generalized in several respects.

1) A moving medium has to be taken into consideration. As electric current – being diffusion flow – is related to a coordinate system moving with mass-average velocity $\mathbf{v}_0(\mathbf{r}, t)$, the intensity $\mathbf{E}(\mathbf{r}, t)$ of the electric field and the intensity $\mathbf{H}(\mathbf{r}, t)$ of the magnetic field have also to be expressed in this system. In a non-relativistic approximation, the intensities $\mathbf{E}^*\mathbf{H}^*$ measured in a system moving at velocity $\mathbf{v}_0$ are related to their "at-rest" values $\mathbf{E}$, $\mathbf{H}$ as follows:

$$\mathbf{E}^* = \mathbf{E} + \mu_0 \mathbf{v}_0 \times \mathbf{H}, \tag{1.3.13}$$

$$\mathbf{H}^* = \mathbf{H} - \varepsilon_0 \mathbf{v}_0 \times \mathbf{E}, \tag{1.3.14}$$

The second term on the right-hand side of Eq. (1.3.13) may be interpreted as an e.m.f induced according to Faraday's Law of Induction in a moving conductor intersecting the lines of a magnetic field. No such interpretation exists for the corresponding term of Eq. (1.3.14) which, moreover, can usually be neglected in comparison with $\mathbf{H}$. Because of the good conductivity of plasma, electric fields are mostly very small while for induced fields $E \sim \mu_0 v_0 H$, hence the magnitude of the second member in Eq. (1.3.14) is

$$\varepsilon_0 v_0 E \sim \varepsilon_0 \mu_0 v_0^2 H \ll H, \tag{1.3.15}$$

where $\varepsilon_0 \mu_0 = \dfrac{1}{c^2}$ is the inverse square of the velocity of light in vacuum.

2) In addition to the electric field, various factors existing in a plasma can generate an electric current. In the stationary state, currents can be generated for instance by a pressure (or density) gradient, a temperature gradient, or magnetic forces $\sim \mu_0(\mathbf{j} \times \mathbf{H})$, cf. Eq. (1.3.25); in the non-stationary state, there are, in addition, the force of inertia $\left(\sim \dfrac{\mathrm{d}\mathbf{v}_0}{\mathrm{d}t}\right)$, and the force generated by the time change $\dfrac{\mathrm{d}\mathbf{j}}{\mathrm{d}t}$ of current density $\mathbf{j}$. Various authors have derived generalized forms of Ohm's Law which take into account all the factors mentioned [Ref. 1.3.3, 1.3.4, 1.3.5, 1.3.6]. The starting point of their derivations is usually the set the of basic equations due to Schlüter [Ref. 1.3.3]. These are actually

equations of motion determining the macroscopic velocities $\bar{\mathbf{v}}_i(\mathbf{r}, t)$ (Eq. 1.2.11). The interaction of the components is described in terms of friction forces.

The various generalized forms of Ohm's Law slightly differ from each other. Here, as an example, we quote the formula which Maecker [Ref. 1.3.3] has derived from Schlüter's equations by the methods of macroscopic thermodynamics of irreversible processes (the formula is valid for a plasma consisting of three components: electrons, ions, and one type of neutral particles):

$$\begin{aligned} \mathbf{j} = \sigma \Bigg\{ & \mathbf{E} + \mu_0(\bar{\mathbf{v}}_z \times \mathbf{H}) - \left[\frac{\varrho_e \varepsilon_{i0} - \varrho_i \varepsilon_{e0}}{e \varrho_z n_z (\varepsilon_{i0} + \varepsilon_{e0})} + \frac{m_i}{e \varrho_z} \right] \mu_0 (\mathbf{j} \times \mathbf{H}) - \\ & - \frac{\varrho_e \varepsilon_{i0} - \varrho_i \varepsilon_{e0}}{e n_z (\varepsilon_{i0} + \varepsilon_{e0})} \mathbf{g} + \left[\frac{\varrho_e \varepsilon_{i0} - \varrho_i \varepsilon_{e0}}{e \varrho_z n_z (\varepsilon_{i0} + \varepsilon_{e0})} + \frac{m_i}{2 e \varrho_z} \right] \operatorname{grad} p_z + \\ & + \left[\alpha_{ei} - \frac{e n_0 \varrho_z}{\varrho_0 + \varrho_e} \left(\frac{\varrho_e \varepsilon_{i0} - \varrho_i \varepsilon_{e0}}{e \varrho_z n_z (\varepsilon_{i0} + \varepsilon_{e0})} + \frac{m_i}{e \varrho_z} \right) \right] \frac{k}{2e} \operatorname{grad} T + \\ & + \frac{\varrho_e \varepsilon_{i0} - \varrho_i \varepsilon_{e0}}{e n_z (\varepsilon_{i0} + \varepsilon_{e0})} \frac{d\bar{\mathbf{v}}_z}{dt} - \frac{m_e m_i}{e^2} \frac{d}{dt} \left(\frac{\mathbf{j}}{\varrho_z} \right) \Bigg\}. \end{aligned} \tag{1.3.16}$$

The designations are:

$\mathbf{E}$, $\mathbf{H}$	intensity of the electric and magnetic fields, respectively
$\mathbf{j}$	current density
e	electron charge
m_i, m_e	mass of ions and electrons respectively
ϱ_i, ϱ_e, ϱ_0, ϱ_z	gas density of ions, electrons, neutral particles and charged particles; $\varrho_z = \varrho_e + \varrho_i$
n_0, n_z	density of neutral and charged particles $n_z = n_e = n_i$
k	Boltzmann's constant
α_{ei}	thermal diffusion ratio
ε_{io}, ε_{eo}	friction coefficients according to Schlüter's plasma dynamics
$\mathbf{g}$	gravitational force per one kg (numerically equal to gravitational acceleration)
σ	conductivity of plasma
T	temperature of plasma

The mass-average velocity of a charged-particle gas $\mathbf{v}_z$ is described by the formula

$$\varrho_z \bar{\mathbf{v}}_z = \varrho_e \bar{\mathbf{v}}_e + \varrho_i \bar{\mathbf{v}}_i, \tag{1.3.17}$$

where $\bar{\mathbf{v}}_e$, $\bar{\mathbf{v}}_i$ are the macroscopic velocities of electron and ion gases, as defined in Eq. (1.2.11). The partial pressure of charged particles p_z equals the sum $p_z = p_e + p_i$. The coefficients α_{ei}, σ, ε_{io}, ε_{eo} can be calculated in accordance with kinetic theory (for details cf. Ref. [1.3.3]).

Ohm's Law is rarely used in the complex generalized form of Eq. (1.3.16). As a rule, we select only those terms which are essential to the given case. The relative magnitudes of the individual terms of Ohm's Law in generalized form are estimated in Ref. [1.3.4]. In most cases, Ohm's Law as written below will suffice:

$$\mathbf{j} = \sigma[\mathbf{E} + \mu_0 \mathbf{v}_0 \times \mathbf{H}], \tag{1.3.18}$$

$\mathbf{v}_0$ being the mass-average velocity of the plasma.

3) The tensorial nature of plasma conductivity in a strong magnetic field is sometimes a source of trouble. In a rough approxim ation we have discussed this property in Sec. 1.2.4.1, where Eq. (1.2.85) was its expression; this formula is often used in magnetohydrodynamics. For a more thorough discussion of the tensorial nature of conductivity see e.g. Ref. [1.3.7].

1.3.2.3 Energy, momentum, and forces in the electromagnetic field

If after scalar multiplication of Eq. (1.3.1)′ by vector $\mathbf{E}$ and of Eq. (1.3.3.)′ by vector $\mathbf{H}$ we subtract the second product from the first one, we obtain

$$-(\mathbf{H} \,.\, \operatorname{curl} \mathbf{E} - \mathbf{E} \,.\, \operatorname{curl} \mathbf{H}) - \left(\mathbf{E} \,.\, \frac{\partial \varepsilon_0 \mathbf{E}}{\partial t} + \mathbf{H} \,.\, \frac{\partial \mu_0 \mathbf{H}}{\partial t}\right) = \mathbf{j} \,.\, \mathbf{E}. \tag{1.3.19}$$

Using formulae of vector analysis [Ref. 1.3.1] we can write Eq. (1.3.19) in the form

$$-\operatorname{div}(\mathbf{E} \times \mathbf{H}) - \frac{1}{2} \frac{\partial}{\partial t}(\varepsilon_0 E^2 + \mu_0 H^2) = \mathbf{j} \,.\, \mathbf{E}. \tag{1.3.20}$$

where $E = |\mathbf{E}|$, $H = |\mathbf{H}|$ are the magnitudes of vectors $\mathbf{E}$ and $\mathbf{H}$. If we introduce the designations

$$\mathbf{S} = \mathbf{E} \times \mathbf{H}, \tag{1.3.21}$$

$$u_{\text{elmg}} = \frac{1}{2}(\varepsilon_0 E^2 + \mu_0 H^2), \tag{1.3.22}$$

we can write Eq. (1.3.20) in the form of a typical balance equation (in local form)

$$-\frac{\partial u_{\text{elmg}}}{\partial t} = \mathbf{j} \cdot \mathbf{E} + \operatorname{div} \mathbf{S}. \tag{1.3.23}$$

All terms of this equation have the dimension of power density [W m^{-3}]. The first member on the right-hand side stands for the input density of the electric forces. The time derivative on the left-hand side indicates the time change in the density–(Eq. 1.3.22)–of the electromagnetic energy. Finally, the last member of Eq. (1.3.23) denotes the divergence of the electromagnetic flow density i.e. the energy radiated by an infinitesimal unit volume. This flow density is described by the Poynting vector $\mathbf{S}$ (Eq. 1.3.21).

Experiments have yielded the following equation for the force $\mathbf{f}$ acting upon a unit volume of conductive medium under current of density $\mathbf{j}$, which carries a space charge of density ϱ_{el} ($\varepsilon_{\text{rel}} = \mu_{\text{rel}} = 1$):

$$\mathbf{f} = \varrho_{\text{el}}\mathbf{E} + \mu_0(\mathbf{j} \times \mathbf{H}). \tag{1.3.24}$$

The first term on the right-hand side is negligible in a high-conductivity plasma (quasi-neutrality; except for the border regions or for strong induced electric fields; strong electric fields rarely occur in plasma). It is the second term called the "magnetic force", that is of prime importance for the entire field of magnetohydrodynamics:

$$\mathbf{f}_m = \mu_0(\mathbf{j} \times \mathbf{H}). \tag{1.3.25}$$

By substituting for current density $\mathbf{j}$ from Maxwell's equation Eq. (1.3.1)' -neglecting the term $\dfrac{\partial \varepsilon_0 \mathbf{E}}{\partial t}$ –Eq. (1.3.25) assumes the following form:

$$\mathbf{f}_m = \mu_0 \operatorname{curl} \mathbf{H} \times \mathbf{H}. \tag{1.3.26}$$

Using the formula of vector analysis

$$\boldsymbol{a} \times \operatorname{curl} \boldsymbol{a} = \frac{1}{2} \operatorname{grad} |\boldsymbol{a}|^2 - (\boldsymbol{a} \,.\, \operatorname{grad})\, \boldsymbol{a},$$

we obtain

$$\boldsymbol{f}_m = -\frac{\mu_0}{2} \operatorname{grad} H^2 + \mu_0 (\boldsymbol{H} \,.\, \operatorname{grad}) \boldsymbol{H}, \tag{1.3.27}$$

where the first member on the right-hand side is the gradient of energy density of the magnetic field.

For the body force $\boldsymbol{f}_m$ we may substitute an equivalent surface force $\boldsymbol{T}$, acting on the surface S of some plasma-filled volume V defined by the relation:

$$\int_{(V)} \boldsymbol{f}_m \, \mathrm{d}V = \int_{(S)} \boldsymbol{T}_n \, \mathrm{d}S. \tag{1.3.28}$$

In Eq. (1.3.28) $\boldsymbol{T}_n \, \mathrm{d}S$ is a force acting in the direction of the normal $\boldsymbol{n}$ on area $\mathrm{d}S$ of surface S. This force depends on the magnitude and direction of the magnetic field $\boldsymbol{H}$ at the place occupied by element $\mathrm{d}S$ and on the direction of its normal $\boldsymbol{n}$ [cf. Refs. 1.3.5, 1.3.8, 1.3.9].

$$\boldsymbol{T}_n = \mu_0 \boldsymbol{H} H_n - \frac{\mu_0}{2} H^2 \boldsymbol{n}. \tag{1.3.29}$$

Two forces act on every point of surface S, according to Eq. (1.3.29). The first one, aiming along the lines of force of the magnetic field, numerically equals $\mu_0 H$ multiplied by H_n, the number of lines of force intersecting the element $\mathrm{d}S$. The second force, directed towards the surface, acts along its normal and can be interpreted as an isotropic magnetic pressure of the magnitude $\frac{n_0}{2} H^2$.

According to Eq. (1.3.28) we can consider $\boldsymbol{f}_m$ as the divergence $\operatorname{Div} \mathsf{T}$ of a second-order tensor (Gauss-Ostrohradsky's theorem). The same applies to the total density of the ponderomotive force $\boldsymbol{f}$ described by Eq. (1.3.24).

$$\boldsymbol{f} = \operatorname{Div} \mathsf{T}, \tag{1.3.30}$$

or written componentwise

$$f_\alpha = \sum_{\beta=1}^{3} \frac{\partial T_{\alpha\beta}}{\partial x_\beta}, \tag{1.3.31}$$

where the components $T_{\alpha,\beta}$ of the so-called Maxwell tensor of electromagnetic stress are defined by the equation

$$T_{\alpha\beta} = \varepsilon_0 E_\alpha E_\beta + \mu_0 H_\alpha H_\beta - \frac{1}{2}\delta_{\alpha\beta}(\varepsilon_0 E^2 + \mu_0 H^2). \qquad (1.3.32)$$

$E_\alpha, E_\beta, H_\alpha, H_\beta$ are components of the vectors **E** and **H**, $\delta_{\alpha\beta}$ is Kronecker's symbol defined by the relationship

$$\delta_{\alpha\beta} = \begin{cases} 0 & \text{for} \quad \alpha \neq \beta \\ 1 & \text{for} \quad \alpha = \beta . \end{cases} \qquad (1.3.33)$$

The equations (1.3.26) to (1.3.31) are valid for a stationary electromagnetic field. In the case of a non-stationary field, the density of the ponderomotive force $\boldsymbol{f}$ is defined by Eq. (1.3.34) which follows from Eq. (1.3.24) and the Maxwell equations (for its derivation see Ref. [1.3.1]):

$$f_\alpha = \sum_{\beta=1}^{3} \frac{\partial T_{\alpha\beta}}{\partial x_\beta} - \frac{\partial}{\partial t} g_\alpha, \qquad (1.3.34)$$

where the vector $\boldsymbol{g}$ has the form

$$\boldsymbol{g} = \varepsilon_0 \mu_0 (\mathbf{E} \times \mathbf{H}). \qquad (1.3.35)$$

Equation (1.3.34) is again a balance equation (in local form); all its terms have the dimension of the time variation of momentum density:

$$\frac{\partial \boldsymbol{g}}{\partial t} = \operatorname{Div} \mathrm{T} - \boldsymbol{f}. \qquad (1.3.34)'$$

We interpret $\boldsymbol{g}$ as the macroscopic – momentum density of the electromagnetic field, and components $-T_{\alpha 1}$, $-T_{\alpha 2}$, $-T_{\alpha 3}$ of tensor T represent three components of the flow density of the αth component of the electromagnetic-field momentum. Compared with the general balance equation (1.2.45), Eq. (1.3.34) also makes it obvious that σ_g – the production of electromagnetic-field momentum – in the unit volume per unit time is given by the density of the ponderomotive force $\boldsymbol{f}$. The law of conservation of momentum (Eq. 1.3.34)′ signifies that the time change of momentum density of the electromagnetic field in any infinitesimal volume is given by the flow of electromagnetic momentum through the surface bounding this volume and by the magnitude of the electromagnetic momentum which inside this volume changes into mechanical momentum (or vice versa) in one second.

1.3.3 Equations of Gas Dynamics

The physical description of a flowing medium hinges on the three fundamental Laws of Conservation discussed in Sec. 1.2.3.2. They are expressed in:

1 – The equation of the conservation of mass (1.2.51), also called the continuity equation; for the individual components of a gas mixture we use the balance equations (1.2.47).

2 – The equation of the conservation of momentum (1.2.49), also called the equation of motion.

3 – The equation of the conservation of energy (1.2.50), also called the energy equation.

Several requirements of a thermodynamic nature are added to these three laws; they are in particular the selection of independent thermodynamic variables of state (cf. Sec. 1.1.1.2) and of the equation of state (Sec. 1.1.1.4); and – if need be – the suitable formulation of deviations from thermodynamic equilibrium.

1.3.3.1 The equation of motion

The law of the conservation of momentum reads:

$$\varrho\left[\frac{\partial \mathbf{v}_0}{\partial t} + (\mathbf{v}_0 \, . \, \mathrm{grad})\, \mathbf{v}_0\right] =$$
$$= -\mathrm{Div}\,(p\mathsf{I} + \mathsf{S}) + \mu_0(\boldsymbol{j} \times \boldsymbol{H}) + \varrho_{\mathrm{el}}\mathbf{E}. \qquad (1.3.36)$$

Equation (1.3.36) differs from Eq. (1.2.49) by the developed form of the total derivative $\frac{\mathrm{d}\mathbf{v}_0}{\mathrm{d}t}$ (cf. Eq. 1.2.46); by the substitution from Eq. (1.3.24) for the resultant external force (i.e. $\Sigma\, \varrho_i \boldsymbol{F}_i = \boldsymbol{f}$); and by the pressure tensor P split into its isotropic part

$$p\mathsf{I} = -p\delta_{\alpha\beta} \qquad (1.3.37)$$

and the tensor of shear stress with components as shown below:

$$S_{\alpha\beta} = \eta \left(\frac{\partial v_{0\alpha}}{\partial x_\beta} + \frac{\partial v_{0\beta}}{\partial x_\alpha} - \frac{2}{3} \delta_{\alpha\beta} \sum_{\gamma=1}^{3} \frac{\partial v_{0\gamma}}{\partial x_\gamma} \right) + \zeta \, \delta_{\alpha\beta} \sum_{\gamma=1}^{3} \frac{\partial v_{0\gamma}}{\partial x_\gamma} . \tag{1.3.38}$$

In the equations (1.3.36) to (1.3.38), **I** is the unit tensor, p the hydrostatic pressure, Div the tensor divergence, $\delta_{\alpha\beta}$ Kronecker's symbol (Eq. 1.3.33), η the coefficient of dynamic viscosity (cf. Sec. 1.2.6.1), ζ the coefficient of volume viscosity [cf. Ref. 1.3.8], and $v_{0\alpha}$, $v_{0\beta}$, $v_{0\gamma}$ are components of the mass-average velocity.

If we carry out the tensor divergence indicated in Eq. (1.3.36), we obtain the well-known form of the Navier-Stokes equation:

$$\varrho \left[\frac{\partial \mathbf{v}_0}{\partial t} + (\mathbf{v}_0 \operatorname{grad}) \mathbf{v}_0 \right] = -\operatorname{grad} p + \eta \, \Delta \mathbf{v}_0 + $$
$$+ \left(\zeta + \frac{\eta}{3} \right) \operatorname{grad} \operatorname{div} \mathbf{v}_0 + \mu_0 \mathbf{j} \times \mathbf{H} + \varrho_{el} \mathbf{E}, \tag{1.3.39}$$

where Δ is the Laplace operator $\left(\Delta = \frac{\partial^2}{\partial x^2} + \frac{\partial^2}{\partial y^2} + \frac{\partial^2}{\partial z^2} \right)$.

In the equation for stationary flow the term $\frac{\partial \mathbf{v}_0}{\partial t}$ is absent.

If a fluid can be considered incompressible, then $\varrho =$ constant, div $\mathbf{v}_0 = 0$ (for velocities not too near to the local velocity of sound in subsonic gas flow), and the term $\left(\zeta + \frac{\eta}{3} \right)$ grad div $\mathbf{v}_0$ is omitted from Eq. (1.3.39).

Provided the flow is irrotational (curl $\mathbf{v}_0 = 0$), the convection term on the left-hand side of Eq. (1.3.39) may be expressed in the form:

$$(\mathbf{v}_0 \operatorname{grad}) \mathbf{v}_0 = \operatorname{grad} \left(\frac{1}{2} v_0^2 \right) . \tag{1.3.40}$$

If the gradients of mass-average velocity in the flow field are small enough, the gas mixture may be considered non-viscous; in this case the terms containing η and ζ vanish from Eq. (1.3.39).

In the simplest case—with the plasma considered non-viscous and flow being irrotational—the equation of motion assumes the following form:

$$\varrho\left[\frac{\partial \mathbf{v}_0}{\partial t} + \operatorname{grad}\left(\frac{1}{2}v_0^2\right)\right] = -\operatorname{grad} p + \mu_0(\boldsymbol{j}\times\boldsymbol{H}) + \varrho_{el}\mathbf{E}. \quad (1.3.41)$$

The effect of the gravitional force is usually neglected in the forced flow of a plasma (e.g. in a plasma torch). In other instances, however, it can make itself felt as uplift; such is, for example, the natural convection of an arc freely burning in the air, where gravity causes the characteristic arc shape in the horizontal position. In these cases the right-hand side of the equation of motion has to be supplemented by the term $\varrho\mathbf{g}_0$, where $\mathbf{g}_0$ is the gravitional force per unit mass (numerically equalling gravitational acceleration).

1.3.3.2 The equation of continuity

The balance equation (1.2.51) is the most frequent form in which the law of conservation of mass is applied to the flow:

$$\frac{\partial \varrho}{\partial t} + \operatorname{div}(\varrho\mathbf{v}_0) = 0. \quad (1.3.42)$$

This equation does not describe how the total density ϱ of the mixture is related to the partial densities ϱ_i of the individual components. It is no more than the mathematical expression of the fact that the total mass contained in any infinitesimal volume cannot change except by the inflow of mass into this volume or its outflow from it.

When the composition of the mixture does not change (frozen chemical equilibrium), Eq. (1.3.42) may be written for each component separately:

$$\frac{\partial \varrho_i}{\partial t} + \operatorname{div}(\varrho_i\mathbf{v}_0) = 0. \quad (1.3.42)'$$

This case rarely occurs in the flow of technical plasmas. As a rule, this flow is accompanied by changes in the temperature of the plasma, and owing to the short relaxation times, a state of local dissociation and ionization equilibrium is established in accordance with the local temperature. The gradients of particle densities thus created cause diffusion flow, with the continuity equation in a local-balance form applying to each of the components:

$$\frac{\partial \varrho_i}{\partial t} = -\operatorname{div} \varrho_i \mathbf{v}_0 - \operatorname{div} \mathbf{J}_i + \Gamma_i . \tag{1.3.43}$$

Eq. (1.3.43) can be derived from Eq. (1.2.47) by expanding the total derivative and transforming its convection term. The notation is the same as in Sec. 1.2.3.2, and Γ_i indicates the production of the ith component due to chemical reactions.

1.3.3.3 The energy equation

For the description of a flowing plasma, the energy equation is essential as the starting point in determining the temperature field of the plasma and hence in describing its local thermodynamic properties. The Heller-Elenbaas equation, well known from electric-arc physics, is a typical instance of an energy balance of a plasma at rest:

$$\operatorname{div}(\varkappa \operatorname{grad} T) + \mathbf{E} \cdot \mathbf{j} = 0, \tag{1.3.44}$$

where the first term describes the heat removal by conduction from the arc body, the second indicates the ohmic heating of the plasma, $\varkappa$ is the coefficient of heat conductivity, and T the temperature.

Energy is also transported by convection (i.e. by flow) in moving plasmas. A general description of the energy balance of flowing plasma is given in Eq. (1.2.50), where we substitute for the external forces $\mathbf{F}_i$:

$$\mathbf{F}_i = \frac{(\varrho_{\text{el}})_i}{\varrho_i} \mathbf{E} + \mu_0 \frac{(\varrho_{\text{el}})_i}{\varrho_i} \bar{\mathbf{v}}_i \times \mathbf{H}; \tag{1.3.45}$$

ϱ_i is the density of ith particles, $(\varrho_{\text{el}})_i$ the corresponding density of their electric charge, $\bar{\mathbf{v}}_i$ their macroscopic velocity (see Eq. 1.2.11). After rewriting, Eq. (1.2.50) assumes the following form:

$$\varrho \frac{\mathrm{d}}{\mathrm{d}t}\left(u + \frac{v_0^2}{2}\right) = -\operatorname{div} \mathbf{Q} - \operatorname{div}(\mathbf{P} \cdot \mathbf{V}_0) + \mathbf{E} \cdot \mathbf{j}; \tag{1.3.46}$$

the notation is the same as in Eq. (1.2.50).

The work performed by the magnetic forces does not appear in the total-energy balance (Eq. 1.3.46), because the increase in the kinetic energy of the plasma due to these forces, is balanced by the corresponding decrease in internal energy. This example shows that the total-energy balance according to Eqs. (1.2.50) or (1.3.46) is not sufficient for a detailed description of the energy conditions in the plasma. Therefore we quote here separately the balances of the total kinetic energy and of internal energy [Ref. 1.3.4]. The scalar multiplication of the equation of momentum conservation (Eq. 1.3.36) by the vector $\mathbf{v}_0$ yields the first of these balances:

$$\frac{1}{2}\varrho\frac{dv_0^2}{dt} = -\mathbf{v}_0 \cdot \mathrm{Div}(p\mathsf{I} + \mathsf{S}) + \varrho_{el}\mathbf{v}_0 \cdot \mathbf{E} + \mu_0(\mathbf{j}\times\mathbf{H}) \cdot \mathbf{v}_0 . \tag{1.3.47}$$

We obtain the balance of internal energy by subtracting Eq. (1.3.47) from Eq. (1.3.46):

$$\varrho\frac{du}{dt} = -\mathrm{div}\,\mathbf{Q} - \mathrm{div}(\mathsf{P}\cdot\mathbf{v}_0) + \mathbf{v}_0 \cdot \mathrm{Div}\,\mathsf{P} + \\ + \mathbf{E}\cdot\mathbf{j} - \varrho_{el}\mathbf{v}_0\cdot\mathbf{E} - \mu_0(\mathbf{j}\times\mathbf{H})\cdot\mathbf{v}_0 , \tag{1.3.48}$$

After rearrangement the equation reads

$$\varrho\frac{du}{dt} = -\mathrm{div}\,\mathbf{Q} - p\,\mathrm{div}\,\mathbf{v}_0 + (\mathbf{E} + \mu_0\mathbf{v}_0\times\mathbf{H})\cdot(\mathbf{j} - \varrho_{el}\mathbf{v}_0) + \mathsf{S} : \mathrm{Grad}\,\mathbf{v}_0 , \tag{1.3.49}$$

where S is the tensor of shear stress (Eq. 1.3.38), Grad $\mathbf{v}_0$ the vectorial gradient (second-order tensor) of the mass-average velocity $\mathbf{v}_0$, and symbol : indicates the binary tensor product; the rest of the notation is the same as in the preceding equations. The term $-p$ div $\mathbf{v}_0$ describes the compression work, and the density $\mathbf{Q}$ of the heat flux comprises the total heat transport by all mechanisms described in Sec. 1.2.7.1.

The last member on the right-hand side of the equation—called the dissipation function—describes the change in internal energy due to internal friction. Except for the factor $\frac{1}{T}$ it is the same term that also appears in the equation for the production of entropy (the penultimate term in Eq. 1.2.56). The dissipation function, usually denoted by Φ, reads in its expanded form as follows (the notation being the same as in Eq. 1.3.38):

$$\varphi = \mathsf{S} : \operatorname{Grad} \mathbf{v}_0 = \frac{1}{2}\eta \sum_{\alpha,\beta}\left(\frac{\partial v_{0\alpha}}{\partial x_\beta} + \frac{\partial v_{0\beta}}{\partial x_\alpha}\right)^2 + \left(\zeta - \frac{2}{3}\eta\right)\left(\sum_\gamma \frac{\partial v_{0\gamma}}{\partial x_\gamma}\right)^2. \tag{1.3.50}$$

The penultimate member of Eq. (1.3.49),

$$(\mathbf{E} + \mu_0 \mathbf{v}_0 \times \mathbf{H}) \,.\, (\mathbf{j} - \varrho_{el}\mathbf{v}_0) = \mathbf{E}^* \,.\, \mathbf{j}^* \tag{1.3.51}$$

stands for the joulean heat; $\mathbf{E}^*$ is the electric-field intensity in a coordinate system moving with the plasma at velocity $\mathbf{v}_0$, and $\mathbf{j}^*$ is the current density in this system.

1.3.3.4 The complete system of equations for the motion of a plasma in an electromagnetic field

Let us now sum up the set of equations obtained for the motion of a plasma in an electromagnetic field. If we consider the transport coefficients σ, $\varkappa$, η, ζ constant—an assumption which is rarely justified—then we have to write a total of 16 unknown functions of coordinates and time:

1 – magnetic-field intensity $\mathbf{H}$ (three scalar equations)
2 – electric-field intensity $\mathbf{E}$ (three scalar equations)
3 – electric-current density $\mathbf{j}$ (three scalar equations)
4 – electric space charge ϱ_{el}
5 – the mass-average velocity $\mathbf{v}_0$ (three scalar equations)
6 – gas pressure p
7 – gas density ϱ
8 – gas temperature T

We have to select 16 equations from the relations quoted so far. The following set of basic equations roughly corresponds to the selection made in Ref. [1.3.11].

The first ten equations (three vectorial, one scalar) are drawn from electrodynamics:

Maxwell's equations (1.3.1)′, (1.3.3)′:

$$\text{curl}\,\boldsymbol{H} = \varepsilon_0 \frac{\partial \boldsymbol{E}}{\partial t} + \boldsymbol{j}, \tag{1.3.52}$$

$$\text{curl}\,\boldsymbol{E} = -\mu_0 \frac{\partial \boldsymbol{H}}{\partial t}. \tag{1.3.53}$$

Ohm's Law supplemented by the convection-current density $\varrho_{el}\mathbf{v}_0$:

$$\boldsymbol{j} = \sigma(\boldsymbol{E} + \mu_0 \mathbf{v}_0 \times \boldsymbol{H}) + \varrho_{el}\mathbf{v}_0. \tag{1.3.54}$$

The equation of continuity of the electric charge (Eq. 1.3.9):

$$\frac{\partial \varrho_{el}}{\partial t} + \text{div}\,\boldsymbol{j} = 0. \tag{1.3.55}$$

The remaining six equations are from continuum mechanics. The law of mass conservation (Eq. 1.3.42):

$$\frac{\partial \varrho}{\partial t} + \text{div}\,(\varrho\mathbf{v}_0) = 0. \tag{1.3.56}$$

The Law of Momentum Conservation (equation of motion 1.3.36):

$$\varrho \frac{\mathrm{d}\mathbf{v}_0}{\mathrm{d}t} = -\text{grad}\,p - \text{Div}\,\mathsf{S} + \mu_0(\boldsymbol{j} \times \boldsymbol{H}) + \varrho_{el}\boldsymbol{E}. \tag{1.3.57}$$

The equation of state for a mixture of ideal gases (Eq. 1.1.9)

$$p = \frac{kT}{V}\sum N_i = \varrho R'T, \tag{1.3.58}$$

where R' – the gas constant of the mixture – depends on the composition.

The law of conservation of internal energy:

$$\varrho \frac{\mathrm{d}u}{\mathrm{d}t} = -\text{div}\,\boldsymbol{Q} - p\,\text{div}\,\mathbf{v}_0 + \boldsymbol{E}^* . \boldsymbol{j}^* + \mathsf{S} : \text{Grad}\,\mathbf{v}_0. \tag{1.3.59}$$

The quantities $\boldsymbol{Q}$, S, u appearing in Eqs. (1.3.57) and (1.3.59) are not independent functions, but related to other quantities:

$$\boldsymbol{Q} = -\varkappa\,\text{grad}\,T, \tag{1.3.60}$$

$$u = u(\varrho, T). \tag{1.3.61}$$

S is given by Eq. (1.3.38), hence

$$\mathsf{S} = \mathsf{S}\left(\varrho, T, \frac{\partial v_{0\alpha}}{\partial x_\beta}\right). \tag{1.3.62}$$

Thorough analyses of the system of equations describing the macroscopic motion of a plasma are found in Refs. [1.3.12, 1.3.4, 1.3.11, 1.3.13].

1.3.4 Fundamentals of Magnetohydrodynamics

1.3.4.1 The law of electromagnetic induction in moving conductive media

In the complete system of equations (1.3.52) to (1.3.59), the mutual effect of flowing plasma and an electromagnetic field was formally described by the introduction of mass-average velocity $\boldsymbol{v}_0$ into the equations of electrodynamics, and vice versa the introduction of the magnetic force $\mu_0(\boldsymbol{j} \times \boldsymbol{H})$ into the equations describing the flow. However, this formal notation gives no true picture of the specific phenomena of magnetohydrodynamics and their physical essence, that is of the fundamental role which Faraday's Law of Electromagnetic Induction plays in the flow of a conductive fluid through a magnetic field. In integral form, the Law of Induction (cf. Eq. 1.3.53) reads as follows, for a medium at rest:

$$\oint_{(l)} \mathbf{E} \,.\, \mathrm{d}\boldsymbol{l} = -\mu_0 \frac{\partial}{\partial t} \int_{(S)} H_n \,\mathrm{d}S, \tag{1.3.63}$$

i.e. the line integral of the electric-intensity vector along some closed curve (l) is proportional to the rate of change of the magnetic flux through any randomly selected continuous surface (S) bounded by curve (l).

With the medium in motion, both the integration curve (l) and the integration surface (S) are carried away by it and deformed in accordance with the field of the macroscopic velocities $\boldsymbol{v}_0$ ("flowing curve" and "flowing surface").

The formulation of the Law of Electromagnetic Induction has to be changed, in this instance:

The line integral of the electric-field intensity $\mathbf{E}^*$ (i.e. of the intensity referred to the moving medium) along the instantaneous position of the "flowing curve" (l^*) is proportional to the instantaneous rate of change of the magnetic flux through any continuous "flowing surface" (S^*) bounded by the curve (l^*)

$$\int_{(l^*)} \mathbf{E}^* . \mathrm{d}\mathbf{l}^* = -\mu_0 \frac{\mathrm{d}}{\mathrm{d}t} \int_{(S^*)} H_n \, \mathrm{d}S^* = -\mu_0 \frac{\mathrm{d}\Phi}{\mathrm{d}t} . \qquad (1.3.64)$$

Electric intensity $\mathbf{E}^*$ (cf. Eq. 1.3.13) is given by:

$$\mathbf{E}^* = \mathbf{E} + \mu_0 \mathbf{v}_0 \times \mathbf{H}. \qquad (1.3.65)$$

The total change of magnetic flux $\frac{\mathrm{d}\Phi}{\mathrm{d}t}$ can be resolved into a local and a convection part:

$$\frac{\mathrm{d}}{\mathrm{d}t} \int_{(S^*)} H_n \, \mathrm{d}S^* = \int_{(S^*)} \frac{\partial H_n}{\partial t} \mathrm{d}S^* + \int_{(S^*)} H_n \frac{\mathrm{d}}{\mathrm{d}t} (\mathrm{d}S^*); \qquad (1.3.66)$$

in the first member on the right-hand side, integration is carried out over a fixed surface and the field $\mathbf{H}$ changes, while in the second term the integration surface (S^*) changes and the field does not. The field $\mathbf{E}^*$ is induced if the moving plasma intersects the magnetic-field tubes (by entering them or emerging from them) or the magnetic field changes in time. No electric field is induced if the magnetic field does not change in time and the plasma flows parallel to the lines of force ($\mathbf{v}_0 \parallel \mathbf{H} \Rightarrow \Rightarrow \mu_0 \mathbf{v}_0 \times \mathbf{H}) = 0$.

In consequence of the induction of an electric field $\mathbf{E}$ or $\mathbf{E}^*$, currents arise in the conductive medium and generate their own magnetic field $\mathbf{H}'$ which combines with the field $\mathbf{H}$ that induced the intensity $\mathbf{E}$ (or $\mathbf{E}^*$); the field $\mathbf{H}' + \mathbf{H}$ induces an intensity $\mathbf{E} + \mathbf{E}'$ (or $\mathbf{E}^*$ and $\mathbf{E}^{*\prime}$), which generates supplementary currents and a supplementary field $\mathbf{H}''$, etc. In this phenomenon, called self-induction, the magnetic fields act as if counteracting the change which has induced them (Lenz' Law; note the sign of the right-hand side of Eq. 1.3.64). Therefore enforced changes of the magnetic field in a highly conductive medium proceed at a slower rate than in a non-conductive one; for instance, in a superconductive conductor the current and the corresponding magnetic field may continue to exist for several months after the electromotive force has been disconnected.

1.3.4.2 Diffusion of the magnetic field into the plasma

If we introduce a cloud of highly conductive plasma into a magnetic field, self-induction causes the field not to appear "instantaneously" in the entire plasma-filled space, but to infiltrate gradually from the periphery of the cloud towards its centre. The change in the field, as the cloud is introduced into it, induces an emf and generates an electric current on the interface between the plasma and the surrounding medium. The magnetic field of this current cancels the original magnetic field inside the conductor. Owing to ohmic losses the induced current weakens, and the boundary between the external field and the field-free plasma gradually shifts into the cloud. Let us assume a homogeneous plasma at rest, and all macroscopic electric fields $\mathbf{E}$ generated by induction only. Let us apply the operator curl to the Maxwell equation (1.3.52) without the displacement current $\frac{\partial \varepsilon_0 \mathbf{E}}{\partial t}$ and substitute then from Eq. (1.3.53) for curl $\mathbf{E}$; we thus obtain

$$\text{curl curl } \boldsymbol{H} = \text{grad div } \boldsymbol{H} - \Delta \boldsymbol{H} = -\Delta \boldsymbol{H},$$

$$-\Delta \boldsymbol{H} = \sigma \text{ curl } \mathbf{E} = -\sigma\mu_0 \frac{\partial \boldsymbol{H}}{\partial t}; \tag{1.3.67}$$

the condition div $\boldsymbol{H} = 0$ has been demanded here, and Ohm's Law has been used in its simple form

$$\boldsymbol{j} = \sigma \mathbf{E}. \tag{1.3.68}$$

The rate of change of the magnetic field $\frac{\partial \boldsymbol{H}}{\partial t}$ is then described by the equation

$$\frac{\partial \boldsymbol{H}}{\partial t} = \frac{1}{\sigma\mu_0} \Delta \boldsymbol{H} = \nu_m \Delta \boldsymbol{H}. \tag{1.3.69}$$

This is a parabolic equation, describing irreversible processes such as heat conduction or diffussion. In this connection we sometimes speak of diffusion of the magnetic field into the plasma, and ν_m is often called the diffusion coefficient of the magnetic field [Ref. 1.3.9] or magnetic viscosity [Ref. 1.3.4].

If we introduce in Eq. (1.3.69) the characteristic values for dimension L and field H, then we obtain the following relationship for the characteristic time t_0 in which the field changes:

$$t_0 = \sigma\mu_0 L^2 = \frac{L^2}{\nu_m}. \tag{1.3.69$'$}$$

As a rule, this time is very short; for instance, for a spherical plasma cloud of 0·1 m diameter, having one thousandth of the conductivity of copper, this time is about 10^{-4} s.

According to Eq. (1.3.69), the increase of the magnetic field in the plasma after the generation of the magnetizing current is decribed in a rough approximation by the equation:

$$\boldsymbol{H} = \boldsymbol{H}_0\left(1 - e^{-\frac{t}{t_0}}\right), \tag{1.3.70}$$

where $\boldsymbol{H}_0$ is the stationary value (saturated value) of the magnetic field due to the same magnetization current.

In a plasma cloud introduced into a magnetic field, we can consider the thickness d of the layer already penetrated by the magnetic field as the characteristic dimension L (cf. Eq. 1.3.69). The rate of diffusion of this field can be estimated according to the following formula:

$$v \sim \frac{d}{t_0} = \frac{1}{\sigma\mu_0 d} = \frac{\nu_m}{d}. \tag{1.3.71}$$

1.3.4.3 Freezing of lines of force into moving plasma

Let us again start from Maxwell's equation (1.3.52) in which we substitute from Eq. (1.3.18) for current density $\boldsymbol{j}$:

$$\boldsymbol{j} = \sigma(\boldsymbol{E} + \mu_0\boldsymbol{v}_0 \times \boldsymbol{H}); \tag{1.3.72}$$

the term in round brackets is the electric-field intensity referred to a coordinate system moving together with the plasma at mass-average velocity $\boldsymbol{v}_0$. Equation (1.3.52) assumes then the following form:

$$\text{curl}\,\boldsymbol{H} = \sigma(\boldsymbol{E} + \mu_0\boldsymbol{v}_0 \times \boldsymbol{H}). \tag{1.3.73}$$

If we apply curl to this equation and substitute for curl $\mathbf{E}$ from Eq. (1.3.53), which is generally valid (with the limitation $\mu_{rel} = 1$), then—after the same rearrangement which yielded Eq. (1.3.69) and with the same notation—we obtain what is called the equation of induction:

$$\frac{\partial \mathbf{H}}{\partial t} = \text{curl}\,(\mathbf{v}_0 \times \mathbf{H}) + \nu_m \Delta \mathbf{H}; \tag{1.3.74}$$

this is the fundamental equation of magnetohydrodynamics.

The term $\nu_m \Delta \mathbf{H}$ in this equation stands for the ohmic losses. It is negligible if conductivity is sufficiently high; the condition under which it may be omitted is formulated by comparing the magnitudes of the two terms on the right-hand side:

$$\frac{|\,\text{curl}\,(\mathbf{v}_0 \times \mathbf{H})\,|}{|\,\nu_m \Delta \mathbf{H}\,|} \sim \frac{v_0 L}{\nu_m} = \sigma \mu_0 v_0 L = \text{Re}_m, \tag{1.3.75}$$

where L is the characteristic dimension and $\mathbf{v}_0$ the flow velocity. In analogy with hydrodynamics, the non-dimensional quantity Re_m is called the magnetic Reynolds number (ν_m corresponds to $\frac{\eta}{\varrho}$ where η is dynamic viscosity and ϱ density).

For

$$\text{Re}_m = \frac{v_0 L}{\nu_m} \gg 1 \tag{1.3.76}$$

i.e. high conductivity, the ohmic losses may be neglected and Eq. (1.3.74) assumes the form

$$\frac{\partial \mathbf{H}}{\partial t} - \text{curl}\,(\mathbf{v}_0 \times \mathbf{H}) = 0, \tag{1.3.77}$$

which permits the physical interpretation presented below.

With the aid of Stokes' theorem we convert the line integral of Eq. (1.3.64) into a surface integral:

$$\oint_{(l^*)} \mathbf{E}^* \,.\, \mathrm{d}\mathbf{l}^* = \int_{(S^*)} \text{curl}_n\, \mathbf{E}^*\, \mathrm{d}S^*. \tag{1.3.78}$$

By substituting from Eq. (1.3.65) for **E*** we obtain

$$\int_{(l^*)} \mathbf{E}^* \,.\, \mathrm{d}\mathbf{l}^* = \int_{(S^*)} \mathrm{curl}_n\,(\mathbf{E} + \mu_0 \mathbf{v}_0 \times \mathbf{H})\, \mathrm{d}S^*; \qquad (1.3.79)$$

using Maxwell's equation (1.3.53), we may re-write this as follows:

$$\int_{(l^*)} \mathbf{E}^* \,.\, \mathrm{d}l^* = \int_{(S^*)} \left[-\mu_0 \frac{\partial \mathbf{H}}{\partial t} + \mu_0 \,\mathrm{curl}\,(\mathbf{v}_0 \times \mathbf{H}) \right]_n \mathrm{d}S^*. \qquad (1.3.80)$$

The line integral on the left-hand side, however, equals $-\mu_0 \dfrac{\mathrm{d}\Phi}{\mathrm{d}t} =$ $= -\mu_0 \dfrac{\mathrm{d}}{\mathrm{d}t} \displaystyle\int_{(S^*)} H_n \,\mathrm{d}S^*$; if Eq. (1.3.77) is valid, then Eq. (1.3.80) yields the following condition for the total change of the magnetic flux:

$$\frac{\mathrm{d}}{\mathrm{d}t} \int_{(S^*)} H_n \,\mathrm{d}S^* = 0. \qquad (1.3.81)$$

Hence if the plasma is sufficiently conductive (i.e. Eq. 1.3.76) is valid), the magnetic flux over any surface chosen at random moving together with the plasma does not change.

In view of the random selection of the fluid surface S^*, the only possible relative position of the flowing medium and the magnetic field is a position in which the lines of magnetic force move simultaneously with the plasma as if frozen into it (similarly to the vortex lines frozen into the barotropic fluids – Helmholtz' theorem). There is no other way of fulfilling the demand of Eq. (1.3.81) that the flowing plasma must not intersect the lines of magnetic force. The mechanism of this phenomenon may be imagined to act in such a way that whenever the flowing plasma intersects the magnetic lines, self-induced currents generate a magnetic field which – superimposed on the original field – causes the resultant lines of force to move simultaneously with the plasma.

Unless the conductivity of the plasma is infinitely high, the entrainment of the lines of force by the plasma and the diffusion of the magnetic field counteract each other. The lines of force are partly entrained by the plasma, but the magnetic field diffuses against the direction of the motion; which of the two phenomena prevails is indicated by the magnetic Reynolds number $\mathrm{Re_m}$ as written in Eq. (1.3.75).

1.3.4.4 Reduced system of equations of magnetohydrodynamics

The complete system of equations introduced in Sec. 1.3.3.4 formally expresses the relationships between electrical, magnetic, and thermodynamic quantities; it tells nothing, however, about the relative importance of the individual quantites or about their causal relationships. The preponderance of self-induction, as formulated in the preceding section, does not allow the electrodynamical equations to be solved by the usual procedure. This procedure starts—as a rule—from the electric field $\boldsymbol{E}$, the intensity of which is determined by the voltage source. The current $\boldsymbol{j}$ and its magnetic field $\boldsymbol{H}$ are then calculated from the known electric resistivity of the medium. From the change of the magnetic field $\boldsymbol{H}$ we next calculate the induced electric field and the induced current. The causal relations may then be represented by diagrams as they are in Ref. [1.3.8]:

$$\begin{array}{c} \varrho_{el} \quad \boldsymbol{v}_0 \\ \swarrow \ \nwarrow \ \swarrow \\ \frac{\partial \boldsymbol{H}}{\partial t} \rightarrow \boldsymbol{E} \rightarrow \boldsymbol{j} \rightarrow \boldsymbol{H}; \\ \uparrow\_\_\_\_\_\_\_\_\_\_\_\_\_\_| \end{array} \tag{1.3.82}$$

the arrows indicate the causalities.

The phenomenon of self-induction interferes with the causal relationship $\boldsymbol{E} \rightarrow \boldsymbol{j}$: the current is determined not exclusively by the superimposed field $\boldsymbol{E}$, but by the induced field $\sim \frac{\partial \boldsymbol{H}}{\partial t}$, too. Thus we have to look for other causalities. As a rule, the magnetic field $\boldsymbol{H}$ and flow velocity $\boldsymbol{v}_0$ are considered the fundamental quantities in magnetohydrodynamics: given the distribution of the magnetic field $\boldsymbol{H}$ in space and time, we can calculate the electric current $\boldsymbol{j}$ from the formula

$$\operatorname{curl} \boldsymbol{H} = \boldsymbol{j}. \tag{1.3.83}$$

The electric field $\boldsymbol{E}$ is then calculated so that the current $\boldsymbol{j}$ corresponds to the given conductivity σ of the plasma. The electric charge ϱ_{el} has to satisfy the continuity equation (1.3.55) and Ohm's Law.

This procedure and the underlying causalities can be represented in the following way:

$$\begin{array}{ccc} \mathbf{v}_0 & \rightleftarrows & \boldsymbol{H} \\ \downarrow & & \downarrow \\ \mathbf{E} & \leftarrow & \boldsymbol{j} \\ \uparrow & & \\ \varrho_{el} & & \end{array} \tag{1.3.84}$$

The pair of arrows $\rightleftarrows$ indicates that the relation between $\mathbf{v}_0$ and $\boldsymbol{H}$ is not in the nature of cause and effect, but of mutual influences.

If the relative importance of the terms appearing in the system of equations (1.3.52) to (1.3.59) is estimated, we find—at sufficiently high conductivity and for phenomena occurring not too rapidly—that in the "magnetohydrodynamic approximation" we can neglect the following terms:

$$\frac{\partial \varepsilon_0 \mathbf{E}}{\partial t} \quad \text{in Eq. (1.3.52)}$$

$$\varrho_{el}\mathbf{v}_0 \quad \text{in Eq. (1.3.54)}$$

$$\varrho_{el}\mathbf{E} \quad \text{in Eq. (1.3.57).}$$

The equation for Joule's heat (Eq. 1.3.51) will then read:

$$\mathbf{E}^* \,.\, \boldsymbol{j}^* = \frac{(j^*)^2}{\sigma} = \frac{j^2}{\sigma} = \nu_m \mu_0 \,|\operatorname{curl} \boldsymbol{H}|^2. \tag{1.3.85}$$

This omission reduces the number of unknown quantities in the equations requiring simultaneous solution to 9 : $\mathbf{v}_0$, $\boldsymbol{H}$, ϱ, p, T. The equation of state (1.3.58) may be explicitly solved with respect to the variable ϱ or p which permits the number of unknown quantities to be reduced to 8.

The equations of the "magnetohydrodynamic approximation" will then have the following forms:

Equation of induction:

$$\frac{\partial \boldsymbol{H}}{\partial t} = \operatorname{curl} \mathbf{v}_0 \times \boldsymbol{H} + \nu_m \Delta \boldsymbol{H}. \tag{1.3.86}$$

Equation of motion:

$$\varrho \frac{d\mathbf{v}_0}{dt} = -\operatorname{grad} p - \operatorname{Div} \mathrm{S} + \mu_0 \operatorname{curl} \boldsymbol{H} \times \boldsymbol{H}. \tag{1.3.87}$$

Equation of continuity:

$$\frac{\partial \varrho}{\partial t} + \operatorname{div}(\varrho \mathbf{v}_0) = 0. \tag{1.3.88}$$

Equation of conservation of internal energy (for calculating temperature):

$$\varrho \frac{\mathrm{d}u}{\mathrm{d}t} = -\operatorname{div}(\varkappa \operatorname{grad} T) - p \operatorname{div} \mathbf{v}_0 + \mathsf{S} : \operatorname{Grad} \mathbf{v}_0 + \nu_{\mathrm{m}}\mu_0 \,|\operatorname{curl} \boldsymbol{H}|^2 \tag{1.3.89}$$

Equation of state:

$$p = \varrho R'T. \tag{1.3.90}$$

The notation is the same as in Sec. 1.3.3.4. Internal energy u is considered a function of T and ϱ ($u = u(T, \varrho)$).

We see that in spite of various simplifying assumptions and omissions the system of equations of magnetohydrodynamics remains very complex; since general functions of variables of state—such as Eqs. (1.3.61), (1.3.62)—appear in it, the system of equations (1.3.86) to (1.3.90) can only be solved in the simplest cases and by using digital computers.

1.3.5 Fundamentals of Similarity Theory in Plasma Physics and Magnetohydrodynamics

It follows from the discussion in earlier Chapters that both the total macroscopic state of a plasma and its detailed microscopic state are determined by a great many different factors. The number of these factors and their complex interaction in every actual technical form of plasma virtually prevent the theoretical equations from being solved exactly. Therefore semi-empirical methods come increasingly to the fore in the physical description of high-pressure plasma. In such descriptions, the main assignment of theory is to interpret the experimental results, to classify them, and to devise new experimental procedures allowing the conclusions drawn from one kind of experiment to be extended to the widest possible range of "similar" phenomena. The criteria of the similarity or dissimilarity of phenomena are the subject of similarity theory. In its application to plasma, this theory is not very elaborate, so far; as a rule it is formulated separately for the description of the microscopic state, where it hinges on the analysis of Boltzmann's kinetic equations, and for the description of the macroscopic state, which is based on the equations of the magnetohydrodynamic approximation. Here we restrict ourselves to a few fundamental concepts of each of these trends.

1.3.5.1 Similarity of microscopic states of plasma

The microscopic state of a plasma is uniquely determined by the distribution functions f_i $(\mathbf{r}, \mathbf{v}_i, t)$ of the component types of particles. These distribution functions in turn, are determined by the system of Boltzmann's equations (1.2.32) or Focker-Planck's equations (1.2.39) in the cases of binary or multiple coulombic collisions, respectively. These equations are supplemented by the boundary conditions for the distribution functions and for the fields. The state is fully defined by the distribution functions $f_i(\mathbf{r}, \mathbf{v}_i, t)$, and the similarity principle of various configurations of plasma may therefore be formulated as follows:

If two plasma configurations are geometrically similar, they are also physically similar provided the distribution functions $f_i(\mathbf{r}, \mathbf{v}_i, t)$ of all types of particles have the same values at corresponding places and times. (For the conduction of electricity in gases the similarity principle has been formulated in Ref. [1.3.14]).

The distribution functions of two configurations are similar if their Boltzmann equations and boundary conditions have the same form, irrespective of transformations:

$$x' = kx, \quad y' = ky, \quad z' = kz, \quad t' = kt \tag{1.3.91}$$

The dashed quantities stand for coordinates and times in one configuration, and the quantities without dash for those in the other; k is the factor of proportionality.

In other words, the quantities $t, \mathbf{v}_i, \mathbf{F}_i, m_i, \left[\frac{\delta f_i}{\delta t}\right]_{\text{coll}}$ in Boltzmann's equation (1.2.27) have to be transformed in such a way that—in accordance with the transformation (1.3.91) and with the physical dimensions of these quantities—all terms of the transformed Boltzmann equation should exhibit the same factor. After reduction by this factor, the Boltzmann equations of both configurations have the same form. If the transformed boundary conditions are also identical with the original ones, the values of the functions $f_i(\mathbf{r}', \mathbf{v}_i', t')$ and $f_i(\mathbf{r}, \mathbf{v}_i, t)$ are equal.

The left-hand sides of Boltzmann equations are comparatively easily brought to the same form; it is the collision term on the right-hand side that causes difficulties. This collision member is determined by various elementary processes, some of which (ionization by stages, spatial recombination, etc.) do not permit the requisite transformation. We therefore speak of similar plasmas only where such processes do not in any considerable degree participate in the formation of the physical state of the plasma.

Some basic results of the similarity theory based on the analysis of Boltzmann's equations are tabulated below. If on the transition from one plasma configuration to another we scale down the dimensions by a factor of k, the remaining quantities have to be reduced according to the following table:

	Scale factor
time intervals and distances (including mean free paths of all elementary processes)	k
energy of particles, velocities, electromagnetic potentials, current, ohmic resistance	1
field strengths ($\mathbf{E}$, $\mathbf{H}$), pressures, surface charge density	k^{-1}
electric-current density, space charge density, density of charged particles	k^{-2}

For a detailed discussion of the "microscopic" similarity of plasma cf. Ref. [1.3.14].

1.3.5.2 Similarity of magnetohydrodynamic phenomena

Within the bounds given by the assumptions of the magnetohydrodynamic approximation, the system of equations (1.3.86) to (1.3.90) is complete and can be used as a starting point for the similarity theory of macroscopic plasma processes. In order to estimate the similarity of two different magnetohydrodynamic processes, we have to transform the

Eqs. (1.3.86) to (1.3.90) by introducting the following dimensionless quantities:

$$\boldsymbol{r} = \frac{\boldsymbol{r}}{L}, \qquad \boldsymbol{v}_0 = \frac{\boldsymbol{v}_0}{V}, \qquad t' = \frac{V}{L}\,t,$$

$$\text{grad}' = L\,\text{grad}, \qquad \text{curl}' = L\,\text{curl},$$

$$\text{div}' = L\,\text{div}, \qquad \Delta' = L^2\,\Delta,$$

$$\varrho' = \frac{\varrho}{\varrho_0}, \qquad p' = \frac{p}{\varrho_0 V^2}, \qquad T' = \frac{T}{T_0},$$

$$\varkappa' = \frac{\varkappa}{\varkappa_0}, \qquad \sigma' = \frac{\sigma}{\sigma_0}, \qquad \mathsf{S}' = \frac{L}{\eta_0 V}\,\mathsf{S}, \qquad \eta' = \frac{\eta}{\eta_0}. \tag{1.3.92}$$

L, V are the characteristic length and velocity, and index 0 denotes the reference value of a quantity.

The transformed equations (expressed in dashed variables) have then the same form as the original equations (1.3.86) to (1.3.90). In front of the individual differential operators there are non-dimensional complexes, called criteria. The equation of induction (1.3.86), for instance, has the following form in dashed quantities:

$$\frac{\partial \boldsymbol{H}'}{\partial t'} = \text{curl}'\,\boldsymbol{v}_0' \times \boldsymbol{H}' + \frac{1}{VL\sigma_0\mu_0}\,\Delta'\boldsymbol{H}'. \tag{1.3.93}$$

The dimensionless parameter (i.e. the criterion)

$$\frac{1}{VL\sigma_0\mu_0} = \text{Re}_\text{m}^{-1} \tag{1.3.94}$$

is the reciprocal value of the magnetic Reynolds number Eq. (1.3.75). The value of Re_m describes the intensity of the entrainment of the magnetic lines of force by the conductive medium.

Other dimensionless complexes are also found in the transformed equations of the magnetohydrodynamic approximation. The most important among them are listed below.

$$\text{Re} = \frac{\varrho_0 VL}{\eta_0} = \text{Reynolds number} \tag{1.3.95}$$

showing the ratio of inertia and internal friction forces, and hence the conditions for the transition to turbulent flow.

$$M = \frac{V}{\sqrt{\frac{C_p}{C_v} R' T_0}} = \frac{V}{a_0} = \text{Mach number} \tag{1.3.96}$$

indicating the ratio of the flow velocity V to the propagation velocity of elastic deformations a_0 (sound velocity). The conditon $M \ll 1$ points out that the gas may be considered incompressible. Under the radical sign there is another dimensionless complex:

$$\gamma = \frac{C_p}{C_v} = \text{Poisson's ratio.} \tag{1.3.97}$$

It indicates the ratio of heat capacities at constant pressure and constant volume.

$$\text{Pr} = \frac{C_p \eta_0}{\varkappa_0} = \text{Prandtl's number} \tag{1.3.98}$$

is the measure of similarity between the velocity field and the temperature field. (In the boundary layer, with $\text{Pr} = 1$ and grad $p = 0$, the fields are similar).

$$R_H = \frac{\mu_0 H_0^2}{\varrho_0 V^2} = \text{the magnetic pressure number} \tag{1.3.99}$$

is the ratio between magnetic pressure $\frac{\mu_0}{2} H_0^2$ and dynamic pressure $\frac{1}{2} \varrho_0 V^2$. Only if $R_H \gtrsim 1$, the magnetic field affects the flow. The number R_H can also be expressed as a velocity ratio

$$R_H = \frac{V_A}{V}, \tag{1.3.100}$$

where

$$V_A = \frac{H_0 \sqrt{\mu_0}}{\sqrt{\varrho_0}} \tag{1.3.101}$$

is the propagation velocity of Alfven's magnetohydrodynamic waves [Ref. 1.3.2] in the direction $\mathbf{H}_0$ in a non-viscous, incompressible, perfectly conductive fluid. The ratio between viscous and magnetic forces in a flowing plasma is called Hartmann's number [Ref. 1.3.9]

$$H_0 L \sqrt{\frac{\mu_0}{\nu_{0\text{m}}\eta_0}} = \text{Hartmann's number} \qquad (1.3.102)$$

where η_0 is viscosity, and $\nu_{0\text{m}}$ magnetic viscosity $\left(\nu_{\text{om}} = \dfrac{1}{\sigma\mu_0}\right)$. Hartmann's number indicates e.g. the conditions for the suppression of turbulence by the magnetic field (dissipation effect of magnetic viscosity).

Having reviewed the principal dimensionless parameters, we can answer the question as to when two magnetohydrodynamic phenomena are similar. This case arises if

a) the non-dimensional differential operators in the transformed equations have the same form as in the original ones, and

b) the dimensionless parameters of the types Eqs. (1.3.94) to (1.3.102) appearing in front of these operators have the same value.

Let us briefly mention similarity theory as formulated by T. Kihara [Refs. 1.2.13, 1.2.14]. He started from the general set of macroscopic equations for the system {plasma + electromagnetic field} and found that the law of similarity is expressed by the invariance of this system with respect to transformations:

$$(\mathbf{r}, t) \to k^{-1}(\mathbf{r}, t)$$

$$\varepsilon, \mu, \varphi, m_j, e_j, \bar{\mathbf{v}}_j \quad \text{do not change}$$

$$(\mathbf{E}, \mathbf{B}, \mathbf{D}, \mathbf{H}) \to k(\mathbf{E}, \mathbf{B}, \mathbf{D}, \mathbf{H})$$

$$(\mathbf{j}, \varrho_{\text{el}}, \varrho, n_j, \mathsf{P}_j, \mathsf{T} \to k^2(\mathbf{j}, \varrho_{\text{el}}, \varrho, n_j, \mathsf{P}_j, \mathsf{T})$$

$$\mathbf{R}_j \to k^3 \mathbf{R}_j$$

where k is a positive number, ε permittivity, μ magnetic permeability, φ the gravitational potential, m_j, n_j, $\bar{\mathbf{v}}_j$, e_j are mass, density, macroscopic velocity and charge of a j-type particle, $\mathbf{E}$ is electric intensity, $\mathbf{H}$ magnetic intensity, $\mathbf{D}$ electric displacement, $\mathbf{B}$ magnetic induction, $\mathbf{j}$ current density, ϱ_{el} electric-charge density, ϱ plasma density, P_j the partial-pressure tensor of the jth component, T the tensor of electromagnetic stress, $\mathbf{R}_j$ the friction force per unit volume of the jth component (in the sense of Schlüter's plasma dynamics; cf. Sec. 1.3.2.2).

1.3.6 The Plasma-to-Solid Interface

Physicists are paying ever increasing attention, in recent years, to the phase boundary on the plasma-to-solid interface, since the control of conditions there is of great immediate importance in the technical application of plasma. This applies to plasma cutting, metallurgical problems, hard-facing by plasma jets, material testing at extremely high temperatures, as well as to electrodes and walls in future MHD generators and similar problems. The physico-mathematical groundwork of these problems is only in the initial stages of its development, we therefore content ourselves with a few general remarks and avoid explicit mathematical formulations which would invariably require extensive supplementary discussions.

1.3.6.1 The interface between a solid wall and stationary plasma

We have to distinguish between current-carrying and currentless phase boundaries, moreover between conductive and nonconductive walls.

a) **Interface between a nonconductive wall and a plasma.** No chemical surface reactions will occur if the wall consists of chemically inert material. High-temperature plasma melts and vapourizes the material of the wall. This evaporation alters the composition of the plasma (contamination with wall material) and shifts the interface in the direction of the wall. The decisive quantity in this case is the density of the heat flow from the plasma into the wall, which is greatly influenced by the evaporating material. The total heat balance at the interface is described by the following equation:

$$Q_{\text{plas}} = Q_{\text{rem}} + Q_{\text{ag}} + Q_{\text{diss}} + Q_{\text{ion}} + Q_{\text{htg}} + Q_{\text{rad}}. \quad (1.3.103)$$

The heat flux Q_{plas} from the plasma to the wall—due to heat transfer—is partly removed by the wall (Q_{rem}), partly consumed for heating, fusing and evaporating the wall material (Q_{ag}) and for dissociation, ionization and heating ($Q_{\text{diss}}, Q_{\text{ion}}, Q_{\text{htg}}$) of the wall-material vapours entering into the plasma; another part (Q_{rad}) radiates back. Where heterogeneous chemical reactions take place on the interface, their reaction heat participates in the heat balance Eq. (1.3.103).

If the wall material is not melted and evaporated, it can affect the plasma as a catalyst for various chemical reactions (particularly in the surface recombination of electric charges).

The wall-to-plasma interface is also the scene of ambipolar diffusion of electric charges. The diffusion flux is determined by the ability of the charged particles to recombine on the wall. A surface charge can be formed on the wall, and its field is added to the electric field of the space charge formed by ambipolar diffusion.

b) **Interface between a conductive wall and a plasma.** The heat balance written in Eq. (1.3.103) continues to be valid. Moreover, metallic walls can be intensively cooled with water, and a considerable temperature gradient between the plasma and the wall may thus be maintained: if the wall is thin enough, its temperature can be kept down to a few hundred degrees while the plasma temperature ranges in the tens of thousands. The cooled wall acts as a good catalyst for recombination reactions, and the ambipolar diffusion of charges can then reach a considerable intensity. The electric field on the interface is greatly influenced by the presence of the highly conductive metal, particularly in a thin layer whose thickness corresponds to the local Debye length. Beyond this layer, the field is determined by ambipolar diffusion. For a structural description of this layer see for instance Ref. [1.3.15], where it is shown to consist of several sub-layers, the one nearest to the metal calling for quantum statistical description.

c) **Current-carrying interface.** The decisive factors are the elementary processes on the interface itself, their sum total determining the density of the emission current (provided the face is the cathode). In complex but technically very important instances, such as the cathode of an electric arc, all known emission processes (thermal emission, field emission, emission due to metastable atoms, photoemission, etc.) may participate in the current balance at the interface. In addition to the terms listed in Eq. (1.3.103), heating due to the passage of an electric current Q_J can also appear in the balance; it may be written in the form

$$Q_J = J(\pm U_{\mathrm{wf}} + U_{\mathrm{dr}}), \qquad (1.3.104)$$

where J is the current, U_{wf} the voltage corresponding to the work function, and U_{dr} part of the voltage drop which arises on the electrodes in the form of directed kinetic energy of electrons or ions.

The current flow through the interface is often accompanied by a characteristic contraction of the current-carrying region ahead of the electrode. Current densities may attain there values of $10^8 - 10^9$ A m^{-2}, and heat flux (Eq. 1.3.104) may range around 10^{11} W m^{-2}. Intensive local heating of the electrode by such a heat flow may lead to "micro-explosions" due to heat stresses, that is to the explosive release of entire microregions from the wall material; they shoot into the plasma at supersonic velocities, evaporate and are ionized there, affecting thereby the properties of the surrounding plasma. A great amount of experimental material related to processes in these constricted regions has been collected and a number of theories partially explain some phenomena observed. However, a systematical theory embracing the entire subject is lacking so far.

1.3.6.2 The interface between a solid wall and a moving plasma

If a plasma flows past a solid wall, a boundary layer always forms in the immediate proximity of the wall. Depending on the actual circumstances, this layer can be either laminar or turbulent (with a laminar sublayer). The velocity boundary layer is accompanied by thermal and diffusion boundary layers. The flow of plasma in the boundary layer is described by balance equations (laws of conservation of mass, momentum and energy) which have been solved in a fairly general form for the flow around a plate and past rotationally symmetrical bodies [cf. e.g. Ref. 1.3.16], cases that are of great importance in rocketry. A thorough analysis of problems involved in boundary layers that are the scenes of chemical reactions (including dissociation), diffusion and heat conduction may, for instance, be found in Refs. [1.2.19, 1.3.17].

For illustration let us consider a comparatively simple case: a boundary layer formed in the flow of a conductive high-pressure plasma – combustion products at ~3,000°K with a slight addition of easily ionized substances such as potassium or caesium – past a well-cooled metal wall. Provided the concentration of the additive is low, it does not affect the dynamic and thermodynamic properties of the combustion products, hence the kinetic and thermal boundary layers remain unchanged and are described by the ordinary system of boundary-layer equations [Ref. 1.3.17]. The ionization degree in the boundary layer is then determined by space ionization and recombination processes as well as by the diffusion of charged particles to the wall.

At some distance from the wall, ionization and recombination proceed at a rapid rate, hence the ionization degree corresponds to the temperature (in accordance with Saha's equation 1.1.79). In the immediate proximity of the wall, the density of charged particles is so low that diffusion makes them reach the wall before their space recombination can take place; they recombine then on the wall. This is why the ionization boundary layer can be divided into an equilibrium sublayer controlled by the temperature field at the interface, and a sublayer with frozen recombination, determined by the ambipolar diffusion of charge carriers to the wall. The internal structure of the sublayer with frozen recombination is similar to that of the interface at rest.

1.4 Radiation of Plasma

1.4.1 Intensity of Spectral Lines Emitted by Plasma

Intensive radiation of light energy is one of the main features of high-temperature plasma. The spectrum emitted by plasma has a line structure underlaid by a weaker continuous background.

Spectral lines are emitted as an electron passes from one energy level to another. The frequency of the light emitted is proportional to the difference between the energy level on which the electron was immediately before the emission and the one on which it is immediately afterwards. An electron cannot pass from a lower level to a higher one unless the atom is supplied with an amount of energy equal to or larger than the excitation energy of the state under consideration. The internal energy of the atom can be increased in several ways:

- By collision of the atom with an electron, an ion or another atom; in this case, the kinetic energy of the particles in collision is converted into excitation energy.
- By the absorption of radiation.
- By the excitation energy of another atom transferred to the atom under consideration.

Collisions with an electron, in which the energy required for excitation is transferred to an atom, may be of two different kinds:

1. Elastic collisions, in which the electron transfers virtually no kinetic energy to the atom. Hence the kinetic energy of the electron after the collision equals its kinetic energy prior to it.

2. Inelastic collisions, in which the electron transfers a part of its kinetic energy to the atom and leaves it in an excited state. An inelastic collision of an electron with an atom occurs when the energy

of the electron is sufficient to transfer it into the next higher state. Some inelastic collisions of atoms with electrons do not result in excitation although the energy of the electron is sufficient. In other words, the transfer of energy in the collision is subject to a specific probability, defined as the ratio between the number of inelastic collisions with electrons and the number of all collisions—elastic and inelastic.

In describing these phenomena, we must recall that the collision of an atom with an electron is not identical with the collision of two spheres; hence it is independent of the geometrical dimensions of the particles in collision, which are governed by the laws of the kinetic theory of gases. The occurrence of a collision resulting in excitation is dependent upon a quantity called the effective (excitation) cross-section for the collision of an atom with an electron of a given velocity.

In the same connection we should also mention the effective radius R, that is the minimum distance at which a moving electron may pass an atom without exciting it. The effective cross-section of the atom is then

$$Q = \pi R^2 \tag{1.4.1}$$

and the quantity Q is a measure of the probability that the collision will cause excitation.

The effective cross-section Q of the atom changes with the velocity of the electron in collision; this dependence is termed the excitation function. In discussing this quantity we have to distinguish two concepts: the function for excitation to a specific energy level, and the excitation function for a specific spectral line. These two functions can only coincide if no more than a single transition to the level under consideration is permissible.

The excitation functions for some hydrogen lines are plotted in Fig. 4. The shape of the function is dependent on the properties of the given energy level, particularly its multiplicity. If the kinetic energy of the electron is lower than the excitation energy for the level under consideration, the effective cross-section is zero and elastic collisions are the only ones to occur. The value of the effective cross-section increases rapidly with the velocity of the electrons and attains its maximum at a velocity slightly higher than is sufficient for excitation to the level under consideration. Beyond this maximum, the value of the effective cross-section slowly decreases.

Fabrikant [Ref. 1.4.1.] has derived the following mathematical expression for the excitation function:

$$Q = Q(U_m)\frac{U - U_b}{U_m - U_b}\, e^{1-\frac{U-U_b}{U_m-U_b}} \tag{1.4.2}$$

where $Q(U_m)$ is the maximum value of the excitation function, U_b the excitation voltage, and U_m the voltage corresponding to the maximum of the excitation function. The quantities of Eq. (1.4.2) can be ascertained by experiment. In the form written above, the equation describes the course of the excitation function for singlet and triplet levels in the neigbourhood of the maximum. Deviations at larger separations from the

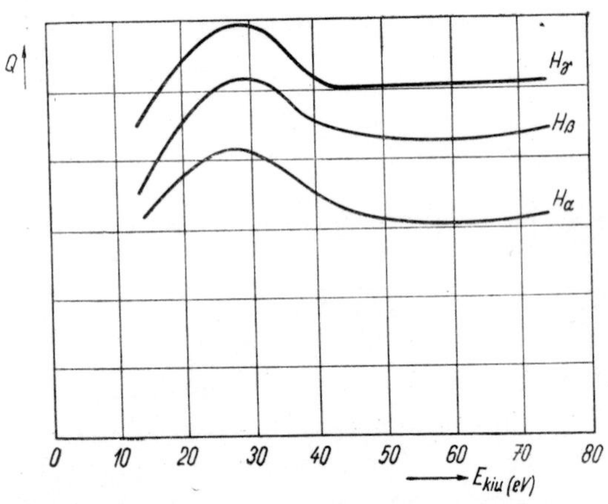

Fig. 4 Excitation functions for the first three Balmer lines for hydrogen

maximum may be corrected by the addition of a constant member Q_∞. We obtain then the following equation for the excitation function:

$$Q = Q(U_m)\frac{U - U_b}{U_m - U_b}\, e^{1-\frac{U-U_b}{U_m-U_b}} + Q_\infty . \tag{1.4.3}$$

This formula yields values in good agreement with those ascertained by experiment.

According to the way in which the electron passes from one level to another, we distinguish three phenomena: Spontaneous emission, absorption and induced emission (cf. diagram Fig. 5).

Spontaneous emission occurs as an atom passes from the mth to the nth energy level. The transition involves the emission of a light quantum

$$h\nu = E_m - E_n. \tag{1.4.4}$$

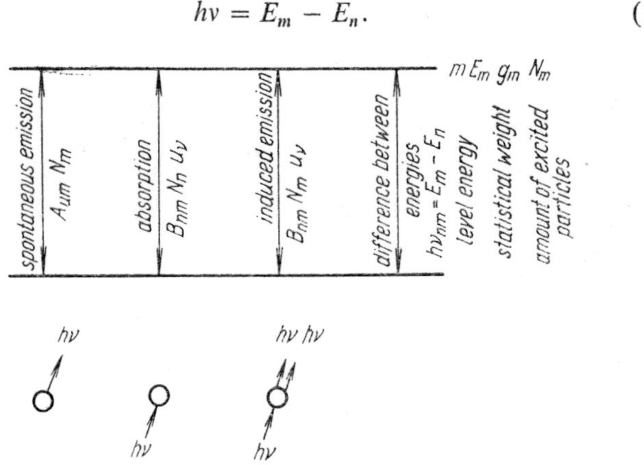

Fig. 5 Transition of electrons between successive energy levels

If at a certain time a number N_m of atoms is in the quantum state m, then—according to Einstein—the number of such transitions will amount to

$$A_{nm}N_m \tag{1.4.5}$$

and the energy emitted per second is

$$I_{nm} = A_{nm}N_n h\nu_{nm}; \tag{1.4.6}$$

the proportionality constant A_{nm} is called *Einstein's transition probability of emission.*

In the absorption of light, the atom passes from a lower quantum state to a higher one. The number of these transitions is proportional to the number N_n of atoms in the initial state n and to the density ϱ_ν of absorbed radiation having the frequency ν, the assumption being that ϱ_ν does not perceptibly vary in the frequency range under consideration. The number per second of these transitions is

$$B_{mn}N_n\varrho_\nu, \tag{1.4.7}$$

where B_{mn} is the transition probability of absorption.

More intricate phenomena are involved in induced emission. According to classical theory, the atom or oscillator need not absorb the energy of the radiation field only, but it may be excited to oscillate and radiate in the frequency range of the absorption. Einstein considers induced emission as an analogy to this phenomena. A light quantum of frequency ν forces the excited atom to emit a light quantum of the same frequency. The number per second of these phenomena is

$$B_{nm}N_m\varrho_\nu, \tag{1.4.8}$$

B_{nm} being the coefficient of the transition probability of induced emission.

We derive the relation existing between the different coefficients of transition according to Einstein. In thermodynamic equilibrium, the number of transitions from level m to level n equals the number of transitions in the inverse direction, therefore

$$(A_{nm} + B_{nm}\varrho_\nu)\, N_m = B_{mn}\varrho_\nu N_n. \tag{1.4.9}$$

The numerical proportion of atoms in the different quantum states—$N_m : N_n$—is, in thermodynamic equilibrium, described by Boltzmann's distribution (cf. Eq. 1.1.7). In the equation quoted below we consider the number of particles in the nth state instead of the number of particles in the ground state:

$$\frac{N_m}{N_n} = \frac{g_m}{g_n}\,\mathrm{e}^{-\frac{E_m - E_n}{kT}} = \frac{g_m}{g_n}\,\mathrm{e}^{-\frac{h\nu}{kT}}, \tag{1.4.10}$$

where g_m and g_n are the statistical weights of the spectral terms indicating their multiplicity. On substituting Boltzmann's distribution into Eq. (1.4.8) we obtain the radiation density of a black body at temperature T

$$\varrho_v = \frac{A_{nm}}{\dfrac{g_n}{g_m}\,B_{mn}\,\mathrm{e}^{\frac{h\nu}{kT}} - B_{nm}}. \tag{1.4.11}$$

In order to facilitate the transition to classical theory, Eq. (1.4.11) must for $\frac{h\nu}{kT} \ll 1$ express the Rayleigh-Jeans Law, hence

$$\varrho_\nu = \frac{A_{nm}}{\frac{g_n}{g_m} B_{mn} - B_{nm} + \frac{g_n}{g_m} B_{mn} \frac{h\nu}{kT}} \rightarrow \frac{8\pi\nu^3}{c^3} kT. \tag{1.4.12}$$

By comparing the last expressions we obtain the following equations expressing the relations between the transition probabilities:

$$g_m B_{nm} = -g_n B_{mn} \tag{1.4.13}$$

and

$$A_{nm} = \frac{8\pi h\nu^3}{c^3} B_{nm} = \frac{8\pi h\nu^3}{c^3} \frac{g_n}{g_m} B_{mn}. \tag{1.4.14}$$

For the transition probability we may substitute the strength of the oscillator, defined as the number of classical oscillators which replace the absorptive activity of one atom in the nth state. Defined in this way, the strength of the absorption oscillator f_{mn} is related to the transition probability by the equation

$$f_{mn} = \frac{g_m}{g_n} \frac{mc^3}{8\pi^2 e^2 \nu^2} A_{nm}. \tag{1.4.15}$$

In analogy to the strength of the absorption oscillator, we may define the strength of the emission oscillator f_{nm}

$$f_{nm} = -\frac{g_n}{g_m} f_{mn} = -\frac{mc^3}{8\pi^3 e^2 \nu^2}. \tag{1.4.16}$$

In computing the intensity of a spectral line, we start from Eq. (1.4.6) in which we replace the number of excited atoms by their density and proceed by substituting for the latter the density of excited atoms according to Boltzmann's Law

$$\frac{n_m}{n} = \frac{g_m}{Z} e^{\frac{E_m}{kT}} = \frac{g_m}{Z} e^{-\frac{E_n}{kT}} e^{-\frac{h\nu}{kT}}. \tag{1.4.17}$$

After substitution into Eq. (1.4.6) we obtain for the intensity radiated by the spectral line

$$I_{nm} = A_{nm} h\nu n \frac{g_m}{Z} e^{-\frac{E_m}{kT}}. \qquad (1.4.18)$$

In some cases it is more convenient to substitute in Eq. (1.4.18) the strength of the absorption oscillator according to Eq. (1.4.15) for the transition probability A_{nm}. Thus we obtain for the intensity of the line

$$I_{nm} = \frac{8\pi^2 e^2 \nu^3 h}{mc^3} f_{mn} \frac{g_n}{Z} n\, e^{-\frac{E_m}{kT}}. \qquad (1.4.19)$$

In discussing the intensities radiated by the lines of one series it seems advantageous to substitute for the excitation energy of the higher energy level the energy of the ground level E_n which is equal for all members of the series. To this end we use the form of Eq. (1.4.17) and obtain thus for the intensity of the line

$$I_{nm} = A_{nm} h\nu n \frac{g_m}{Z} e^{-\frac{E_n}{kT}} e^{-\frac{h\nu}{kT}}. \qquad (1.4.20)$$

We obtain an analogous form by substituting the absorptive strength of the oscillator for the transition probability. We quote this form for the lines of the Balmer series, and simultaneously we pass from the frequency scale to the scale of wavelengths. We use the Boltzmann distribution (Eq. 1.1.7) adjusted to the form

$$\frac{N_n}{n_H} = \frac{g_m}{Z} e^{-\frac{eU_m}{kT}} = \frac{g_n}{Z} e^{-\frac{E_n}{kT}} \frac{g_m}{g_n} e^{-\frac{c_1}{\lambda T}}, \qquad (1.4.21)$$

where the statistical weight $g_m = 2m^2$, and n_H is the number of all excited hydrogen atoms. After substitution we obtain for the intensity of the Balmer lines

$$I_{nm} = \frac{2\pi e^2 h f_{mn}}{m\lambda^2} n_H e^{-\frac{E_n}{kT}} e^{-\frac{c_1}{\lambda T}}. \qquad (1.4.22)$$

In this equation the main quantum number n of the lower-order term was equated to 2 and the statistical weight of the ground term of the Balmer series was substituted for the sum of state Z. At temperatures below 11,000°K the higher members of the sum of state may be neglected.

In the equation for the intensity of a spectral line emitted by a plasma, the particle density occurs as a multiplication factor. We determine it by solving a system of equations which take into consideration the chemical composition of the plasma, its degree of ionization, and the partial pressures of the individual particles.

As a general example we compute in the following paragraphs the intensity of a line emitted by a plasma according to Eq. (1.4.2). We first assume that the plasma consists of atoms and ions of a single gas; the extension to mixtures of gases will be carried out in Div. 1.5 where the composition of plasma in water—and air-stabilized arcs is being computed.

For the calculation of the intensity emitted we have available:

1. The equation for the intensity of a spectral line:

$$I = \frac{A_{nm}}{4\pi} h\nu \frac{G_n}{Z} n_e \, e^{-\frac{eU_m}{kT}} \tag{1.4.23}$$

2. We shall use Saha's equations to take into consideration the degree of ionization:

$$\frac{n_{i+1}}{n_i} = S_i(T), \tag{1.4.24}$$

where $i = 1, 2, 3 \ldots z$, and $S_i(T)$ is the right-hand side of Eq. (1.1.80) divided by the electron density n_e.

3. For consideration of partial pressures we shall use the equations of state

$$p = [n_0 + 2n_1 + 3n_2 + \ldots + (Z + 1)\, n_z]\, kT, \tag{1.4.25}$$

$$p_e = [n_1 + 2n_2 + \ldots + Zn_z]\, kT \tag{1.4.26}$$

First of all we use Eqs. (1.4.24) to (1.4.26) to determine the density of i times ionized atoms. From (1.4.25) and (1.4.26) we calculate n_{i+1} and substitute into Eq. (1.4.24). Then we obtain for $i = 0$

$$n_0 = \frac{p - p_e}{kT} \frac{1}{1 + \sum\limits_{i=1}^{z} \prod\limits_{m=0}^{i-1} \frac{S_m}{p_e}}. \tag{1.4.27}$$

In a similar way, we obtain for $i = 1, 2, \ldots 1 \ldots z$

$$n_i = \frac{p - p_e}{kT} \frac{\sum_{m=0}^{i-1} \frac{S_m}{p_e}}{1 + \sum_{i=1}^{z} \prod_{m=0}^{i-1} \frac{S_m}{p_e}}. \quad (1.4.28)$$

By substituting the results calculated for n_0 and n_i into Eq. (1.4.26), we arrive at the equation for calculating the electron pressure p_e as a function of pressure and temperature

$$p_e \left[1 + \sum_{i=1}^{Z} (i+1) \prod_{m=0}^{i-1} \frac{S_m}{p_e} \right] = p \sum_{i=1}^{i} \prod_{m=0}^{i-1} \frac{S_m}{p_e}. \quad (1.4.29)$$

We substitute the electron pressure thus computed into Eqs. (1.4.27) and (1.4.28), and the result of the calculation is then the density of ionized particles depending on plasma temperature and pressure.

Since Eq. (1.4.29) is not generally solvable, we use graphical or other suitable methods. Where only two degrees of ionization occur at a time, this equation is sufficiently accurately reduced to a quadratic equation.

If the plasma consists of neutral atoms and single ions only, then

$$p_e = S_0 \left(\sqrt{1 + \frac{p}{S}} - 1 \right). \quad (1.4.30)$$

After substitution into Eqs. (1.4.26) and (1.4.27) we obtain

$$n_0 = \frac{p}{kT} \left[1 - \frac{S_0}{p} \left(\sqrt{1 + \frac{p}{S_0}} - 1 \right) \right]. \quad (1.4.31)$$

and

$$n_1 = \frac{p}{kT} \frac{S_0}{p} \left(\sqrt{1 + \frac{p}{S_0}} - 1 \right). \quad (1.4.32)$$

The intensity of the spectral line first increases and then—having reached its maximum—decreases with increasing temperature. We shall need the temperature at which the line emits the maximum of light energy when we discuss temperature measurements (Chap. 1.5.6). In calculating this temperature we start from the equation for the intensity of the line and introduce into it the degree α of the ionization of the emitting particles. The equation for the intensity of the line has then the following form:

$$I = \frac{A}{4\pi} h\nu \frac{g_n}{g_o} \frac{1-\alpha}{1+\alpha} \frac{p}{kT} e^{-\frac{eU_m}{kT}}. \tag{1.4.33}$$

We differentiate this equation with respect to kT, equate the derivative to zero, and from the nullified equation we calculate the temperature at which the line reaches its maximum intensity. We denote this temperaturc by T'.

In the equation for T' there is a derivation with respect to kT. We calculate its value from Saha's equation, and after substitution into the equation for T' we obtain the following expression from which T' can be calculated:

$$\frac{1}{2} \frac{h^3}{(2\pi m)^{5/2} e^{5/2}} \frac{g_m}{g_0} \frac{p}{U_j^{5/2}} \left(\frac{eU_j}{kT'}\right)^{5/2} e^{\frac{eU_j}{kT}} = \left[\frac{\frac{eU_j}{kT} + \frac{5}{2}}{\frac{eU_m}{kT'} - 1}\right]^2 = 1. \tag{1.4.34}$$

The equation is transcendental and we solve it graphically. We therefore substitute

$$\frac{eU_j}{kT} = \xi \quad \text{and} \quad \frac{U_m}{U_j} = \eta \qquad [1.4.35)$$

and introduce into the constant coefficient on the left-hand side a dimensional factor which permits us to insert the pressure in atmospheres; in other words we set

$$\frac{g_m}{g_n} \frac{p}{U_j^{5/2}} = K. \qquad [\text{at, V, at} . \text{V}^{-5/2}] \tag{1.4.36}$$

Eq. (1.4.32) is thus for convenience transformed into

$$1.05 . 10^{-4} \xi^{5/2} e^{\circ} = \left[\frac{\xi + \frac{5}{2}}{\eta\xi - 1}\right]^2 - 1. \tag{1.4.37}$$

This formula is graphically represented in Fig. 6. The logarithms of both sides are plotted on the axes, and the curves show the corresponding values of parameters K and η. The point of intersection of any specific pair of parameters determines the solution for that pair by its abscissa ξ. The ranges of both parameters are sufficient for almost every case.

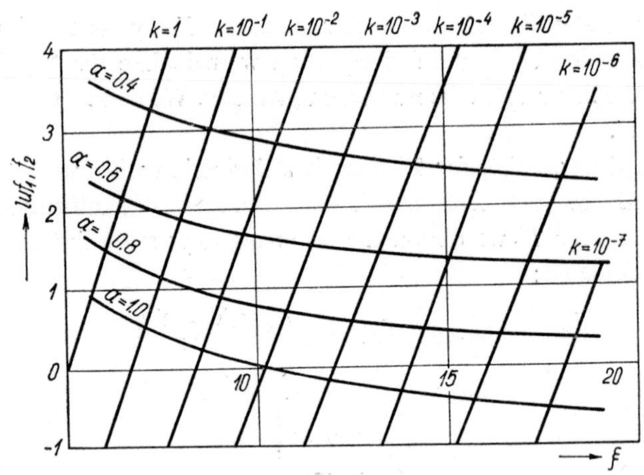

Fig. 6 Graphical solution of Eq. (1.4.36)

The range $10^{-7} < K < 1$ is sufficient if we consider that increasing pressure causes the self-absorption of the line; η cannot exceed the value 1 nor drop below 0.4.

In a similar way we can calculate the lines emitted by ionized atoms and obtain as the final result that the temperatures at which the line emits the maximum intensity are proportional to the ionizing voltages:

$$T'_0 : T'_1 : T'_2 : \ldots = eU_{j0} : eU_{j1} : eU_{j2} \ldots . \tag{1.4.38}$$

1.4.2 The Intensity of Continuous Radiation Emitted by Plasma

According to Finkelnburg and Maecker [Ref. 1.4.3] we define the emission coefficient of continuous radiation as the output emitted by a unit volume into a unit solid angle in the frequency interval $d\nu = -c\, d\lambda/\lambda^2$

$$\varepsilon_\nu = \frac{dE}{dt\, d\tau\, d\nu\, d\omega}. \tag{1.4.39}$$

In the emission from a thick layer, the intensity is proportional to the thickness. Unless absorption occurs,

$$I_\nu = \varepsilon_\nu l. \tag{1.4.40}$$

The radiated intensity dimensionally equals the intensity of a surface radiator.

The emission coefficient of continuous radiation has been numerically determined by Kramers. He used the assumption of classical physics that an electron deflected by an ion emits an output $\mathrm{d}R/\mathrm{d}t$ proportional to the acceleration b of the electron:

$$\frac{\mathrm{d}R}{\mathrm{d}t} = \frac{2}{3}\frac{e^2}{c^3}b^2. \tag{1.4.41}$$

The radiated output appears in the spectrum as so-called free continuum. Kramers derived the intensity curve of the free continuum by applying Fourier's analysis to the radiation pulse generated as particles pass each other. To include emission due to electron capture, hence to the limit continuum, he applied a semi-quantum treatment based on the correspondence principle; the results of the reasoning were later verified by quantum-mechanical computations. Unsöld [Ref. 1.4.4] extended this theory to the continuous spectrum of hydrogen-like elements, and later, when Finkelnburg had interpreted the continuous spectrum of intensive discharges as interaction between electrons and ions, to these phenomena too. The results of these computations may be summarized as follows:

At low frequencies, the total continuum, free continuum and limit continuum have a constant intensity independent of the frequency. At higher frequencies, where limit continua of lower terms come into play at the end of line series, only the mean intensity remains constant. Sawtooth oscillations of amplitude increasing with frequency are superimposed on this constant intensity, and the last sawtooth oscillation corresponds to the limit continuum of the principal series. Since the last lines of a series broaden until they merge in a continuum, the individual limit continua are already formed before the theoretical limits. At the end of a series, where lines are still visible, the wings of the lines overlap, hence the intensity close to the limit actually does not alter by jumps but continuously.

In the ultraviolet and visible ranges, the curve of the continuum plotted against frequency is either constant or rapidly decreases depending on the spacing of the terms.

The mathematical development of these theories is briefly described in the following paragraphs.

Let the limit frequency ν_g be defined as a frequency whose h-fold multiple is the energy difference between the spectral term for ionization energy and the term which terminates a sufficiently dense sequence of terms. At a frequency $\nu \geqq \nu_g$ the intensity of the continuum is

$$\varepsilon_\nu = \frac{32\pi^2}{3\sqrt{3}} \frac{e^6}{c^3} \frac{(Z+s)^2}{(2\pi m)^{3/2}} \frac{n_e n_i}{\sqrt{kT}} =$$

$$= 6.36 . 10^{-47} (Z+s)^2 \frac{n_e n_i}{\sqrt{kT}} \tag{1.4.42}$$

The intensity of this continuum is independent of the frequency. Between consecutive maxima, the intensity decreases according to the relation

$$\varepsilon_\nu \sim \frac{n_e n_i}{\sqrt{kT}} e^{-\frac{h\nu}{kT}}. \tag{1.4.43}$$

For atomic hydrogen there applies

$$\varepsilon_{\nu'H} = 6.36 . 10^{-47} \frac{n_i n_e}{\sqrt{kT}} e^{-\frac{h\nu}{kT}} \left\{ e^{\frac{h\nu_g}{kT}} + 2 \frac{E_{iH}}{kT} \sum \frac{1}{n^3} e^{\frac{E'_n}{kT}} \right\}, \tag{1.4.44}$$

where n is the quantum number and E_n the energy of the lower term of the limit continuum boundary of ionization. The summation must be carried out over the limit continua which contribute from the range of larger than limit frequencies.

The numerical value of $(Z + s)^2$ is derived from the fact that it must lie between the value computed from the differences of the terms according to the equation

$$(Z+s)^2 = n^2 \frac{E_j - E_n}{kT} \tag{1.4.45}$$

and Z^2. This is borne out by experimental results which confirm that $(Z + s)^2 \approx Z^2$.

It follows from the equations presented in this Chapter that by measuring the absolute intensity we can ascertain the electron density which, in turn, enables us to determine the temperature, with the aid of Saha's equation. The effect of the denominator $\sqrt{kT}$ may be taken into account by estimation or iteration.

Let us supplement our discussion by computing the intensity of a continuum radiated by the plasma of a water-stabilized arc with 3,500 Å and 5,600 Å wavelengths; of a continuum formed by the blur of the higher members of the Balmer series; and of a limit continuum of the Balmer series.

The intensity of the continuum emitted in the wavelengths of 3,500 and 5,600 Å consists of three components:

- the intensity of the pertinent part of the Balmer series ($I_{\nu B}$);
- the contribution of the limit continuum of the oxygen atom and of continuous radiation as far as oxygen participates in it ($I_{\nu 0}$);
- the contribution of the limit continuum of the hydrogen atom and of continuous emission as far as hydrogen participates in it ($I_{\nu H}$).

Since the wavelength $\lambda = 3{,}500$ Å lies between the ninth and tenth lines of the Balmer series, we calculate $I_{\nu B}$ as the intensity radiated in the frequency range $\Delta\nu$ between the pertinent lines, and by dividing $\Delta\nu$ we assign the calculated radiation to the frequency-scale unit. We have then

$$I_{\nu B} = \frac{A_{nm}}{4\pi} \frac{1}{\Delta\nu} n_H \frac{g_1}{Z_H} e^{-\frac{eU_m}{kT}} h\nu. \tag{1.4.46}$$

We obtain the sum of the limit-continuum contributions and the continuous radiation of the hydrogen atom by multiplying the pertinent coefficient of absorption with the Planck function. This sum has been computed by Unsöld and Maue [Ref. 1.4.5]. The effect of Paschen's continuum was taken into consideration, and the remainder of the sum is replaced by an integral. We thus obtain for the second part of the intensity

$$I_{\nu k} = \frac{64\pi^4 m e^{10}}{3\sqrt{3} c h^6 \nu^3} \overset{+}{g_2}(\nu)\, n_H \frac{2}{Z_H} e^{-\frac{eU_{jH}}{kT}} \times$$
$$\times \frac{1}{8} e^{\frac{1}{4}\frac{eU_{jH}}{kT}} \frac{2h\nu^3}{c^2} e^{-\frac{h\nu}{kT}}. \tag{1.4.47}$$

The factor g^+ is the wave-mechanical correction to Kramer's formula for the coefficient of absorption. According to Maue [Ref. 1.4.6] $g^+ = 0{\cdot}89$ for $\lambda = 3{,}600$ Å, and $g^+ = 0{\cdot}95$ for $\lambda = 5{,}600$ Å.

The third component equals the sum of the given coefficient of absorption and Planck's function. For spectra of other than hydrogen type, the coefficient of absorption can only be estimated in a rough approximation. For the "coefficient of absorption" $\varkappa_k$ ... per ion and electron in the continuum emitted by partly ionized gas we introduce

$$\varkappa_k = \frac{16\pi^2 e^6 Z^2}{3\sqrt{3}chv^3}(2\pi m)^{-3/2}(kT)^{-3/2}\,e^{\frac{hv}{kT}}. \qquad (1.4.48)$$

For Z we insert here the charge of the ions (without considering the effective charge). In calculating the contribution of oxygen to the continuous emission, we have to consider the part which twice charged ions contribute to the value of the coefficient of absorption. This part is only effective at high temperatures, but it is important because it limits the temperature range in which the computed intensity is usable for temperature measurements. We obtain then for the oxygen part

$$I_{v0} = \frac{16\pi^2 e^6}{3\sqrt{3}ch(2\pi m)^3 v^3}(kT)^{-1/2}\,e^{\frac{hv}{kT}}(n_{0+} + 4n_{0+})\,n_e\,\frac{2hv^3}{c^2}\,e^{-\frac{lv}{kT}} =$$

$$= \frac{32\pi^2 e^6}{3\sqrt{3}c^3(2\pi m)^{3/2}}(n_{0+} + 4n_{0++})\,n_e(kT)^{-1/2}. \qquad (1.4.49)$$

This part is independent of the frequency and identical with the formula derived by Maecker [Ref. 1.4.7] from Unsöld's premises for the coefficient of absorption of a partly ionized gas, with the densities of electrons and ions replacing their pressures. The densities of particles n_H, n_{0+} and n_e are computed in Chap. 1.5.3.

The calculated intensities of the continuum together with partial intensities are plotted in Figs. 7 and 8. In both cases these quantities are related to their maximum values.

A continuum formed by the blurring and merging of the higher members of the Balmer series passes at the wavelength of $\lambda = 3{,}647$ Å into a true limit continuum, owing to the microfields of ions and electrons.

Based upon the principle of correspondence, Kramers has calculated a formula for the intensity of this continuum:

$$I = \frac{14\pi e^2 h}{R_H m_0 \lambda^5} e^{\frac{eU_2}{kT}} e^{-\frac{c_2}{\lambda T}} \cdot n_H l \tag{1.4.50}$$

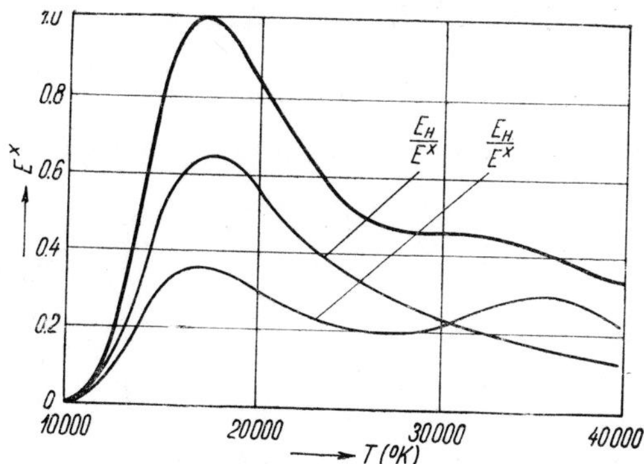

Fig. 7 Intensity of continuum radiated by arc plasma at the 6,500-Å wavelength

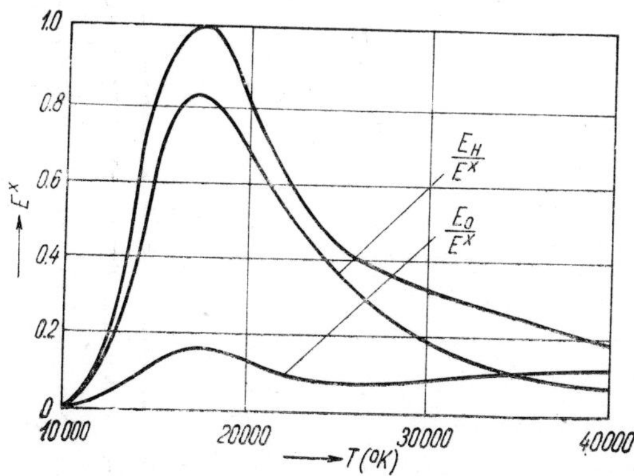

Fig. 8 Intensity of continuum radiated by arc plasma at the 3,500-Å wavelength

Gaunt [Ref. 1.4.8] and Maue [Ref. 1.4.9] have later verified it by quantum mechanical calculations.

In calculating the intensity of the limit continuum of the Balmer series, we start from the "continuous – discrete" coefficient of absorption per atom in the nth quantum state according to Kramers

$$\varkappa = \gamma_n \frac{64\pi^4 m\, e^{10}}{3\sqrt{3}ch^6 n^5 \nu^3}; \tag{1.4.51}$$

γ_n is the coefficient and its numerical value has been ascertained by Gaunt, Maue and Menzel-Pekeris [Ref. 1.4.9]. The multiplication of Eq. (1.4.51) with Boltzmann's formula yields the coefficient of absorption $\varkappa_\nu$. Before the Kirchhoff law is applied, $\varkappa_\nu$ has to be reduced – according to Unsöld – by the factor $1 - e^{h\nu/kT}$, because the effect of induced emission is not considered in Eq. (1.4.51). The intensity of the limit continuum is then

$$I = \gamma_n \frac{128\pi^4 m\, e^{10}}{3\sqrt{3}n^3 c^2 h^5 \lambda^2} e^{-\frac{eU_m}{kT}} e^{-\frac{c^2}{\lambda T}} n_{\text{H}} l. \tag{1.4.52}$$

The quantum state denoted n has to be considered the state in which the electron is captured.

In order to determine the effect of the limit continuum of the Balmer series on the emitted radiation, we have to summarize all transitions fulfilling the condition n = 2 and all free transitions. We obtain then for the "continuous – discrete" contribution

$$I = \gamma_2 \frac{128\pi^4 m\, e^{10}}{3\sqrt{3}c^2 h^5 \lambda^2} e^{-\frac{c^2}{\lambda T}} n_{\text{H}} l \times$$

$$\times \left\{ \sum_{n=2}^{4} \frac{\gamma_n}{\gamma_2 n^3} e^{-(u_1 - u_n)} + \sum_{n=5}^{\infty} \frac{\gamma_n}{\gamma_2 m^3} e^{-(u_1 - u_n)} \right\}, \tag{1.4.53}$$

where for the sake of simplicity we have substituted

$$u_n = \frac{Rhc}{n^2 kT} \quad \text{and} \quad u_1 - u_n = \frac{eU_n}{kT}. \tag{1.4.54}$$

If in the second summation we consider $\Delta n = 1$ as the summation interval, we may approximate the summation by an integral

$$\sum_{n=5}^{\infty} \frac{\Delta n}{n^3} e^{-(u_1 - u_n)} \to -\frac{1}{2} \int_{n=5}^{\infty} e^{-(u_1 - u_n)}\, d\frac{1}{u^2} =$$

$$= \frac{e^{-u_1}}{2u_1} \int_{u_n = u_5}^{u_2 = 0} e^{u_n}\, du_n. \tag{1.4.55}$$

The contribution of the continuous transitions is taken into consideration, according to Unsöld, by integrating in Eq. (1.4.55) through all states with positive energies, i.e. up to the limit $u_n \to \infty$. The summation (1.4.55) then assumes the form $e^{-(u_1-u_5)} : 2u_1$. At temperatures lower than 14,000 °K, the contribution of the higher continua is small in comparison with that of the Palmer continuum; hence in the first member of the summation we can equate $\frac{\gamma_n}{\gamma_2} = 1$. Thus from Eq. (1.4.52) we obtain the expression

$$I = \gamma_2 \frac{16\pi^4 e^{10}}{3\sqrt{3}c^2h^5\lambda^5} e^{-\frac{eU_2}{kT}} e^{-\frac{c_2}{kT}} n_H l \times$$
$$\times \left\{1 + \sum_{n=3,4} \frac{2^3}{n^3} e^{-(u_2-u_n)} + \frac{2^3}{2u_1} e^{-(u_2-u_5)}\right\}. \qquad (1.4.56)$$

The member preceding the brackets—a joint coefficient dependent upon the wavelength—equals the intensity of the Balmer continuum.

The second member of the summation in the brackets stands for the contribution of the Paschen and Bracket continua and is independent of the wavelength. The last member, independent of the wavelength, represents the contribution of the higher continua for $n > 4$.

1.4.3 Self-inversion of Lines Radiated from a Plasma

So far, we have considered the emission from an optically thin layer, i.e. from a medium where self-inversion does not occur. In using the individual equations, we have to check to what extent this assumption is correct.

Absorption processes are characterized by the coefficient of absorption, defined as the probability that a radiation will be absorbed by an atom or ion. The coefficient of absorption $\varkappa_\nu$ indicates the magnitude of the relative decrease in the intensity of radiation, which is dependent on the thickness of the absorbing medium:

$$\varkappa_\nu = -\frac{dI_\nu}{I_\nu\, dl}. \qquad (1.4.57)$$

After the passage of the ray through the homogeneous absorbing layer of thickness l, the initial intensity is reduced to

$$I_\nu = I_{\nu 0}\, \mathrm{e}^{-\varkappa_\nu l}. \tag{1.4.58}$$

The product of the coefficient of absorption and the thickness of the absorbing layer is called the optical depth of the layer

$$\tau = \varkappa_\nu l. \tag{1.4.59}$$

If $\tau \ll 1$, the absorption of radiation is negligible, whereas almost all the radiation is absorbed if $\tau \gg 1$. The coefficient of absorption changes with the absorbing matter, with temperature, pressure, density of radiation and above all with frequency.

At thermal equilibrium, Kirchhoff's law must hold good:

$$B_\nu = \frac{\varepsilon_\nu}{\varkappa_\nu}, \tag{1.4.60}$$

where B_ν is the intensity of black-body radiation; its magnitude is given by Planck's law. In a medium whose emission is not described by Planck's law, Kirchhoff's law may be used if Boltzmann's energy distribution applies to the excitation. Symbol ε_ν is here the total coefficient of emission, composed of the coefficient of spontaneous emission ε'_ν and a contribution due to induced emission. According to Einstein

$$\varepsilon_\nu = \frac{\varepsilon'_\nu}{1 - \mathrm{e}^{-\frac{h\nu}{kT}}}. \tag{1.4.61}$$

After substitution into Eq. (1.4.59) we obtain Kirchhoff's law in the form

$$\varepsilon'_\nu = \varkappa_\nu \left(1 - \mathrm{e}^{-\frac{h\nu}{kT}}\right) B_\nu = \varkappa'_\nu B_\nu. \tag{1.4.62}$$

If we consider a unit volume of the plasma, intensity increases in it by ε_ν, and simultaneously it is reduced by a value of $I_\nu \varkappa_\nu$ owing to absorption. The change in intensity in any place selected at random is then

$$\frac{\mathrm{d}I_\nu}{\mathrm{d}l} = \varepsilon_\nu - I_\nu \varkappa_\nu = \varkappa_\nu (B_\nu - I_\nu). \tag{1.4.63}$$

The integration of this equation yields

$$I_\nu = B_\nu - C\,\mathrm{e}^{-\varkappa_\nu l}. \tag{1.4.64}$$

The constant C is determined from the initial conditions. For $l = 0$, there holds $C = B_\nu$, hence

$$I_\nu = B_\nu(1 - \mathrm{e}^{-\varkappa_\nu l}). \tag{1.4.65}$$

For small optical depths $-\tau \ll 1-$ the intensity rises proportionally with the depth of the layer; at large optical depths ($\tau \gg 1$) the intensity is

$$I_\nu = B_\nu \tag{1.4.66}$$

It follows from the last equation that the plasma can at most emit black radiation corresponding to its temperature.

The self-reversal of lines emitted from plasma has been thoroughly studied by Bartels [Ref. 1.4.10, 1.4.11, 1.4.12]. We shall discuss the results of his computations in Chap. 1.5.6.

1.4.4 Broadening of Lines Emitted from Plasma

A variety of factors broadens spectral lines emitted by plasma. Most important is pressure broadening which includes—according to Finkelnbusch and Maecker [Ref. 1.4.3] and Unsöld [Ref. 1.4.13]—also broadening due to the Stark effect of microfields of electrons and ions. Collision damping and the Doppler effect are the other causes of line broadening.

As the electric fields of ions and electrons affect terms with higher energies, the lines are subject to pressure broadening. The magnitude of line broadening is described by the half width δ, which is the width of the line at half its maximum intensity.

The accurate calculation of line broadening is very difficult. The half-width of a line is proportional to the number of collisions, hence the number of perturbing particles, i.e. ion or electron density. More distant particles also contribute towards broadening, and there is no saying what phase changes must be caused by the encounter of two excited particles if it is to be qualified as collision. As a rule, an encounter is qualified as a collision if the frequency change $\Delta\omega$ multiplied by the duration of the "disturbance" t_p is far larger than 1:

$$t_p \Delta\omega \gg 1. \qquad (1.4.67)$$

The duration of the disturbance is the ratio of the minimum distance of the particles involved to their relative velocity. The condition expressed in Eq. (1.4.67) is simultaneously the criterion for the fact that the line is broadened by pressure.

Pressure broadening of spectral lines is particularly conspicuous where ions and electrons affect lines that are sensitive to Stark's effect (e.g. Balmer lines).

The dependence between the line width in the Balmer series and the ion density may be found as follows [Ref. 1.4.14]: The lines broaden in linear proportion to the intensity of the electric field F. For a point charge of the ions, F is proportional to $\frac{e}{r^2}$, and the mean distance between ions is proportional to $n^{-1/3}$. Therefore

$$F \approx en^{2/3}. \qquad (1.4.68)$$

Since processes in a plasma are of a static nature, the Stark effect does not split the line but causes its characteristic broadening.

In order to be independent of the ion density in our further calculations, we introduce, for mathematical reasons [Ref. 1.4.14] the normal field intensity F_0

$$F_0 = 2{\cdot}61en^{2/3} \qquad (1.4.69)$$

and relate all intensities F to this value. The relation of the intensity under consideration to the normal intensity is the reduced intensity

$$\beta = \frac{F}{F_0}. \qquad (1.4.70)$$

We use the reduced intensity to calculate the probability distribution $W(\beta)\, d(\beta)$ of intermolecular fields. The curve of this probability according to Holtsmark [Ref. 1.4.14] is plotted in Fig. 9. Very weak and very strong

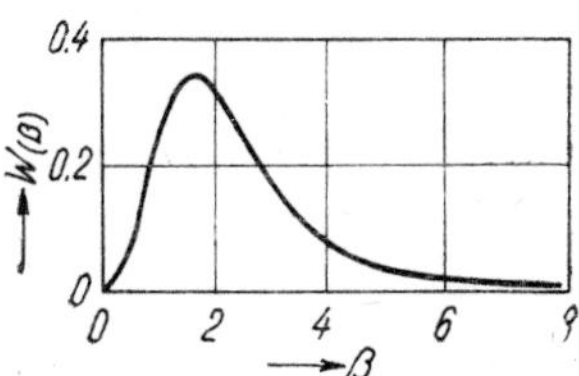

Fig. 9 Function $W(\beta)$

fields occur rather seldom, hence the distribution must have its maximum in medium fields. For hydrogen, the shift of the individual Stark components is symmetrical and proportional to the instantaneous field intensity.

The mathematical form of $W(\beta)$ is

$$W(\beta) = \frac{2}{\pi\beta} \int_0^\infty v \sin v \, e^{-(v/3)^2} \, dv, \tag{1.4.71}$$

where v is the mean relative velocity of the radiating and perturbing particles. The displacement law applying to the individual components of the line is

$$\Delta\lambda = C_i F, \tag{1.4.72}$$

where C_i is the shift of component i. By introducing another variable

$$\alpha = \frac{\Delta\lambda}{F_0} \tag{1.4.73}$$

which indicates the line broadening at unit intensity—i.e. at $F_0 = 1$— we obtain for the profile of the line

$$S(\alpha)\, d\alpha = \sum I_i \frac{1}{C_i} W\left(\frac{\alpha}{C_i}\right) d\alpha, \tag{1.4.74}$$

where C_i and I_i are the shift and intensity respectively of the individual components. The computed curves for the first four pairs of the Balmer series are plotted in Figs. 10 to 13. The profiles are normalized to unity.

$$\int_{-\infty}^{+\infty} S(\alpha)\, d\alpha = 1. \tag{1.4.75}$$

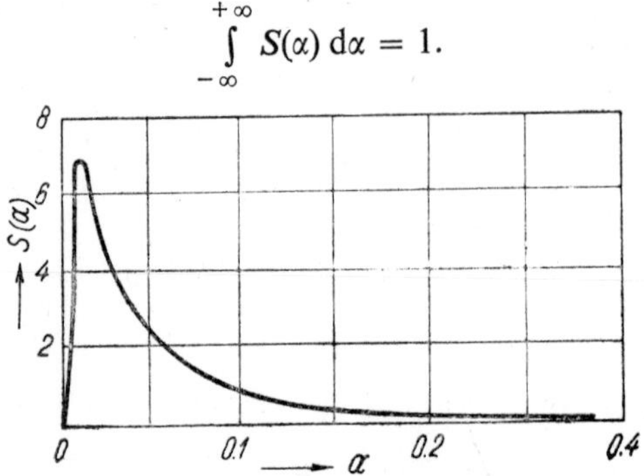

Fig. 10 Profile of Balmer line H_α

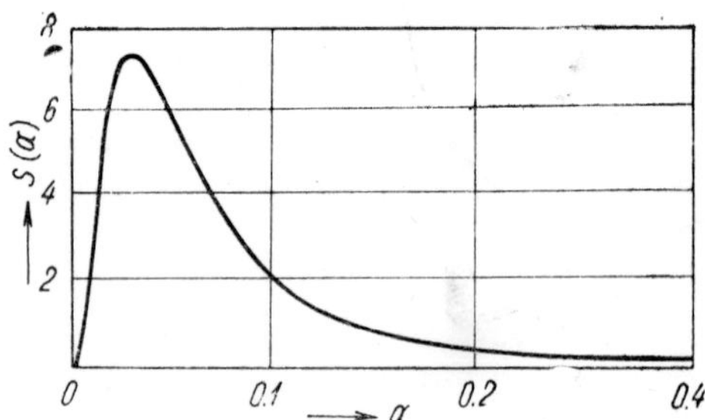

Fig. 11 Profile of Balmer line H_β

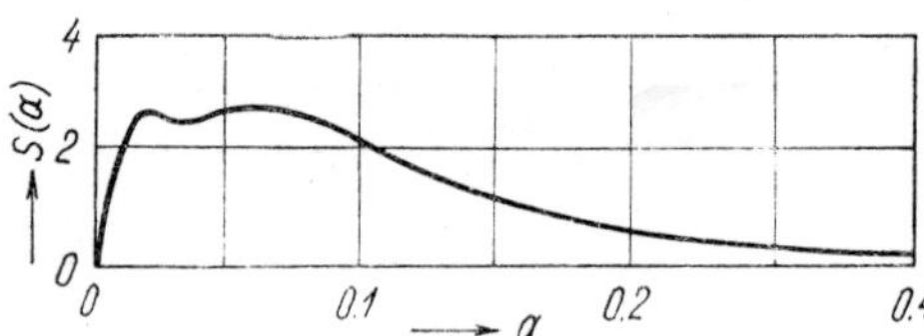

Fig. 12 Profile of Balmer line H_γ

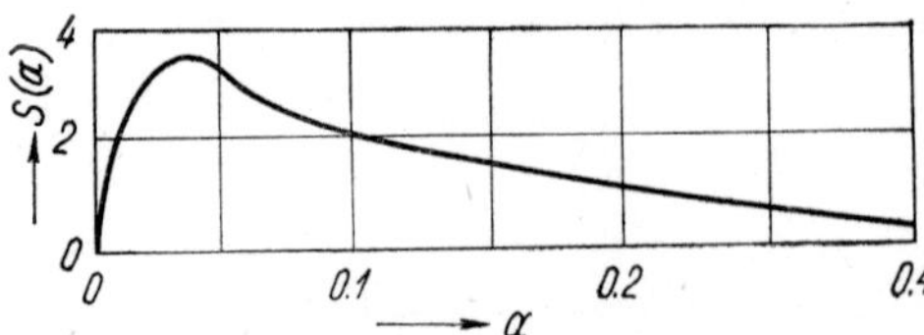

Fig. 13 Profile of Balmer line H_δ

The lines of the remaining elements are broadened by a quadratic Stark effect. Since in these cases we know neither the shift nor the intensity of the individual components, normalized profiles cannot be plotted as they were for hydrogen lines.

The half width has been computed by Lindholm [Ref. 1.4.15]. He laid down the condition that an encounter causing a change in frequency $\Delta\omega \geqq 0{\cdot}64$ is considered a collision. For the half width of a line broadened by collisions with electrons or ions he derived the equation

$$\delta = 50.3 \,.\, 10^{-10} T^{1/6} C^{3/2} \left(\frac{1}{\mu_1} + \frac{1}{\mu_2} \right)^{1/6} n, \tag{1.4.76}$$

where C is the constant of the quadratic Stark effect defined by the equation

$$\Delta\nu = \frac{C}{c^2} F^2, \tag{1.4.77}$$

μ_1 is the atomic mass of the radiating atom, μ_2 the mass of the perturbing electron or ion, and n the electron (or ion) density.

Collisional damping is the most frequent cause for the broadening of spectral lines. In calculating the half width of the line we start according to Lorentz' dispersion theory from the equation of force of an oscillating electron

$$m_2 \frac{\mathrm{d}^2 x}{\mathrm{d}t^2} + m_e \omega_0^2 x + m_e \gamma \frac{\mathrm{d}x}{\mathrm{d}t} = -e\bar{E}, \tag{1.4.78}$$

where $m_e \omega_0^2 x$ is the reverse force expressed by its self-frequency, $m_e \dfrac{\mathrm{d}x}{\mathrm{d}t}$ friction with the damping constant γ, and $e\bar{E}$ is the electric force of the radiating field.

The solution of this equation with the incident wave $E = E_0\, \mathrm{e}^{i\omega t}$ and under the condition of $x = x_0\, \mathrm{e}^{i\omega t}$ leads via the polarizability of the refracting and absorbing media to the curve of the coefficient of absorption as a function of the distance from the centre of the line

$$\varkappa_\nu = \frac{\pi e^2}{mc} \bar{N} f_{mn} \frac{2\gamma}{(\Delta\omega)^2 + \left(\dfrac{\gamma}{2}\right)^2}; \tag{1.4.79}$$

$\Delta\omega = 2\pi\,\Delta\nu$, $\bar{N}$ is the number of particles which emit the line under consideration, f_{mn} the intensity of the line oscillator, $\Delta\nu$ the frequency interval from the centre of the line. The second fraction in the last member is the weight function whose integral equals unity. In the centre of the line, where $(\Delta\omega)^2 \ll \left(\frac{\gamma}{2}\right)^2$, $\varkappa_\nu$ is constant, as expected; with increasing distance from the centre of the line it drops rapidly. For $(\Delta\omega)^2 \gg \left(\frac{\gamma}{2}\right)^2$ it decreases in proportion to $\frac{1}{(\Delta\omega)^2} \sim \frac{1}{(\Delta\lambda)^2}$; this decrease is characteristic of line broadening due to damping.

From Eq. (1.4.78) we can calculate the half width of the line

$$\delta = 2(\Delta\omega)_{1/2} = 2\pi(2\Delta\omega)_{1/2} = \frac{2\pi c}{\lambda_0^2}\,2(\Delta\lambda)_{1/2} = \gamma. \qquad (1.4.80)$$

The half width in angular frequency units thus equals the damping constant which is identical with the reciprocal time value of the reverberating line.

For undisturbed atoms, energy losses by radiation of an accelerated electron are the only causes of damping to be considered. In this case

$$\gamma = \frac{2e^2}{3m_e c^3}\,\omega_0^2 = \frac{8\pi e^2 \nu_0^2}{3m_e c^3} = \frac{8\pi e^2}{mc\lambda_0}, \qquad (1.4.81)$$

and the natural line width

$$\delta = 2(\Delta\lambda)_{1/2} = \frac{4\pi e^2}{3mc^3} = 1.18\,.\,10^{-4}\,\text{Å} \qquad (1.4.82)$$

is independent of the wavelength and is so insignificant that in the plasma it is overshadowed by other kinds of broadening.

The effect of damping on the line width may be imagined in such a manner that the frequency of the radiating electron is reduced by a decrease in amplitude or phase jumps owing to collisions. We then obtain the frequency distribution by Fourier's analysis.

Another cause of line broadening in the plasma is the Doppler effect, caused by the random motion of the radiating atoms and ions. If the particles move at a velocity of v_x in the direction of observation, the number of particles moving at a velocity of $v_x = dv_x$ is given by the equation

$$\frac{dn}{n} = \frac{1}{\sqrt{\pi}} e^{-\frac{v_x^2}{v_{x,0}}} \frac{dv_x}{dv_{x,0}}, \tag{1.4.83}$$

where $v_{x,0} = \sqrt{2kT : m_a}$.

The Doppler effect is described by the proportions

$$\frac{\Delta\nu}{\nu} = \frac{\Delta\lambda}{\lambda} = \frac{v_x}{c}. \tag{1.4.84}$$

For the line broadening of an atom or ion moving at the mean velocity of $v_{x,0}$

$$\frac{\Delta\nu}{\nu} = \frac{\Delta\lambda_0}{\lambda} = \frac{v_{x,0}}{c}. \tag{1.4.85}$$

From Eqs. (1.4.84) and (1.4.85) we obtain

$$\frac{\Delta\nu}{\Delta\nu_0} = \frac{\Delta\lambda}{\Delta\lambda_0} = \frac{v_x}{v_{x,0}}, \tag{1.4.86}$$

Since the line intensity I_l grows proportionately to the number of radiating particles

$$\frac{dn}{n} = \frac{dI_l}{I_l}; \tag{1.4.87}$$

at the same time

$$\frac{dI_l}{d(\Delta\nu)} = \frac{dI_l}{d\nu} = I_\nu. \tag{1.4.88}$$

The intensity distribution, due to the Doppler effect only, has then the shape of a Gaussian error-distribution curve

$$I_\nu = I_l \frac{1}{\sqrt{\pi}\,\Delta\nu_D} e^{-\left(\frac{\Delta\nu}{\Delta\nu_D}\right)^2}. \tag{1.4.89}$$

If the exponential member drops to half its value, so does the intensity of the line. Thus we obtain

$$\delta = 2(\Delta\nu)_{1/2} = 2\sqrt{\ln 2}\,\Delta\nu_D = 1.665\,\Delta\nu_D. \tag{1.4.90}$$

Line broadening due to the Doppler effect always occurs together with natural line width, and in plasma its magnitude is of the order of 0·1 Å. The distinction between natural and Doppler broadening is facilitated by the fact that the former is independent of wavelength, while Doppler broadening is proportional to it.

Doppler broadening is comparable to collision broadening which in some cases predominates. In order to differentiate between the two phenomena, we keep in mind that Doppler broadening predominates in the centre of the line and drops exponentially towards the wings, while collision broadening is prevalent at the wings where it drops as $\frac{1}{(\Delta\lambda)^2}$.

1.4.5 Radiation of a Non-Homogenenous Plasma Layer

In a plasma with non-homogeneous temperature distribution, the change in intensity in any random location is described by the equation [Ref. 1.4.56]

$$\frac{dI_\nu}{dl} = \varepsilon_\nu - \varkappa_\nu I_\nu \tag{1.4.91}$$

Except in special instances this equation is not integrable, because $\varkappa_v$ and ε_v are dependent on temperature which changes from point to point.

In an optically thin layer

$$I_\nu = \int \varepsilon_\nu(T_l)\, dl. \tag{1.4.92}$$

For a rotationally symmetrical plasma, Eq. (1.4.90) may be integrated. Since in the technical utilization of plasma this case occurs relatively frequently, we shall discuss it in greater detail.

As we observe the radiation of a cylindrical plasma, a ray at a distance x from the centre line represents the summarized radiation of all elements placed on it. For discussing the properties of the plasma, we need the intensity radiated by one of its points lying at a distance r from its centre line. The conversion of longitudinal into radial distribution is carried out in the following way:

In order to derive the requisite equation, we assume the cylindrical symmetry of the plasma and observe the radiation on rays in a plane perpendicular to the centre line. We relate the points of this plasma to a system of coordinates having its origin in the centre line. The x axis lies in the direction of the observation, and the y axis is perpendicular

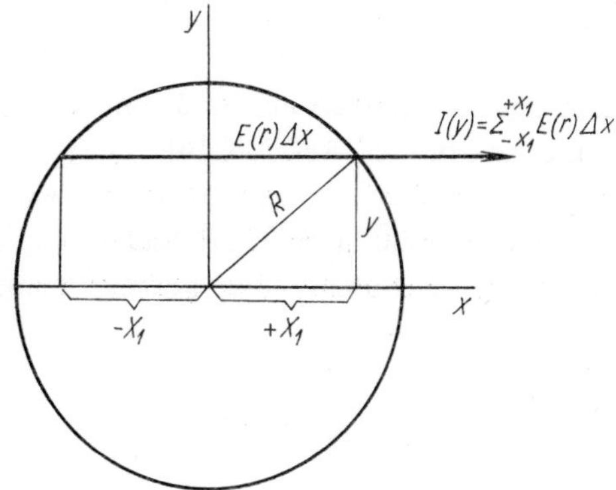

Fig. 14 Conditions for derivation of Abelian integral equation

to it. We designate the radial distance from the centre line by r, and the radius of the plasma by R. The relations are schematically represented in Fig. 14.

We designate by $I(y)$ the radiation density emitted on a ray at a distance y from the centre line. The radial course of the radiation intensity is denoted $E(r)$. Every element Δx of plasma in the plane x, y emits a radiation element $E(r)\,\Delta x$. The total intensity of the radiation is defined by the sum of the individual elements in the interval $-x_1$, $+x_1$. Hence

$$I(y) = \sum_{-x_1}^{+x_1} E(r)\,\Delta x. \tag{1.4.93}$$

After replacing the summation by the integral we obtain

$$I(y) = 2 \int_0^{\sqrt{R^2 - y^2}} E(\sqrt{x^2 + y^2})\,\mathrm{d}x, \tag{1.4.94}$$

where $\sqrt{R^2 - y^2} = x$, and $E(\sqrt{x^2 + y^2})\,\mathrm{d}x = E(r)$ is the value of the energy radiated at $P(x, y)$. The radial distribution of intensity $E(r)$

is calculated from the measured intensity $I(y)$ – the longitudinal intensity distribution – by solving Eq. (1.4.94). After substituting radius r for the variable x, we obtain the equation

$$I(y) = 2 \int_{y}^{R} \frac{E(r)\, r\, \mathrm{d}r}{\sqrt{r^2 - y^2}} . \tag{1.4.95}$$

This is an Abelian equation and its accurate solution is rather difficult. For an approximate solution cf. Ref. [1.4.16].

In cour consideration, we can at every wavelength replace a non-homogeneous cylindrical layer of $2R$ diameter by a homogeneous layer with an equivalent thickness l_q, this quantity being defined by the equation

$$l_q = \frac{1}{I(r)} \frac{\int I(r)\, \mathrm{d}r}{I(0)} . \tag{1.4.96}$$

If l_q is known, we obtain from the intensity $I = I(0)\, l_Q$, as ascertained by the absolute measurement, the intensity in the centre line of the cylindrical plasma. It follows from Eq. (1.4.95) that for determining l_q no more than one relative measurement of the intensity of the line $I(r)$ is required. In the case of lateral observation, we calculate $I(r)$ by solving the Abelian integral equation (1.4.95), whereas in the case of observation in the direction of the plasma axis we obtain $I(r)$ directly. Since l_q depends on the wavelength, we have to repeat this procedure for every line whose intensity we measure.

1.5 Plasma Diagnostics

1.5.1 Calculating the Plasma Temperature of a Water-Stabilized Electric Arc

The electric arc is probably the most common source of high-pressure plasma. In this Chapter we shall calculate the plasma temperature as it depends on the parameters that determine the properties of the arc, using for this demonstration the plasma of a torch with a water-stabilized arc. The design of such a torch is described in Sec. 2.3.2.2.

The parameters determining the properties of the arc in the plasma torch are the power input (given by the product of the arc voltage and the current flowing through the arc) and the diameter of the water vortex channel in which the arc is burning.

We begin the calculation from the current density in the arc plasma which is

$$i = \sigma E, \tag{1.5.1}$$

σ being the conductivity of the plasma, and E its electric-field intensity. We obtain the total current flowing through the arc by integrating the current density over the radius R

$$I = 2\pi E \int_0^R j(T)\, r\, \mathrm{d}r. \tag{1.5.2}$$

Electric conductivity varies with temperature; hence by substituting the first of these equations into the second, we obtain

$$I = 2\pi \int_0^R \sigma(T)\, r\, \mathrm{d}r. \tag{1.5.3}$$

For computing the integral in Eq. (1.5.3), we have to determine the dependence of conductivity upon temperature. This calculation is carried out in Chap. 1.5.6, and in our present computation we shall use Eq. (1.5.88) derived there. We substitute this equation in Eq. (1.5.3) and obtain

$$I = \frac{2\pi e^2}{\sqrt{3mk}} E \int_0^R \frac{n_e}{\sqrt{T}(n_0 Q_0 + n_+ Q_+ + n_{++} Q_{++})} r \, dr. \tag{1.5.4}$$

As it is difficult to compute the integral for the whole temperature range, we first assume that the plasma contains only singly ionized atoms and neglect neutral and multiply-ionized atoms. This simplifies the equation and we can write

$$I = \frac{2e^2\pi}{\sqrt{3mk}} E \int_0^R \frac{n_e}{T n_+ Q_+} r \, dr. \tag{1.5.5}$$

Since the arc plasma is quasineutral, n_e must equal n_+, and we can therefore cancel these quantities. After substituting (according to Gvosdover, cf. Eq. 1.5.63) for the effective cross-section Q_+

$$Q_+ = \frac{\pi}{2} \frac{e^4}{(kT)^2} \ln \frac{3kT}{2e^2 n_i^{1/3}}, \tag{1.5.6}$$

we obtain for the total current

$$I = \frac{4k^{3/2}}{e^2 \sqrt{3m_e}} E \int_0^R \frac{T^{3/2}(r)}{\ln \dfrac{3kT}{2e^2 n_e^{1/2}}} r \, dr. \tag{1.5.7}$$

The logarithmic member has at various temperatures an almost constant value of approximately 4, and we can therefore put it in front of the integration sign.

In Eq. (1.5.7) temperature occurs as a function of the radius. Measurements of the radial temperature distributions make it obvious that a parabolic shape of their curves can be assumed with a sufficient degree of accuracy, i.e.

$$T(r) = T_m \left(1 - \frac{r^2}{R^2}\right); \tag{1.5.8}$$

T_m is the temperature in the centre line of the arc, and R the radius where the plasma ceases to be electrically conducting.

After integration and solving for T_m, we obtain for the temperature in the centre line of the plasma

$$T_m^{3/2} = \frac{5e^2\sqrt{3m}\ln\dfrac{3kT}{2e^2 n\dfrac{1}{3}}}{4k^{3/2}} \frac{I}{R^2 E}. \qquad (1.5.9)$$

When we substitute numerical values, we have for the temperature along the axis of a plasma in which singly ionized atoms preponderate

$$T_m = 1.041 \cdot 10^3 \left[\frac{I}{R^2 E}\right]^{2/3}. \qquad (1.5.10)$$

Although multiply ionized atoms are neglected in Eq. (1.5.10), it is valid with a sufficient degree of accuracy throughout the whole temperature range. At temperatures above 25,000°K, where multiple ionization has to be taken into account, we can start from Spitzer's equation for the conductivity of a fully ionized gas [Ref. 1.5.1]. Then we obtain for the temperature along the stabilization axis

$$T_m = 1.082 \cdot 10^3 \left[\frac{I}{R^2 E}\right]^{2/3}. \qquad (1.5.11)$$

This equation yields slightly higher temperature values than Eq. (1.5.10).

The temperature in the centre line of the plasma thus depends on three parameters: the current flowing through the arc, the arc voltage, and the radius of the water channel in the plasma torch. The temperature increases with rising arc voltage and with decreasing radius and electric-field intensity. The current and arc radius can be chosen independently, whereas the intensity of the electric field depends on them.

Burnhorn and Maecker [Ref. 1.5.2] have empirically derived the following formula for the gradient as a function of current and radius:

$$\pi R^2 = 0.007\,26 \frac{I + 190}{E - 42}. \qquad (1.5.12)$$

The current-dependence of the measured and calculated temperature values in the centre line of the plasma is compared in Fig. 15.

According to Burnhorn, Maecker and Peters [Ref. 1.5.26], the results of theoretical reasoning about the plasma temperature and of the measurements taken by various authors can be summarized in an isothermal plasma diagram which permits conclusions on the temperatures achievable in plasma torches.

The isothermal diagram represents the dependence of temperature in the centre line of the plasma on the parameters determining the arc (cf. Fig. 16). In the coordinate plane whose axes indicate the current and arc gradient on a logarithmic scale, the curves of equal inputs supplied to the plasma appear as straight lines. At low inputs (hence low temperatures) these straight lines represent points of equal temperature, while at high inputs (and consequently high temperatures) the

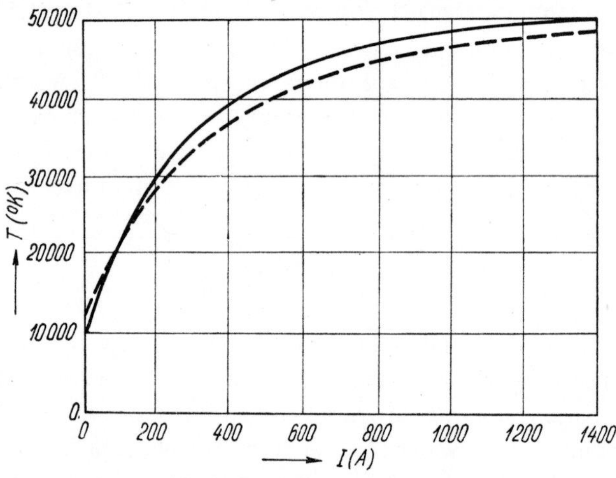

Fig. 15 Measured and calculated temperature values in centre line of water-stabilized arc vs. current

isotherms are curved. Temperatures measured in the plasma of arcs burning at various currents and in plasma of different diameters are also plotted in the diagram.

Conclusions on the temperature in the centre-line of the plasma can be based on the isothermal diagram. The plasma temperature rises, as expected, with increasing current, and at high current intensities and small stabilization radii it approaches a limit value of 60,000°K. Generally speaking, the higher the current intensity and the smaller the radius—the higher the plasma temperature.

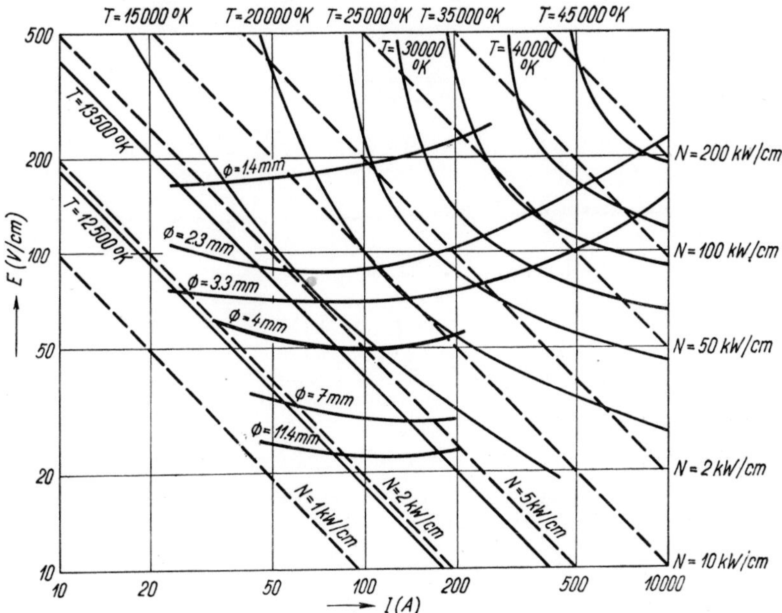

Fig. 16 Isothermal diagram of plasma in a water-stabilized arc

1.5.2 Computing the Composition of the Plasma in a Water-Stabilized Arc

For this calculation we have available: Saha's equations for hydrogen and oxygen atoms (ions)

$$\frac{n_{H^+}}{n_H} n_e = \frac{S_H(T)}{p_e}, \tag{1.5.13}$$

$$\frac{n_{O^+}}{n_O} n_e = \frac{S_O(T)}{p_e}, \tag{1.5.14}$$

$$\frac{n_{O^{++}}}{n_{O^+}} n_e = \frac{S_O(T)}{p_e}. \tag{1.5.15}$$

Moreover, we have the equations of state:

$$p = (n_H + n_O + 2n_{H^+} + 2n_{O^+} + 3n_{O^{++}})\,kT, \tag{1.5.16}$$

$$p_e = (n_{H^+} + n_{O^+} + 2n_{O^{++}})\,kT. \tag{1.5.17}$$

From the chemical composition of water follows

$$\frac{n_H + n_{H^+}}{n_O + n_{O^+} + n_{O^{++}}} = 2. \tag{1.5.18}$$

We first calculate the densities of O, O^+, O^{++}. By subtracting Eq. (1.5.17) from Eq. (1.5.16) we obtain

$$\frac{p - p_e}{kT} = n_H + n_{H^+} + n_O + n_{O^+} + n_{O^{++}}. \tag{1.5.19}$$

We substitute for $n_H + n_{H^+}$ from Eq. (1.5.18) and obtain

$$\frac{p - p_e}{kT} = 3(n_O + n_{O^+} + n_{O^{++}}). \tag{1.5.20}$$

We find from Saha's equation that in the temperature range below 40,000°K, $n_{O^{++}}$ is negligible in comparison with $n_O + n_{O^+}$.

This simplifies Eq. (1.5.20) to

$$\frac{p - p_e}{kT} = 3(n_O + n_{O^+}). \tag{1.5.21}$$

After substitution for n_{O^+} from Eq. (1.5.14) we have

$$\frac{p - p_e}{kT} = 3n_O\left(1 + \frac{S_O(T)}{p_e}\right), \tag{1.5.22}$$

and hence

$$n_O = \frac{1}{3}\,\frac{p - p_e}{kT}\,\frac{1}{1 + \dfrac{S_O(T)}{p_e}}. \tag{1.5.23}$$

From Eqs. (1.5.13) and (1.5.23) follows

$$n_{O^+} = n_O\frac{S_O(T)}{p_e} = \frac{1}{3}\,\frac{p - p_e}{kT}\,\frac{\dfrac{S_O(T)}{p_e}}{1 + \dfrac{S_O(T)}{p_e}}. \tag{1.5.24}$$

In Eqs. (1.5.23) and (1.5.24) we have n_O and n_{O^+} as functions of pressure, electron pressure and temperature. Pressure and temperature are known, we must therefore calculate the temperature dependence of electron pressure. By substituting from Eq. (1.5.18) into Eq. (1.5.17) we arrive at

$$\frac{p_e}{kT} = 3n_{O^+} + 4n_{O^{++}} = 4(n_{O^+} + n_{O^{++}}) - n_{O^+}. \qquad (1.5.25)$$

We factor out on the right-hand side and write

$$\frac{p_e}{kT} = (n_{O^+} + n_{O^{++}})\left(4 - \frac{n_{O^+}}{n_{O^+} + n_{O^{++}}}\right); \qquad (1.5.26)$$

$n_{O^+} + n_{O^{++}}$ is given by Eq. (1.5.21), we calculate $n_{O^+} : (n_{O^+} + n_{O^{++}})$ from Eq. (1.5.15) and obtain

$$\frac{n_{O^+}}{n_{O^+} + n_{O^{++}}} = \frac{1}{1 + \dfrac{S_O(T)}{p_e}}. \qquad (1.5.27)$$

We substitute from this equation into Eq. (1.5.26) and after re-arrangement we can write

$$p_e = p\,\frac{3 + 4\dfrac{S_O(T)}{p_e}}{6 + 7\dfrac{S_{O^+}(T)}{p_e}}. \qquad (1.5.28)$$

This equation makes it obvious that the dependence of p_e upon p is not very marked. If on the right-hand side we substitute $p_e = \infty$, we obtain $p_e = \frac{p}{2}$. After substitution for $p_e = 0$, we have $p_e = \frac{8}{7} \cdot \frac{p}{2}$. Therefore

$$\frac{p}{2} < p_e < \frac{8}{7}\frac{p}{2}. \qquad (1.5.29)$$

The validity of these inequalities obviates the necessity to solve Eq. (1.5.18). Thus for p_e we obtain with a sufficient degree of accuracy

$$p_e = \frac{p}{2}\,\frac{3 + 8\dfrac{S_O(T)}{p}}{3 + 7\dfrac{S_{O^+}(T)}{p}}. \qquad (1.5.30)$$

In computing p_e we have used relations for n_{O^+} and $n_{O^{++}}$; Eq. (1.5.30) therefore holds good for temperatures at which both these ions occur, that is the range above 20,000°K.

For lower temperatures we have to derive a different equation. We neglect $n_{O^{++}}$ in Eq. (1.5.25)—which is permissible because at the temperatures covered by our computation, the number of such ions is almost zero—and substitute in it for n_{O^+} from Eq. (1.5.23). After re-arrangement we obtain for p_e the quadratic equation

$$p_e^2 = 2S_O(T)\, p_e - pS_O(T) = 0. \tag{1.5.31}$$

By solving this equation we have for p_e

$$p_e = S_O(T)\left[\sqrt{1 + \frac{p}{S_O(T)}} - 1\right]. \tag{1.5.32}$$

The value thus calculated for p_e is substituted in Eq. (1.5.23) and after re-arrangement we write

$$n_O = \frac{1}{3kT}\, p\left[1 - 2\,\frac{S_O(T)}{p}\left\{\sqrt{1 + \frac{p}{S_O(T)}} - 1\right\}\right]. \tag{1.5.33}$$

By a simple manipulation of Eq. (1.5.27) and substitution from Eq. (1.5.21) we obtain for n_{O^+}

$$n_{O^+} = \frac{1}{3kT}\,\frac{p - p_e}{1 + \dfrac{S_O(T)}{p_e}}. \tag{1.5.34}$$

We substitute for p_e from Eq. (1.5.28), calculate $p - p_e$, and after re-arrangement we obtain

$$n_{O^+} = \frac{p}{kT}\,\frac{1}{6 + 7\,\dfrac{S_{O^+}(T)}{p_e}}. \tag{1.5.35}$$

On substitution for p_e from Eq. (1.5.28) and after re-arrangement this becomes

$$n_{O^+} = \frac{1}{3}\,\frac{p}{kT}\,\frac{3 + 8\,\dfrac{S_{O^+}(T)}{p}}{18 + 90\,\dfrac{S_{O^+}(T)}{p} + 98\left(\dfrac{S_{O^+}(T)}{p}\right)^2}. \tag{1.5.36}$$

We calculate $n_{O^{++}}$ from Eq. (1.5.15) into which we substitute from Eq. (1.5.36), and after re-arrangement we get

$$n_{O^{++}} = \frac{1}{3}\frac{p}{kT}\frac{6\dfrac{S_{O^+}(T)}{p} + 14\left(\dfrac{S_{O^+}(T)}{p}\right)^2}{18 + 90\dfrac{S_{O^+}(T)}{p} + 98\left(\dfrac{S_{O^+}(T)}{p}\right)^2}. \qquad (1.5.37)$$

In a similar way we calculate the density of hydrogen atoms and ions. By substituting for n_{H^+} from Eq. (1.5.18) and for n_{O^+} from Eq. (1.5.23) into Eq. (1.5.16), we determine the electron pressure, and after a simple manipulation, we obtain the quadratic equation

$$p_e^2 + 2S_H(T)\,p_e - S_H(T)\,p = 0. \qquad (1.5.38)$$

From this equation we calculate p_e

$$p_e = S_H(T)\left[\sqrt{1 + \frac{p}{S_H(T)}} - 1\right]. \qquad (1.5.39)$$

The value thus computed is substituted in an equation which we get by solving the system of equations (1.5.13), (1.5.16), (1.5.17) and (1.5.18) for n_H and n_{H^+} by the method described in determining n_O and n_{O^+}; after a simple manipulation we have then for the density of neutral hydrogen atoms

$$n_H = \frac{2}{3}\frac{1}{kT}\,p\left[1 - \frac{2S_H(T)}{p}\sqrt{1 + \frac{p}{S_H(T)}} - 1\right]. \qquad (1.5.40)$$

The same procedure yields for the density of hydrogen ions

$$n_{H^+} = \frac{2}{3}\frac{1}{kT}\,S_H(T)\left[\sqrt{1 + \frac{p}{S_H(T)}} - 1\right]. \qquad (1.5.41)$$

The electron density is computed from the known electron pressure

$$n_e = \frac{S_H(T)}{kT}\left[\sqrt{1 + \frac{p}{S_H(T)}} - 1\right]. \qquad (1.5.42)$$

The calculated particle densities in the plasma of a water-stabilized arc are plotted against the plasma temperature in Fig. 17.

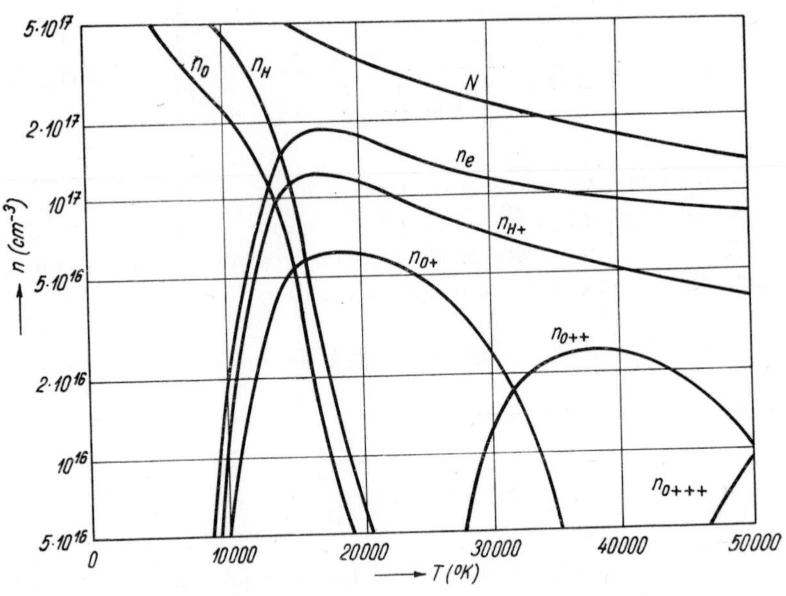

Fig. 17 Particle densities in plasma of a water-stabilized arc

1.5.3 Computing the Composition of the Plasma in an Arc Freely Burning in the Air

The plasma of the arc under consideration consists—according to Maecker [Ref. 1.5.3]—of a mixture of 70% of air and 30% of carbon. The maximum temperatures occuring in the plasma are in the range of magnitude of 10,000°K; molecules are fully dissociated into atoms, and these are substantially ionized at this temperature. The gas of the arc is thus composed of several kinds of particles. In our case we also have to consider molecules, atoms and ions of nitrogen, oxygen and carbon, apart from the transient formation of CN, NO and CO molecules. For the partial pressures of the individual constituents in the temperature range between 1,000 and 10,000°K we have the following system of equations:

$$n_{N_2} + p_{O_2} + p_{C_2} + p_{CO} + p_{CN} + p_{NO} + p_O + p_{N^+} + \\ + p_{O^+} + p_{C^+} + p_e = p = 10{\cdot}16 \text{ kp/cm}^2 \tag{1.5.43}$$

$$p_{N^+} + p_{O^+} + p_{C^+} = p_e \tag{1.5.44}$$

$$\frac{p_N^2}{p_{N_2}} = S_{N_2}(T) \tag{1.5.45}$$

$$\frac{p_O^2}{p_{O_2}} = S_{O_2}(T) \tag{1.5.46}$$

$$\frac{p_C^2}{p_{C_2}} = S_{C_2}(T) \tag{1.5.47}$$

$$\frac{p_C p_O}{p_{CO}} = S_{CO}(T) \tag{1.5.48}$$

$$\frac{p_N p_O}{p_{NO}} = S_{NO}(T) \tag{1.5.49}$$

$$\frac{p_C p_N}{p_{CN}} = S_{CN}(T) \tag{1.5.50}$$

$$\frac{p_{N^+} p_e}{p_N} = S_N(T) \tag{1.5.51}$$

$$\frac{p_{O^+} p_e}{p_O} = S_O(T) \tag{1.5.52}$$

$$\frac{p_{C^+} p_e}{p_C} = S_C(T). \tag{5.1.53}$$

Saha's equations $S_i(T)$ have the form

$$S_i(T) = \frac{p_i p_{i+1}}{p_{i+2}} = \frac{Z_i Z_{i+1}}{Z_{i+2}} \frac{1}{h^3} \left(2\pi \frac{m_i m_{i+1}}{m_{i+2}}\right)^{3/2} kT^{5/2} e^{-\frac{eU_j}{kT}}, \tag{1.5.54}$$

where Z_i is the partition function and U_i the dissociation or ionization energy.

The partition functions are described by the relation

$$Z_i = \sum_{n=0}^{\infty} g_n \, e^{-\frac{eU_n}{kT}}, \tag{1.5.55}$$

where g_n is the statistical weight. The molecular partition functions have been determined by the procedure described in Ref. [1.5.14].

The following values have been used for dissociation and ionization potential: $D_{N_2} = 9{\cdot}76$ eV, $D_{O_2} = 5{\cdot}12$ eV, $D_{C_2} = 3{\cdot}6$ eV, $D_{CO} = 9{\cdot}14$ eV, $D_{CN} = 6{\cdot}23$ eV, $D_{NO} = 6{\cdot}48$ eV, $U_N = 14{\cdot}51$ eV, $U_O = 13{\cdot}57$ eV, and $U_C = 11{\cdot}24$ eV.

Two further relations, in which the composition of the plasma is taken into account—55·3% N, 14·7% O, 30% C—have to be added to the system of equations quoted above:

$$\frac{2p_{N_2} + p_N + p_{NO} + p_{CN} + p_{N^+}}{2p_{O_2} + p_O + p_{NO} + p_{CO} + p_{O^+}} = \frac{55.3}{14.7} \tag{1.5.56}$$

$$\frac{2p_{N_2} + p_N + p_{NO} + p_{CN} + p_{N^+}}{2p_{C_2} + p_C + p_{CO} + p_{CN} + p_{C^+}} = \frac{55.3}{30} \tag{1.5.57}$$

Together, all these equations constitute a non-linear system with thirteen unknown quantities, which cannot be solved by elementary means.

The entire system was solved after division into several temperature ranges and under the assumption that no ions and electrons occur between 1,000 and 6,000°K, and no molecules above 9,000°K. This assumption reduced the number of unknown quantities, but the system still remained unsolvable by elementary means. It was reduced to three or four equations by the method of elimination and then solved by the iterative method. The result of the computation is represented in Fig. 18, where the densities of the various particles are plotted against temperature.

1.5.4 Calculating the Thermal Conductivity of the Plasma in an Arc Freely Burning in the Air

At high temperatures, the molecules of the gases constituting the arc plasma are dissociated into atoms, and these are partly ionized. Consequently, the plasma is the scene of chemical processes such as

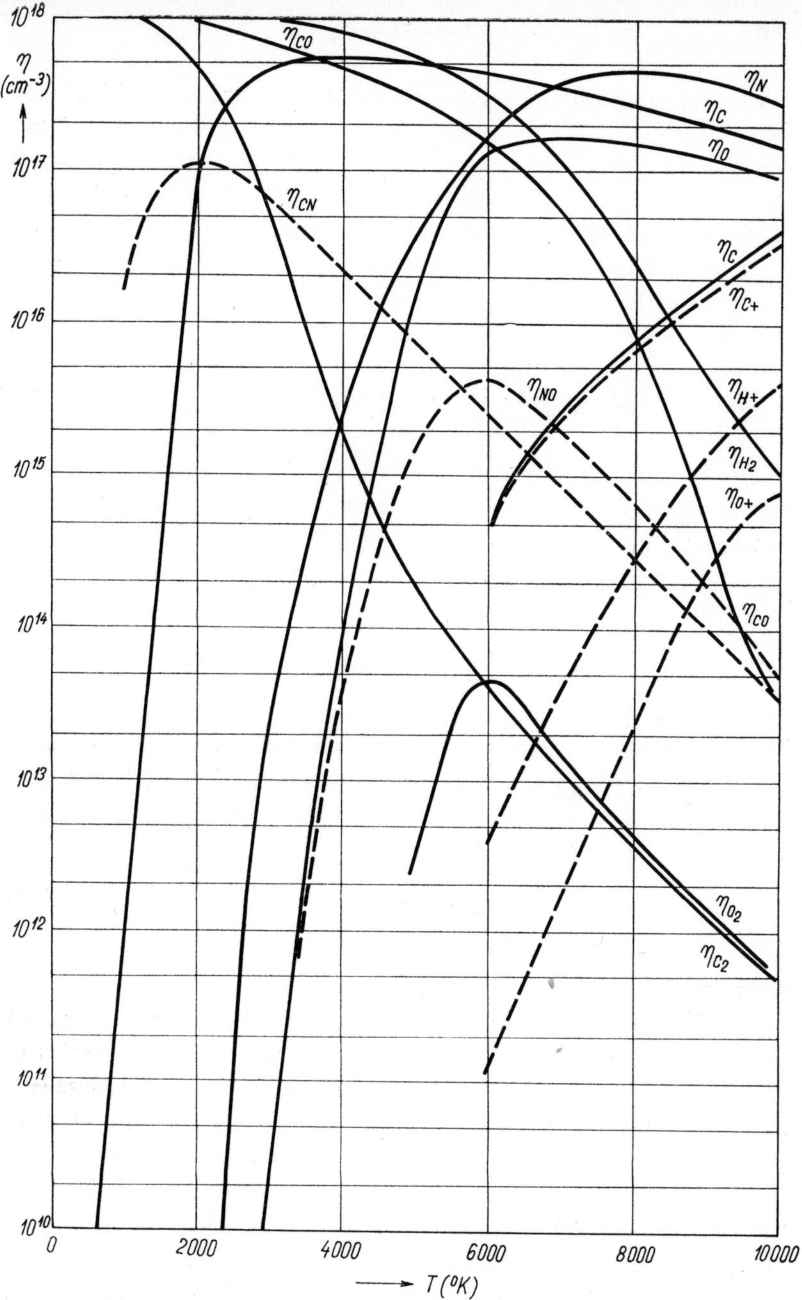

Fig. 18 Particle densities in plasma of an air-stabilized arc

dissociation, ionization and recombination. According to Wiennecke [Ref. 1.5.5] the thermal conductivity of such a gas consists of two components: normal thermal conductivity which causes heat conduction by contact; and thermal conductivity associated with diffusion.

Normal thermal conductivity corresponds to the classical conductivity dealt with in the elementary kinetic theory of gases. Eq. (1.2.129) is valid for it, i.e.

$$\varkappa_i(T) = \sum_{i=1}^{k} \frac{n_i}{3} \overline{c_i l_i} \left(\frac{5}{2} c_{i\,trans} + c_{i\,\text{int}} \right), \tag{1.5.58}$$

where $c_{k\,\text{trans}}$ and $c_{k\,\text{int}}$ are the specific heats of translational and internal degrees of freedom per k-type particle, n_k is the particle density, $\bar{c}_k$ the mean thermal velocity, and $\bar{l}_k$ the mean free path.

The mean free path of k-type particles is described by

$$\frac{1}{\bar{l}_k} = \sum_{i=1}^{k} \left(1 - \frac{1}{2} p_{ki} \right) n_i Q_{ki} \sqrt{1 + \frac{m_k}{m_i}}. \tag{1.5.59}$$

The summation is carried out over all kinds of particles under consideration. Q_{ki} is the effective cross-section of the kth type with respect to the i-th type, p_{ki} is the persistance described by the equation

$$p_{ki} = \frac{m_i - 0.2m_i}{m_k + m_i}, \tag{1.5.60}$$

where m are the masses of the particles.

For the effective cross-section we insert values that depend on the kind of the impinging particles. For molecules we take the effective cross-section from the kinetic theory of gases, for atoms Ramsauer's and for ions and electrons Gvosdover's cross-sections. The effective cross-section of molecules, atoms and ions occuring in the plasma differ by a value which is smaller than the measuring error. In the calculation we can therefore assume that only one kind of particle exists in the plasma; the error due to this assumption is negligible.

We substitute the following values for the effective cross-sections:

Cross-section according to the kinetic theory of gases:

$$Q_k = \left(1 + \frac{113}{T}\right) 8 \,.\, 10^{-16} \qquad [\mathrm{cm}^2] \tag{1.5.61}$$

Ramsauer's cross-section:

$$Q_e = 20 \,.\, 10^{-16} \qquad [\mathrm{cm}^2] \tag{1.5.62}$$

Gvosdover's cross-section:

$$Q_G = \frac{e^4}{(kT^2)} \ln \frac{3kT}{2e^2 n_e^{1/3}} \qquad [\mathrm{cm}^2] \tag{1.5.63}$$

In calculating the thermal conductivity by diffusion of reaction energy we make use of the considerations pointed out in Sec. 1.2.7.3.

In a molecular gas, the density of molecules decreases and the density of atoms increases with rising temperature. Density gradients arise, and they cause diffusion currents J_i in which pairs of atoms diffuse into cooler regions where they recombine. On the other hand, molecules migrate into warm regions, where they dissociate. The sum of the diffusion currents must be zero, in the stationary case. To every mass flow corresponds a flow of energy W_i which equals the product of the mass flow $J_i = \varrho_i v_i$ and the enthalpy h_i

$$W_i = h_i \varrho_i v_i = h_i J_i, \tag{1.5.64}$$

where ϱ_i is the partial density of the ith component, and v_i its velocity.

In a system with k components we obtain the total flow of energy by the summation of all partial energy flows.

$$W = \sum_{i=1}^{k} W_i = \sum_{i=1}^{k} h_i J_i = -\varkappa_r \operatorname{grad} T; \tag{1.6.65}$$

according to the definition, this sum equals the product—taken negative—of thermal conductivity associated with chemical reaction multiplied by the thermal gradient.

For determining the velocity v_i we use the procedure suggested by Schlüter [Ref. 1.5.6] which conforms with the thermodynamics of reversible processes, as proved by Maecker and Peters [Ref. 1.5.7].

For the ith mass current in a gas mixture of k components we can therefore write

$$\varrho_i \frac{dv_i}{dt} + n_1 \sum_{i=1}^{k} n_i \varepsilon_{il}(v_i - v_1) = F_i \varrho_i - \text{grad } p_i - \varrho_i x_i \frac{\text{grad } T}{T}; \tag{1.5.66}$$

F_i are the external forces acting on the ith particles, p_i their partial pressure, and the coefficients ε are defined by the equations

$$\varepsilon_{il} = \varepsilon_{li} = \frac{8}{3} \sqrt{\frac{2}{\pi} kT \frac{m_i m_l}{m_i + m_l}} Q_{il}. \tag{1.5.67}$$

On the right-hand side of Eq. (1.5.66) we have the forces which cause the mass flow J_i, and also the gradient of the partial pressure and the thermal diffusion. On the left-hand side we have, in addition to the force of inertia, friction proportional to the differences between the velocities of the various mass flows. The coefficients of friction ε_{il} are taken from the kinetic theory of gases. Analogous equations are valid for the remaining components.

If we neglect the external forces and thermal diffusion, we obtain for the stationary case under consideration, in which $\frac{dv_i}{dt}$ vanishes, a system of k equations

$$n_m \sum_{l=1}^{k} n_l \varepsilon_{ml}(v_m - v_1) = -\text{grad } p_m, \tag{1.5.68}$$

where $m = 1, 2, 3, \ldots, k$.

To this system are added the conditions that the sum of the mass currents be equal to 0 and the total pressure constant

$$\sum_{m=1}^{k} J_m = \sum_{m=1}^{k} \varrho_m v_m = 0 \tag{1.5.69}$$

$$\sum_{m=1}^{k} \text{grad } p_m = \text{grad } p = 0 \tag{1.5.70}$$

In the temperature interval for which we make our calculation, diffusion conduction is virtually limited to dissociation heat, and ionization can be neglected.

In solving the system of equations for v_i we apply the assumption made on p. 181 that all molecules and atoms have the same masses and effective cross-sections. This assumption reduces the number of factors ε_{il} to three: ε_1 for mutual collisions of molecules, ε_2 for atoms in collision with molecules, and ε_3 for mutual collisions of atoms.

The system of equations can then be solved after the insertion of numerical values.

We obtain for the velocities of molecules

$$v_i = \frac{1}{A}\left\{\frac{1}{\varrho}\operatorname{grad} p_M\left(m_M - \frac{\varepsilon_1}{\varepsilon_2} m_A\right) - \frac{\operatorname{grad} p_i}{m_i}\right\}, \tag{1.5.71}$$

and the velocities of atoms are

$$v_i = \frac{1}{B}\left\{\frac{1}{\varrho}\operatorname{grad} p_M\left(m_M \frac{\varepsilon_2}{\varepsilon_3} - m_A\right) - \frac{\operatorname{grad} p_i}{m_i}\right\}; \tag{1.5.72}$$

p_M and p_A are here the partial pressures of molecules and atoms respectively, $p = p_M + p_A$ is the total pressure, m_M and m_A are the mean masses of molecules and atoms. $A = n_M\varepsilon_1 + n_A\varepsilon_2$, and $B = n_M\varepsilon_2 + n_A\varepsilon_3$.

By multiplying the velocities v_i with the corresponding partial pressures, we obtain the mass flows J_i. The product of the mass flows and the corresponding enthalpies is the energy flow W_i. Summation over all flows W_i then yields the total heat flow:

$$\begin{aligned} W = \sum_{i=1}^{k} W_i = \Bigg[\sum_{i=1}^{k} \frac{h_i\varrho_i}{A}\left\{\frac{\partial p_M}{\partial T}\left(m_M - m_A\frac{\varepsilon_1}{\varepsilon_2}\right) - \frac{1}{n_i}\frac{\partial p_i}{\partial T}\right\} + \\ + \sum_{i=1}^{k} \frac{h_i\varrho_i}{B}\left\{\frac{\partial p_M}{\partial T}\left(m_M\frac{\varepsilon_1}{\varepsilon_2} - m_A\right) - \frac{1}{n_i}\frac{\partial p_i}{\partial T}\right\}\Bigg]\operatorname{grad} T = \\ = -\varkappa_r \operatorname{grad} T. \end{aligned} \tag{1.5.73}$$

The factor $\varkappa$—the expression in square brackets before grad T—is the thermal conductivity associated with diffusion of dissociation, taken negative.

In addition to enthalpy and densities, we need for our calculation the derivative of the partial pressures with respect to the temperature.

These derivatives may be determined graphically. The values of the effective cross-sections are substituted from Eqs. (1.5.61), (1.5.62), and (1.5.63). The coefficients ε_i have the values $\varepsilon_1 = 9{\cdot}65 \,.\, 10^{-35}T$, $\varepsilon_2 = 7{\cdot}15 \,.\, 10^{-35}T$, and $\varepsilon_3 = 5{\cdot}64 \,.\, 10^{-35}T$.

To calculate the enthalpy of the arc plasma, we multiply the enthalpy of the individual particles with their instantaneous mass fraction c_i:

$$c_i = \frac{\varrho_i}{\varrho} \text{ considering that } \sum_i^k c_i = 1; \tag{1.5.74}$$

these products are then summed up over all types of particles

$$h_{\text{total}} = \sum_{i=1}^{k} c_i h_i = \frac{1}{\varrho} \sum_{i=1}^{k} \varrho_i h_i, \tag{1.5.75}$$

The enthalpies of the individual kinds of particles have the following values:

For N_2, O_2, C_2 molecules

$$h_i = \frac{1}{n_i}\left[\frac{5}{2} kT + (kT)^2 \frac{\partial lnZ_i}{\partial T}\right]. \tag{1.5.76}$$

Excited molecules are taken into account in the last member in brackets.

For N, O, C atoms

$$h_i = \frac{1}{m_i}\left(\frac{5}{2} kT + \frac{D_i}{2}\right). \tag{1.5.77}$$

For N^+, O^+, C^+ ions

$$h_i = \frac{1}{m_i}\left(\frac{5}{2} kT + \frac{D_i}{2} + U_j\right). \tag{1.5.78}$$

For electrons

$$h_e = \frac{1}{m_i} \frac{5}{2} kT. \tag{1.5.79}$$

In these equations m_i are the masses of the various particles, z_i their internal partition functions, and U_j ionization energy.

When molecules dissociate, we ascribe half the dissociation energy to each of the resultant atoms. We ascribe the entire ionization energy to the ion which is formed in ionization, and only the translational energy to the electron; the excitation energy of atoms and ions is neglected.

Enthalpy computed from equations (1.5.75)–(1.5.79) is plotted against temperature in Fig. 19. The irregular course of the curve proves

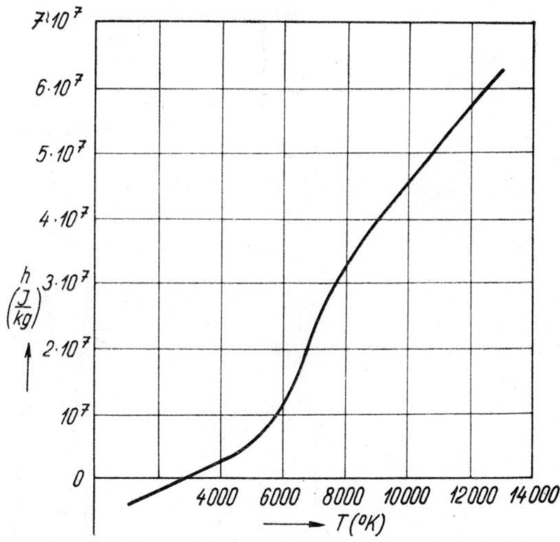

Fig. 19 Enthalpy of air as a function of temperature

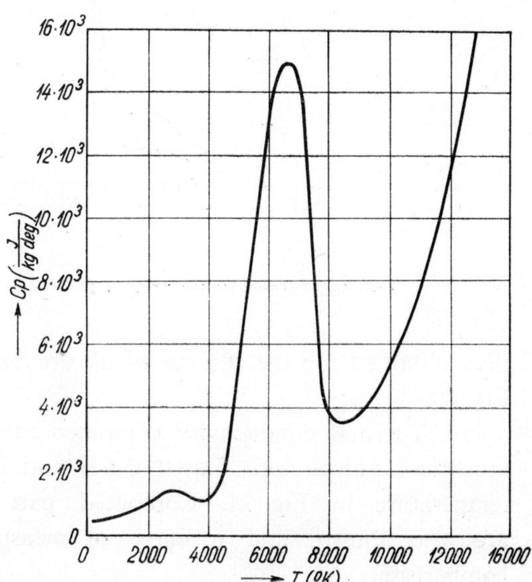

Fig. 20 Temperature dependence of specific heat of plasma burning in air

how various dissociation and ionization processes contribute to the increase of energy. By graphical differentiation of enthalpy with respect to temperature we arrive at the specific heat of the plasma. The curve

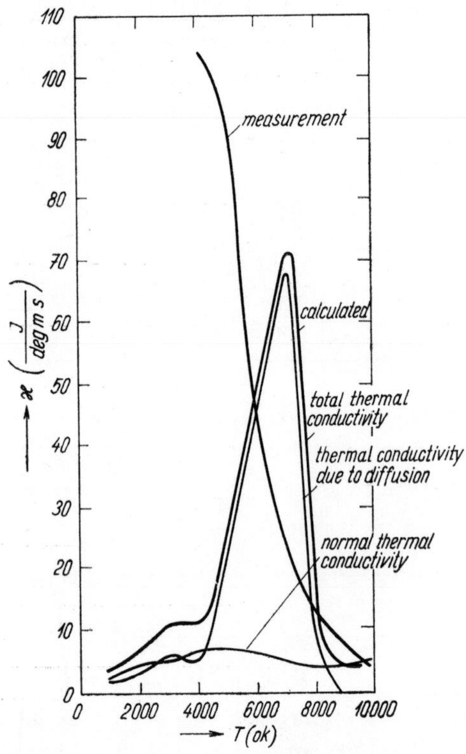

Fig. 21 Thermal conductivity of plasma in an air-stabilized arc

thus obtained for the plasma of an arc burning in air is plotted in Fig. 20.

Thermal conductivity computed after the substitution of the numerical values into Eqs. (1.5.60) and (1.5.73) is plotted against temperature in Fig. 21. Computed partial thermal conductivities are also shown, and a curve of measured values is added for comparison.

1.5.5 Calculating the Electric Conductivity of the Plasma in a Water-Stabilized Arc

The high density of charged particles of both signs produces the relatively high electric conductivity of the plasma. Owing to this high conductivity, plasma left to itself balances out all potential differences that are not kept up from some external source. In calculating the conductivity in this Chapter, we neglect the contribution of positive ions to the electric current in a plasma *.

The electric conductivity of the plasma is then given by the relation

$$\sigma = en_e\mu_e, \tag{1.5.80}$$

where n_e is the electron density and μ_e the mobility of the electrons; the contribution of the ions is neglected. According to the kinetic theory of gases we obtain

$$\mu_e = \alpha \frac{el_e}{m_e\bar{c}_e}, \tag{1.5.81}$$

where

$$\bar{c}_e = \sqrt{\frac{3kT}{m_e}} \tag{1.5.82}$$

is the effective thermal velocity; l_e is the mean free path of the electron

$$l_e = \frac{1}{n_O Q_O + n_i Q_i + n_e Q_e}; \tag{1.5.83}$$

α is the coefficient of uncertainty (ranging between 0·75 and 1·38), n_O the density of neutral particles, and Q_O their effective cross-section. At low ion density,

$$l_e = 4\sqrt{2}l_O. \tag{1.5.84}$$

For computing electric conductivity, it is more convenient to calculate l_e from Eq. (1.5.83). If we are faced with a gas mixture, $n_o Q_o$ has to be divided into its components

$$n_o Q_o = n_1 Q_1 + n_2 Q_2 + \dots. \tag{1.5.85}$$

* For the elementary theory of plasma conductivity, taking into account the influence of both electrons and ions, cf. Sec. 1.2.4.1.

Q_i, the effective cross-section of ions with respect to electrons, may attain high values. Gvosdover [Ref. 1.5.8] and other authors [Ref. 1.5.9, 1.5.10, 1.5.11] quote for this effective cross-section the relation

$$Q_i = \frac{\pi}{2} \frac{e^4}{(kT)^2} \ln \frac{3kT}{2e^2 n_i^{1/3}} . \tag{1.5.86}$$

This equation yields too high values, according to Schulz' experiments [Ref. 1.5.12, 1.5.13]. Maecker and Peters have proved [Ref. 1.5.14] that the coefficient $\frac{\pi}{2}$ has to be reduced to a value of 0·87 if the coefficient of uncertainty in Eq. (1.5.81) is equated to unity. For z-fold ions, the multiplication factor Z^2 is added to the cross-section of the ions. Mutual collisions of electrons do not influence the electric current.

The electron density can be determined from Saha's equation (1.4.9) which at low ionization levels and under the assumption of quasineutrality takes the simplified form

$$n_e = p^{1/2} \left(\frac{Z_i}{Z_o} \right)^{1/2} \frac{(2\pi m_e)^{3/4} (kT)^{1/4}}{h^{3/2}} \, e^{-\frac{eUj}{kT}} \tag{1.5.87}$$

The ionization energy is ascertained in spectroscopic measurements; it equals the work required for removing an electron from the atom to infinity. The mean separation of atoms or ions in the plasma is proportional to $\frac{1}{n_i^{1/3}}$. If the orbit of an electron exceeds the space defined by this limit, its state is no longer discrete and we can consider it to be free. This causes the transitions from such terms to radiate no spectral lines but a continuum which joins up with the discrete end of the limit continuum in the spectrum. This is revealed by the reduced number of lines in a series and the premature ionization, i.e. the reduced ionization potential. This reduction can be evaluated using Eq. (1.4.22); in low-temperature plasma the calculated values are negligible, and only under extreme conditions they equal fractions of electron volts.

In dealing with gas mixtures, we use Saha's equations separately for each component; but we must keep in mind that n_e is identical in all of Saha's equations, and therefore the ion density of one component is not identical with the electron density.

Let us apply equations (1.5.80)–(1.5.87) to the computation of electric conductivity in the plasma of a water-stabilized arc. As earlier stated, our calculation is based on the assumption that the electric current is almost exclusively due to electrons, while ions (because of their relatively large mass) hardly participate.

The plasma of the arc under consideration consists of a mixture of neutral atoms, to which hydrogen and oxygen ions as well as electrons may be added. A procedure for calculating the density of these particles is described in Sec. 1.5.2.

After the substitution of the pertinent values in Eqs. (1.5.80), (1.5.81), (1.5.82), and (1.5.83) we have

$$\sigma = \frac{e^2}{\sqrt{3mkT}} \frac{n_e}{nQ + n_+Q_+ + n_{++}Q_{++}}. \tag{1.5.88}$$

Since the neutral particles in the plasma of a water-stabilized arc are oxygen and hydrogen atoms, the total cross-section of neutral particles in 1 cm^3 is

$$nQ = n_H Q_H + n_O Q_O. \tag{1.5.89}$$

Quantum mechanics offers the following relation for the effective cross-section of the hydrogen atom with respect to electrons;

$$Q_H = 4\pi a_o^2 = 3{\cdot}5 \,.\, 10^{-16}\ \text{cm}^2, \tag{1.5.90}$$

where $a_o = 5{\cdot}3 \,.\, 10^{-9}$ cm is the radius of the first Bohr orbit. For the effective cross-section of the oxygen atom we substitute according to Schulz

$$Q_O = 3 \,.\, 10^{-16}\ \text{cm}^2. \tag{1.5.91}$$

The cross-section of neutral particles need not be more accurate because at $T > 10{,}000$°K it is of minor importance.

As multiple ions are of no consequence up to $T = 25{,}000$°K (cf. Fig. 18), Eq. (1.5.88) is simplified, and after substituting for Q_1 from Eq. (1.5.86) we have

$$\sigma = \frac{2(kT)^{3/2}}{\pi e^2 \sqrt{3m} \ln \dfrac{3kT}{2e^2 n_+^{1/3}}} \frac{1}{1 + \dfrac{nQ}{n_+Q_+}} \tag{1.5.92}$$

The equation can also be obtained from Eq. (1.2.66) by substituting for the effective cross-section from Eq. (1.5.46) and inserting the effective velocity of the electron instead of the thermal velocity $\bar{c}_j$.

The difference between unity and the proportion $\dfrac{nQ}{n_+Q_+}$ is negligible at temperatures above 14,000°K; therefore in the temperature range between 14,000 and 25,000°K, conductivity rises with $T^{3/2}$ independently of electron density.

In the temperature range between 25,000 and 50,000 °K we can ignore neutral atoms, and doubly and trebly charged ions are added to the single ones. In this range

$$\sigma = \frac{2(kT)^{3/2}}{\pi e^2 \sqrt{3m} \ln \dfrac{3kT}{2e^2 n_i^{1/3}}} \frac{n_e}{n_{O^+} + 4n_{O^{++}} + 9n_{O^{+++}}}. \tag{1.5.93}$$

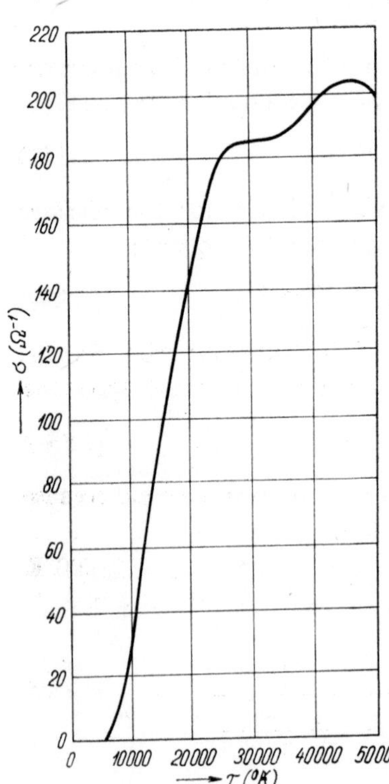

Fig. 22 Electric conductivity of plasma in a water-stabilized arc

Conductivity calculated from Eqs. (1.5.92) and (1.5.93) is about double the value found by experiment, which is due to the inaccurate value for Gvosdover's cross-section. According to the measurements made by Maecker and Peters [Ref. 1.5.14] the formulae for conductivity have to be used with a correction factor of 0·55 to bring the theoretical values into agreement with those measured.

The calculated values for the temperature dependence of conductivity have been plotted in Fig. 22. Above 25,000°K the conductivity of the plasma in a water-stabilized arc is almost constant. This phenomenon is due to the large effective cross-sections of multiple ions and has been borne out by the measured characteristics of water-stabilized arcs [Ref. 1.5.2].

1.5.6 **Measuring Plasma Temperatures**

The spectral methods employed for these measurements are based on either molecular or atomic spectra. In this Chapter we shall discuss the most important of these methods.

Ornstein and Brinkmann [Ref. 1.5.15] were among the first to use molecular spectra for temperature measurements. They measured the temperature of electric-arc plasma and determined it from the relative intensities of the cyanogen lines. For vibrational lines

$$I_{v'v''} = \text{const}\ \nu^4 p(v'v'')\, e^{-\frac{Ev'}{kT}}, \tag{1.5.94}$$

where v' is the vibration quantum number of the upper state and v'' of the lower one, $p(v'v'')$ is the transition probability and ν the wave number of the zero line or limit of the band. Plotting $\frac{\ln I}{\nu^4}$ against E_v, we obtain a straight line, and the temperature can be ascertained from its slope.

Other methods are based on the equation for the intensity of the rotational line

$$I = C(2J + 1)\, i\, e^{-\frac{BJ(J+1)hc}{kT}}, \tag{1.5.95}$$

where J is the rotational quantum number of the upper state, B the rotation constant for the main vibrational state, ν the wave number of the spectral line under consideration, i the intensity factor, $C = g\nu^4$ for emission lines, g the constant corresponding to the statistical weight, which is equal for all bands of singlet electron states.

Ornstein and Wijk [Ref. 1.5.16] based their measurements on the fact that by taking the logarithm of Eq. (1.5.96) we obtain

$$\ln \frac{I}{C_i} = -\frac{hcB}{kT} J(J + 1). \tag{1.5.96}$$

From this equation it follows that the graphical representation of $\ln \frac{I}{C_i}$ as a function of $\frac{J'}{J'+1}$ is a straight line with the slope $\frac{-hcB'}{kT}$. The temperature can be determined from this slope.

If the spectrograph resolves the rotational structure of at least one branch of the vibrational band, then—according to Knaus and Cay [Ref. 1.5.17]—the rotational line with the highest intensity can be utilized for determining the plasma temperature. For this line we ascertain the corresponding J_{max}. The differentiation of Eq. (1.5.95) for the intensity of the rotational line yields

$$\frac{dI}{dJ} = i\,\frac{Bhc}{kT}(2J + 1) + \frac{di}{dJ} = 0. \tag{1.5.97}$$

After substitution of the measured value J_{max} into the nullified Eq. (1.5.97), we can derive the temperature from it, always provided the change of the wave number inside the vibrational band does not affect the intensity curve close to the maximum.

In ascertaining the temperature of a plasma in whose spectrum the components of the various branches are not resolved we set out—according to Janin and Eyraud [Ref. 1.5.18]—from the equation for the intensity of unresolved branches. We have

$$I = \nu^4(i_{1n} + i_{2n} + i_{3n})\, e^{-\frac{F_2(J_{2n})hc}{kT}}. \tag{1.5.98}$$

For individual components of the summation:

$$i_{kn} = e^{-\frac{F_k(J_{kn})hc}{kT}}. \tag{1.5.99}$$

In Eq. (1.5.98) we can cancel out the exponential terms, because for high quantum numbers they have the same values.

After taking the logarithm of Eq. (1.5.98), we have

$$\ln \frac{1}{\nu^4(i_{1n} + i_{2n} + i_{3n})} = \frac{hc}{kT} F_2(J_{2n}). \qquad (1.5.100)$$

The graphic representation of $1/\nu^4(i_{1n} + i_{2n} + i_{3n})$ as a function of $F_2(J_{2n})$ is a straight line from the slope of which we can ascertain the temperature.

We discuss only a few of the numerous methods using atomic spectra for measuring temperatures (cf. Ref. [1.5.19] for a fuller treatment). Most of these methods are based on Eq. (1.4.48) which indicates the temperature dependence of the intensity of spectral lines. Related to the unit solid angle, this equation has the following form:

$$I_{nm} = \frac{1}{4\pi} A_{nm} h\nu \frac{g_m}{Z_l} n_l \, e^{-\frac{eU_m}{kT}} l, \qquad (1.5.101)$$

where l is the thickness of the homogeneous plasma layer emitting the radiation. The number of particles radiating the given line $-n_l-$ and to a lesser degree the partition function Z_l change with the temperature. The dependence of the number of particles in the plasma upon the temperature of the water- and air-stabilized arc is calculated in Chapters 1.5.5 and 1.5.4.

Jürgens [Ref. 1.5.20] has measured the absolute value of intensity I radiated by a specific line and has calculated the temperature by solving Eq. (1.5.101) for T. This method has the disadvantage of requiring the knowledge of the transition probabilities A_{nm} and of the thickness l of the layer. The latter value can be calculated fairly accurately, but the transition probabilities are only known for some of the lines.

Larenz [Ref. 1.5.21] has derived a method which avoids these drawbacks. In applying it we also set out from Eq. (1.5.101). In order to discuss the temperature dependence of the intensity, we replace Boltzmann's factor by the ratio of excited particles to the number of particles in the ground state according to Eq. (1.4.9). We have then for the intensity

$$I_{nm} = \frac{A_{nm}}{4\pi} h\nu \frac{n_{la}}{n_{lo}} n_{lo}. \tag{1.5.102}$$

The relation $n_{la} : n_{lo}$ decreases monotonically and approaches a limiting value at a temperature approaching infinity. If the radiation is emitted by an atom in the ground state [$l = 0$], then n_{l0} decreases monotonically with increasing temperature and approaches zero at a temperature approaching infinity. This decrease is caused by the density of particles in the ground state decreasing owing to excitation, and the number of radiating particles decreasing owing to ionization. If the radiation carrier is an ion, n_{l0} first rises with temperature, and after passing through its maximum it drops monotonically to zero, because simultaneously with the same factors causing the reduction in the number of radiating particles, it is affected by the increase in the number of newly formed ions. In both cases there is a definite temperature at which the radiated intensity passes through a maximum. It is this maximum temperature on which the measurement is based. We normalize the theoretical curve for the radiated energy to the value of the intensity radiated at maximum temperature so that this dimensionless energy has a maximum value equalling unity. The normalized curve is designated $E^+(T)$.

The maximum of function $E^+(T)$ appears as a maximum in the radial distribution of the measured intensity $E^+(r)$. If r' is the radius at which the measured intensity reaches its highest value, then at the distance r' from the axis of the measured arc the temperature is T'. Thus by comparing measured and calculated intensities, we can uniquely assign a temperature T' to a given radius r'.

The determination of temperature and its radial distribution includes then two operations: measuring the radial distribution $E^+(r)$ of the radiated intensity, and calculating $E^+(T)$.

The computation of $E^+(r)$ has been treated in Chap. (1.4.5) when we discussed the radiation of a non-homogeneous plasma layer, and the calculation of $E^+(T)$ is described in Chap. 1.4.1 on the intensity of lines emitted by plasma. We obtain the explicit expressions for the intensity of the oxygen lines by substituting the density values from Eqs.

(1.5.36) and (1.5.37) into Eq. (1.5.101). Thus we have (at atmospheric pressure):

For the relative intensity of the line of the neutral oxygen atom

$$E_{O}^{+}(T) = \frac{T'}{T} \frac{\left\{1 - S_O(T)\left[\sqrt{1 - \frac{1}{S_O(T)}} - 1\right]\right\} e^{\frac{eU_m}{kT}}}{\left\{1 - S_O(T')\left[\sqrt{1 - \frac{1}{S_O(T)}} - 1\right]\right\} e^{\frac{eU_m}{kT}}}; \tag{1.5.103}$$

For the relative intensity of the line of singly ionized oxygen

$$E_{O}^{+}(T) = \frac{T'}{T} \frac{\dfrac{3 + 8S_{O^+}(T)}{18 + 90S_{O^+}(T) + 98[S_{O^+}(T)]^2} e^{\frac{eU_m}{kT}}}{\dfrac{3 + 8S_{O^+}(T')}{18 + 90S_{O^+}(T') + 98[S_{O^+}(T)]^2} e^{\frac{eU_m}{kT}}}; \tag{1.5.104}$$

For the relative intensity of the line of doubly ionized oxygen

$$E_{O^{++}}(T) = \frac{T'}{T} \frac{\dfrac{6S_{O^+}(T) + 14[S_{O^+}(T)]^2}{18 + 90S_O(T) + 98[S_{O^+}(T)]^2} e^{\frac{eU_m}{T'}}}{\dfrac{6S_{O^+}(T') + 14[S_{O^+}(T')]^2}{18 + 90S_O(T') + 98[S_{O^+}(T)]^2} e^{\frac{eU_m}{kT}}}. \tag{1.5.105}$$

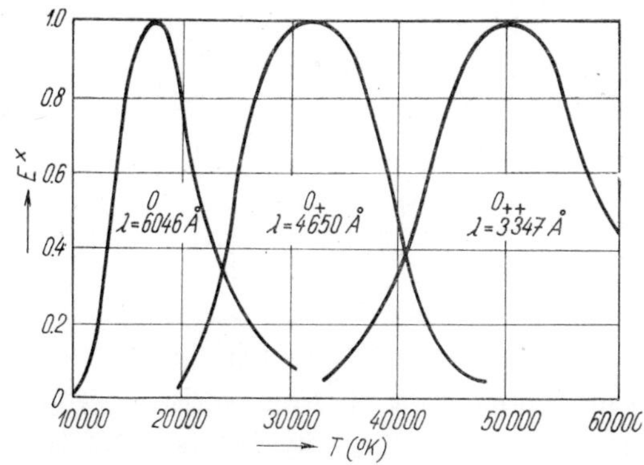

Fig. 23 Relative intensity of spectral lines emitted by oxygen as a function of temperature

Fig. 23 plots the curves of relative intensity calculated from Eqs. (1.5.103) to (1.5.105) for the oxygen lines Oλ = 6,046 Å, O$^+\lambda$ = 4,650 Å, and O$^{++}\lambda$ = 3,341 Å.

If the particle density in the plasma cannot be accurately determined, we apply [according to Maecker, Ref. 1.5.22] the Larenz method as follows:

We start again from Eq. (1.5.101) for the intensity of the line. After normalization to the maximum value we obtain

$$E^{+}(T) = \frac{n_{\emptyset}(T)}{n_{\emptyset}(T')} \frac{Z(T')}{Z(T)} \frac{\mathrm{e}^{\frac{eU_m}{kT'}}}{\mathrm{e}^{\frac{eU_m}{kT}}} . \qquad (1.5.106)$$

We determine the density of the plasma component under consideration by using Saha's equation applied to this component:

$$\frac{n^{+}}{n_{\emptyset}} n_{\mathrm{e}} = S(T), \qquad (1.5.107)$$

and assuming a constant partial pressure of the component

$$p_i = (n_{\emptyset} + n_{+})\, kT = \text{const} \qquad (1.5.108)$$

we obtain for the density

$$n_0 = \frac{p}{\left[1 + \frac{S(T)}{n_{\mathrm{e}}}\right] kT} . \qquad (1.5.109)$$

It follows from this equation that in the temperature range in which we can determine the electron density, constant pressure can be included among the proportionality factors in Eq. (1.5.106).

If the kind of atoms under consideration is the only source of electrons, we can in Saha's equation equate $n_{\mathrm{e}} = n_{+}$ and calculate from it the required density n_0. This case arises where elements having a low ionization voltage occur in a sufficient quantity in the plasma.

Figs. 24 and 25 represent the relative intensities of sodium and strontium lines as functions of temperature. The computation from Eqs. (1.5.106) and (1.5.109) has been performed under the assumption that the partial pressure is 10^{-1} N/m^2.

For measuring relatively low plasma temperatures, the method using the relative intensity of the spectral lines is most convenient. The equation of the line intensity (1.5.101) is again the starting point.

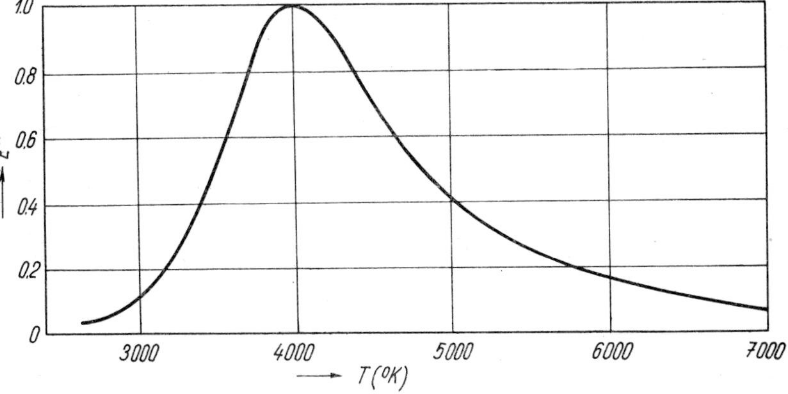

Fig. 24 Relative intensity of spectral line with 3,302·32 Å and 3,202·99 Å wavelength, emitted by sodium

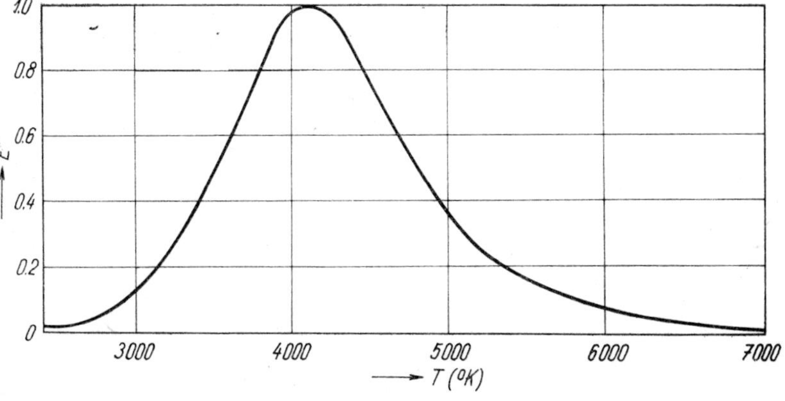

Fig. 25 Relative intensity of spectral line with 4,607·33 Å wavelength, emitted by strontium

For the relation between the intensities of two lines we obtain

$$\frac{I_1}{I_2} = \frac{\lambda_2^3}{\lambda_1^3} \frac{(f_{mn})_1}{(f_{mn})_2} e^{-\frac{eU_{m_1} - eU_{m_2}}{kT}} . \qquad (1.5.110)$$

The number of radiating particles and the thickness of the radiating layer are reduced, because both lines are emitted at equal temperature and from a layer of equal thickness. After taking the Briggs logarithm and solving for T, we obtain the required temperature

$$T = -\frac{eU_{m_1} - eU_{m_2}}{\log \frac{I_1}{I_2} - \log \frac{\lambda_2^3}{\lambda_1^3} \frac{(f_{mn})_1}{(f_{mn})_2}} 5040. \qquad (1.5.111)$$

For using the Balmer lines, we can give the equations of the line intensity a form which is more convenient for the numerical calculation. We substitute in Eq. (1.5.101) the excitation voltage of the lower term for that of the upper term, i.e.

$$eU_m = h\nu + eU_n, \qquad (1.5.112)$$

and obtain

$$T = \frac{c_2 \left(\frac{1}{\lambda_2} - \frac{1}{\lambda_1} \right) 0.4343}{\log \frac{I_1}{I_2} - \log c}; \qquad (1.5.113)$$

here

$$c = \frac{\lambda_2^3}{\lambda_1^3} \frac{(f_{mn})_1}{(f_{mn})_2} . \qquad (1.5.114)$$

After substituting numerical values for lines H_β, H_γ, and H_δ we have then Eq. (1.5.113).

For the ratio $I_{H\beta}/I_{H\gamma}$:

$$T = \frac{1.542 \cdot 10^3}{\log \frac{I_{H\beta}}{I_{H\gamma}} - 0.2843} \qquad (1.5.115)$$

For the ratio $I_{H\beta}/I_{H\delta}$:

$$T = \frac{2.379 \cdot 10^3}{\log \frac{I_{H\beta}}{I_{H\delta}} - 0.5114} \qquad (1.5.116)$$

The relation between the intensities of the lines and of the continuum can be used to measure the temperature by the procedure described below.

Eq. (1.4.22) for the intensity of the Balmer lines can be written as follows

$$\frac{m_0\lambda^3}{2hf_{mn}} I_{nm} = e^{-\frac{C_2}{\lambda T}} e^{-\frac{eU_2}{kT}} n_H 4\pi e^2. \tag{1.5.117}$$

Eq. (1.4.22) for the intensity of the limit continuum in the Balmer series may be given a similar form:

$$\frac{3\sqrt{3}e^2h^5\lambda^2}{\gamma_2 4\pi^3 me^8 l\left\{1+\sum\frac{2^3}{n^3}e^{-(u_2-u_n)}+\frac{2^3}{2u_1}e^{-(u_2-u_5)}\right\}} =$$

$$= e^{-\frac{C_2}{\lambda T}} e^{-\frac{eU_2}{kT}} n_H 4\pi e^2. \tag{1.5.118}$$

The same applies to Eq. (1.4.49) for the intensity of the continuum resulting from the blurring of the higher members in the Balmer series:

$$\frac{R_H m\lambda^5}{3\cdot 5hl} I_\lambda = e^{-\frac{C_2}{\lambda T}} e^{-\frac{eU_2}{kT}} n_H 4\pi e^2. \tag{1.5.119}$$

On the right-hand side of these equations we have constants common to all of them, or functions independent of wavelength, and the factor $e^{-C_2/\lambda T}$.

By multiplying the measured intensity values I_λ with the factors F_λ determined from the left-hand sides of the equations, and plotting $\log F_\lambda I_\lambda$ as a function of $1/\lambda$, we obtain a straight line with the equation

$$\log F_\lambda I_\lambda = -\frac{0.4343c_2}{T}\frac{1}{\lambda} + \text{const} \tag{1.5.120}$$

and from its slope we can determine the temperature.

The method using the relative intensity of the spectral lines of the neutral and the ionized atom [as described by Maecker in Ref. 1.5.22] is convenient for measuring the temperature of a plasma of unknown composition. We start again from the equation for the intensity of the line. For the ratio between the line intensities of the ionized and the neutral atoms we obtain

$$\frac{I_+}{I_0} = \frac{A_{nm}^+ n^+ g_m^+ Z^0 \nu^+}{A_{nm}^0 n^0 g_m^0 Z^+ \nu^+} e^{-\frac{eU_m{}^+ - eU_m{}^0}{kT}}. \tag{1.5.121}$$

We solve this equation for n_+/n_0 and substitute into Saha's equation applied to one component. After adjustment we have

$$\frac{I_+}{I^0} = \frac{2A_{nm}^+ g_m^+ \nu^+}{A_{nm}^0 g_m^0 \nu^0} \frac{(2\pi mkT)^{3/2}}{h^3} e^{-\frac{eU_j + eU_m{}^+ - eU_m{}^0}{kT}}. \tag{1.5.122}$$

We know all quantities on the right-hand side of this equation, hence we can calculate it as a function of temperature.

We determine the relation I_+/I_0 and the electron density by this method and calculate the temperature from Eq. (1.5.122).

A very convenient method for quick temperature estimations, using the measurement of spectrally broad lines, has been described by Schnautz [Ref. 1.5.23]. It is based on the fact that spectrally broad lines emitted by a medium have the same intensity as the radiation of a blackbody of the same temperature on the same wavelength.

This method is applicable on the following conditions, which have to be checked in every single case:

1. The width of the spectral line used for the measurement must be well within the resolving power of the spectrograph.

2. The intensity of radiation in the middle of the line must be constant on a measurable length.

3. The line must not be subject to self-reversal.

In measuring the given line we ascertain its absolute intensity and—using Planck's formula—derive the temperature from it. We want to know the standard radiation in order to determine the radiated intensity and for this Wien's simpler formula may be used instead of Planck's. For the radiation intensity of the standard at temperature T_n and wavelength λ

$$I_n = \frac{c_1}{\lambda^5} e^{-\frac{c_2}{\lambda T_n}} \tag{1.5.123}$$

For the intensity of the line used for measuring at the same wavelength and temperature T_p we write

$$T_p = \frac{c_1}{\lambda^5} e^{-\frac{c_2}{\lambda T_p}}. \tag{1.5.124}$$

Expressing the unknown temperature as a function of ratio $I_r = \frac{I_p}{I_n}$ – i.e. intensity of the measured line over intensity of the standard – we obtain

$$T_p = \frac{\frac{c_2}{\lambda}}{\frac{c_2}{\lambda}\frac{1}{T_n} - \ln I_r}. \tag{1.5.125}$$

All the methods discussed so far are only applicable if the spectral lines used for measuring are not subject to self-reversal. For measuring ihe temperature of plasma whose spectra include lines with self-reversal, Bartels [Ref. 1.5.24] has worked out a method based on the properties of radiation emitted by a nonhomogeneous layer; these properties are discussed in Ref. [1.5.25].

The temperature of an axially symmetrical plasma is determinable, according to Bartels [Ref. 1.5.21], from the contours of lines with self-reversal. The temperature is derived from the peak intensities emitted in the immediate proximity of the self-reversal minimum.

The measuring process is relatively simple: the peak intensity has only to be measured on a single ray, and no other quantities have to be known; hence results are obtained with a minimum of qualifications.

We only require that the plasma to be measured be in thermal equilibrium and the decrease in the number of atoms due to ionization be negligible. This is justified by a theoretical computation taking ionization into account. It can be proved that a 10% ionic concentration does not affect the result and the higher members of the partition function may be neglected in the given range. We set the upper limit permitting the application of this method at a temperature for which the degree of ionization is $\alpha = 0{\cdot}1$. The temperature is derived from the equation

$$T(r) = \left[\frac{T_W(v)}{1 + \frac{kT_W}{hv} \ln M_\infty(v)\, Y_{\max}[p_\infty(v)]}\right]_{v=r}, \tag{1.5.126}$$

where T_W is an auxiliary quantity defined by the equation

$$I_v = \frac{2hv^3}{c^2}\, e^{-\frac{hv}{kT_W}}. \tag{1.5.127}$$

Bartels introduces for it the designation of "Wien's temperature". I_v is the peak intensity defined as the intensity emitted by the peaks of lines in the close proximity of the minimum self-reversal.

The quantities $M_\infty(v)$ and $p_\infty(v)$ are determined by the equations

$$M_\infty(v) = \frac{\displaystyle\int_0^R \left[\frac{T(v)}{T(v')}\right]^{l+1} e^{-\frac{1}{\Theta_m}\left[\frac{T(v)}{T(v')}-1\right]} \frac{v'\,dv'}{\sqrt{v'^2 - v^2}}}{\displaystyle\int_0^R \left[\frac{T(v)}{T(v')}\right]^{l+1} e^{-\frac{1}{\Theta_n}\left[\frac{T(v)}{T(v')}-1\right]} \frac{v'\,dv'}{\sqrt{v'^2 - v^2}}} \tag{1.5.128}$$

and

$$p_\infty(v) = 3\int_r^R \frac{\displaystyle\int_r^R \left[\frac{T(v)}{T(v'')}\right]^{l+1} e^{-\frac{1}{\Theta_m}\frac{T(v)}{T(v'')}} \frac{v'\,dv'}{\sqrt{v'^2 - v^2}}}{\displaystyle\int_r^R \left[\frac{T(v)}{T(v'')}\right]^{l+1} e^{\frac{1}{\Theta_n}\frac{T(v)}{T(v'')}} \frac{v'\,dv'}{\sqrt{v' - v^2}}}\, dv''.$$

$$\cdot\frac{\displaystyle\int_r^R \left[\frac{T(v)}{T(v')}\right]^{l+1} e^{-\frac{1}{\Theta_m}\left[\frac{T(v)}{T(v')}-1\right]} \frac{v'\,dv}{\sqrt{v'^2 - v^2}}}{\displaystyle\int_r^R \left[\frac{T(v)}{T(v')}\right]^{l+1} e^{-\frac{1}{\Theta_n}\left[\frac{I(v)}{I(v')}-1\right]} \frac{v'\,dv}{\sqrt{v'^2 - v^2}}}. \tag{1.5.129}$$

where R is the radius of the plasma to be measured, and l is determined by the form of the line; for broadening by collisions with electrons,

$l = 5/12$. Θ_m and Θ_n are functions of the excitation- and ionization-potentials defined by the equations

$$\Theta_m = \frac{kT_W(r)}{eU_m + \frac{1}{2}\, eU_j} \qquad \Theta_n = \frac{kT_W(r)}{eU_n + \frac{1}{2}\, eU_j}. \qquad (1.5.130)$$

The calculation of temperatures from peak intensities is based upon Eqs. (1.5.126) to (1.5.129). After elimination of $M_\infty(v)$ and $p_\infty(v)$, the system yields the equation for $T(r)$ – the radial temperature distribution.

For easier solution of Eqs. (1.5.126) to (1.5.129), let us consider the magnitudes of $M_\infty(v)$ and $p_\infty(v)$. In the neighbourhood of the maximum temperature, it is the relative shape of the temperature distribution which exerts the decisive influence on the magnitude of these quantities. If this curve approaches the parabolic shape (as it does almost always) then we can determine the limit values $M_\infty(v)_G$ and $Y_{\max G}$ as functions of excitation- and ionization-potentials. $M_\infty(v)$ and $Y_{\max}$ differ from the limit values by a certain small value. The system of equations is solved by the iteration method. We substitute the limit values for the functions M and Y in Eq. (1.5.126). We then obtain $T(r)$ as the first approximation of the radial temperature distribution. The value $T(r)$ thus calculated is used to compute new values M' and Y' from Eqs. (1.5.128) and (1.5.129). After their substitution in Eq. (1.5.126) we obtain the second approximation for $T'(r)$.

In order to give a clear view of the factors affecting the degree of approximation, we calculate $T(r)$:

$$T'(r) = \left[\frac{T_W(v)}{1 - \frac{kT_W(v)}{hv} \ln_{\infty G} Y_{\max G}}\right]_{v=r}. \qquad (1.5.131)$$

For broadening by electron collisions, we replace the factor $kT_W(v)/h$ by the quantity

$$\Theta_{Wh} = \frac{kT_W(v)}{hv}. \qquad (1.5.132)$$

Then we have

$$\frac{kT_W(v)}{hv}\ln\left(M_{\infty G}Y_{\max G}\right)=$$

$$=\Theta_{Wh}\frac{1}{1-\dfrac{\Theta_m}{\Theta_n}}\ln\left\{\left(\frac{\Theta_m}{\Theta_n}\right)^{1/2}Y_{\max}\left(\frac{6}{4}\arctan\frac{\dfrac{\Theta_m}{\Theta_n}}{1+2\dfrac{\Theta_m}{\Theta_n}}\right)\right\}=$$

$$=\Theta_{Wh}\pi\left(\left[\frac{\Theta_m}{\Theta_n}\right]^{1/2}\right)=\Theta_{Wh}\pi(\beta) \qquad (1.5.133)$$

For simplification, we substitute Θ from Eq. (1.5.130) into Eq. (1.5.133) and obtain

$$T'(r)=\left[\frac{T_W(v)}{1-\Theta_{Wh}(v)\,\pi(\beta)}\right]_{v=r}. \qquad (1.5.134)$$

Function $\pi(\beta)$ is plotted in Fig. 26. The error resulting from the omission of the higher members of the iteration amounts to less than 50°K.

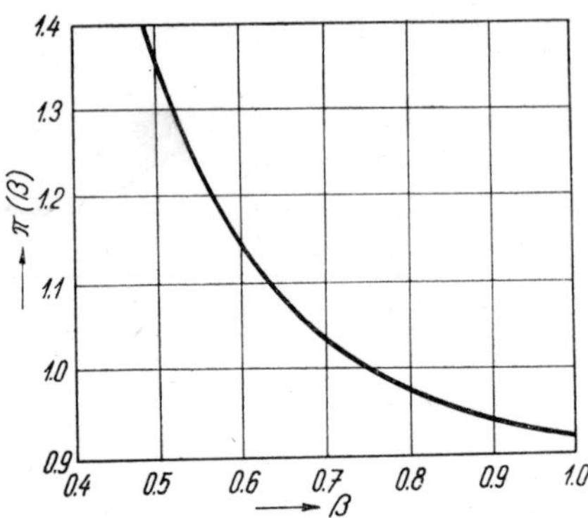

Fig. 26 Function $\pi(\beta)$

1.5.7 **Measuring Electron Density**

The electron density of a plasma is one of the factors determining its electrical conductivity. Various methods of optical plasma diagnostics can be used for ascertaining it. The three principal ways of estimating electron density derive its value from the intensity of the continuum, from the relative intensities of lines emitted by the neutral and the ionized atom, and from the broadening of spectral lines, respectively. The last method is the one used most frequently.

In Chap. 1.4.4 we have calculated the contours of the Balmer lines as a function of ion density. Since the plasma is quasineutral, the number of ions is identical with that of electrons, and electron density can be derived from the measured contours.

We use the H_β line for measuring, because its profile is most convenient for the purpose. We normalize the contours of the line, measured in arbitrarily chosen units, so that

$$\int_{-\infty}^{+\infty} I(\lambda)\,\mathrm{d}\lambda = 1; \tag{1.5.135}$$

we compare the experimental curve with the normalized theoretical curves for various values of F_0 until we find among them the one conformable to the experimental curve. The value F_0 of the theoretical curve conforming to the experimental one is the required value, from which we calculate the electron density according to Eq. (1.4.69).

A simpler way of evaluating the measured contours is by plotting in logarithmic coordinates both the theoretical and experimental contours of the line H_β. The normalization corresponds then to the shift in the vertical direction. We designate the shifted curve $\mathrm{H}\beta$, and for determining F_0 we move it along a straight line inclined at 45° to the vertical axis, until we attain the best possible coincidence with the theoretical curve. Because of the proportionality existing between the broadening $\Delta\lambda$ at a given point on the curve and the mean intensity of the field, the distance by which we shift the normalized experimental curve in the direction of the horizontal axis is the required value F_0 (on a logarithmic scale. Cf. Fig. 27).

$$\Delta\lambda_m : \Delta\lambda_{\mathrm{theor.}} = F_0 : 1 \tag{1.5.136}$$

or

$$\log F_0 = \log \Delta\lambda_m - \log \Delta\lambda_{\mathrm{theor.}} \tag{1.5.137}$$

With F_0 known, we calculate the electron density from Eq. (1.4.57).

Electron density can also be derived from the broadening of a spectral line due to the collision of the emitting particle with electrons. For this broadening we have (in Chap. 1.4.4) derived the equation

$$\delta = 50.3 \,.\, 10^{-10} T^{1/6} C^{3/2} \left(\frac{1}{\mu_1} + \frac{1}{\mu_2} \right)^{1/6} n_e, \qquad (1.5.138)$$

where C is the constant of the quadratic Stark effect, μ_1 the atomic mass of the radiating atom, μ_2 the "atomic mass" of the electron, and n_e electron density.

The estimation of density n_e is based on the measured halfwidth δ. Although the temperature is required for calculating the density, a rough

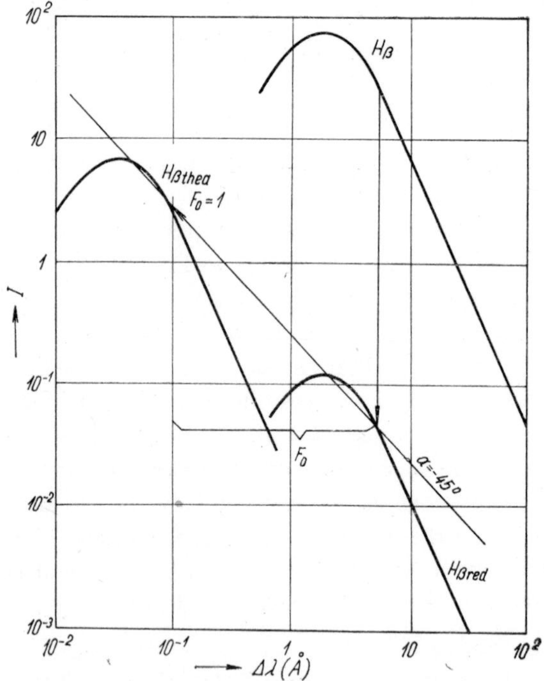

Fig. 27 Determining F_0 from the contours of line H_β

estimation is sufficient because in Eq. (1.5.138) this quantity figures in the sixth radical. From the density thus estimated we can, by using

Saha's equation, obtain the temperature and improve the accuracy of the density value.

In deriving the electron density from the intensity of the continuum, we start from the equations written in Chap. 1.4.2. Since it follows from them that the intensity of the continuum emitted from a plasma is proportional to n_e^2, we may write

$$I_\nu = \frac{C_k n_e^2}{\sqrt{T}}. \tag{1.5.139}$$

By applying this equation, we can estimate the radial distribution of electron density in a cylindrically symmetrical plasma. The procedure for evaluation is similar to that for measuring the radial temperature distribution. Since the computation of constant C_k is relatively difficult, we conveniently use Eq. (1.5.139) for estimating the relative density distribution, and calculate the absolute value from the contour of the H_β line.

In deriving the electron density from the ratio between the spectral intensities of the neutral and the ionized atom, we use Eq. (1.5.122) and obtain from it

$$n_e = 2 \frac{I^+}{I^0} \frac{\nu^{+3}}{\nu_0^3} \frac{g^+}{g^0} \frac{f_{mn}^+}{f_{mn}^0} \left(\frac{2\pi mkT}{h^2}\right)^{3/2} e^{-\frac{1}{kT}(eU_j{}^0 - \Delta eU_j{}^0 + eU_m{}^+ - eU_m{}^0)}. \tag{1.5.140}$$

With temperature known, we can derive the electron density from this equation. For the ionization voltage, we must insert the voltage reduced by the value ΔU_j calculated in Eq. (1.1.92).

References

1.1.1 Finkelnburg, W.—Maecker, H.: Elektrische Bögen und thermisches Plasma (Electric Arcs and Thermal Plasma) *Handbuch d. Physik*, Bd. XXII, Berlin, Springer Verlag, 1956, pp. 254—444

1.1.2 Stupochenko, E. V. et al.: Termodinamicheskie svoistva vozdukha etc. (Thermodynamic Properties of Air in the Temperature Interval 1,000 to 12,000°K and Pressure Interval 0.001 to 1,000 atm.) *Fizicheskaya gazodinamika.* Moscow, IAN SSSR, 1959

1.1.3 Zeise, H.: Thermodynamik (Thermodynamics), Bd. III/1 Tabellen. Leipzig, S. Hirzel Verlag 1954

1.1.4 Levič, V. G.: Úvod do statistické fiziky (Introduction to Statistical Physics). Prague, ČSAV 1954

1.1.5 Hilsenrath, J. et al.: Mechanical Computation of Thermodynamic Tables etc. (*Thermodynamic and Transport Properties of Gases, Liquids and Solids* pp. 416—437) N. York, McGraw-Hill

1.1.6 Kroepelin, H.—Winter, E.: Equilibria in CH_2 and $C + 2H_2$ Systems at Temperatures between 1,000 and 6,000 °K. *ibid.* pp. 438—452

1.1.7 Kroepelin, H.—Neumann, K. K.: Equilibria in a Thermal Plasma Composed of $C + H_2$ and $C + 2\,H_2$ in a Temperature Range from 5,000 to 50,000 °K at a Total Pressure of 1 Bar. *ibid.* pp. 453—464

1.1.8 Martinek, F.: Thermodynamic and Electrical Properties of Nitrogen at High Temperatures *ibid.* pp. 130—156

1.1.9 Denbigh, K.: The Principles of Chemical Equilibrium. Cambridge University Press, 1955

1.1.10 Gross, B.: Měření vysokých teplot (Measuring High Temperatures) Prague SNTL, 1962

1.1.11 Hála, E.—Reiser, A.: Fyzikální chemie 1. (Physical Chemistry, Part 1). Prague, ČSAV, 1960

1.1.12 Unsöld, A.: Physik der Sternatmosphären (Physics of Stellar Atmospheres) Berlin, Springer Verlag, 1955, pp. 79—105

1.1.13 Harris, G. M.—Trulio, J.: *Thermonuclear Research* **2** (1961), pp. 224—234

1.1.14 Saha, M. N.: *Z. f. Physik* **6** (1921), p. 40

1.1.15 EGGERT, J.: *Physik Z.* **20** (1919), p. 570
1.1.16 LEVINE, H. B.: *Plasma Physics* **2** (1961), pp. 206—217
1.1.17 GREEN, H. S.: *Nuclear Fusion* **1** (1961), pp. 69—77
1.1.18 THEIMER, O.—GENTRY, R.: *Annals of Physics* **17** (1962), pp. 93—113
1.1.19 VEDENOV, A. A.—LARKIN, A. J.: *ZhETF* **36** (1959), pp. 1132—1142
1.1.20 ECKER, G.—WEIZEL, W.: *Ann. d. Phys.* **17** (1956), pp. 126—140
1.1.21 BRUNNER, J.: *Z. f. Phys.* **159** (1960), pp. 280—310
1.1.22 VEIS, S.: *Čs. Čas. Fys.* **10** (1960), pp. 398—403
1.1.23 BURNHORN, F.—WIENECKE, R.: *Z. f. Phys. Chem.* **213** (1960) pp. 37—43
1.1.24 BURNHORN, F.—WIENECKE, R.: Spezifische Wärmen von Gasen im Plasmazustand (Specific Heats of Gases in the Plasma State) in Landolt-Börnstein, Zahlenwerte und Funktionen, Bd. 2 (6th ed.) Berlin, Springer Verlag, 1961
1.1.25 KÜMMEL, F.: *Beitr. a. d. Plasmaphysik* **1** (1960/61), pp. 94—106
1.1.26 GROSS, P. A.—EISEN, C. L.: Some Properties of a Hydrogen Plasma (Quoted from the Russian Collection Ionnie, plazmennie i dugovie raketnie dvigateli, Moscow 1961).
1.1.27 RAJZER, J. P.: *ZhETF* **36** (1959), pp. 1583—1585
1.1.28 FAST, J. D.: *Philips Res. Rept.* **2** (1947) pp. 382—390
1.1.29 GRANOVSKI, V. L.: Elektricheski tok v gaze (Electric Current in Gas) Moscow-Leningrad, 1952
1.1.30 KUDRIN, L. P.: *ZhETF* **40** (1961) pp. 1134—1139
1.2.1 FINKELNBURG, W.—MAECKER, H.: Elektrische Bögen und thermisches Plasma (Electric Arcs and Thermal Plasma) in *Handbuch der Physik,* Bd. XXII, Berlin, Springer Verlag, 1956, pp. 254—444
1.2.2 EDELS, H.: The Determination of the Temperatures of an Electrical Discharge in Gas. *Techn. rep. ERA*, ref. L/T 230, 1950
1.2.3 SHULER, K. E.: On the Kinetics of Elementary Gas Phase Reactions at High Temperatures Fifth International Symposium on Combustion, Reinhold Publ. Corp., 1955, pp. 56—74
1.2.4 DELCROIX, J. L.: Introduction a la théorie des gaz ionisés (Introduction to the Theory of Ionized Gases), Paris, 1959
1.2.5 DEWAN, E. M.: *Phys. Fluids* **4** (1961), pp. 759—764
1.2.6 MEIXNER, J.: Thermodynamik der irreversiblen Prozesse (Thermodynamics of Irreversible Processes), Handbuch d. Physik, Bd. III/2, Berlin, Springer Verlag, 1959, pp. 413—505
1.2.7 CHAPMAN, S.—COWLING, T. G.: The Mathematical Theory of Non-Uniform Gases, Cambridge University Press, 1939
1.2.8 HIRSCHFELDER, J. O.—CURTISS, CH. F.—BIRD, R. B.: Molecular Theory of Gases and Liquids, London, Chapman and Hall, 1954
1.2.9 LOEB, L. B.: Basic Processes of Gaseous Electronics, Los Angeles, 1955
1.2.10 GRAD, H.: *Communications on Pure and Applied Mathematics* **2**, 331 (1949)
1.2.11 LINHART, J. G.: Plasma Physics. Amsterdam, 1960
1.2.12 DE GROOT, S. R.: *Journal Nucl. Energy*, Part C **2** (1961), pp. 188—194

1.2.13 KIHARA, T.: *J. Phys. Soc. Japan* **13** (1958), p. 473
1.2.14 KNOFF, J. B.: *Phys. Letters* **1** (1962), pp. 229—230
1.2.15 ENGEL, A. VON: Ionized Gases. Oxford, 1955
1.2.16 LOCHTE-HOLTGREVEN, W.: *Repts. Prog. Physics* **21** (1958), p. 312
1.2.17 FROST, L. S.: *J. Appl. Physics* **32** (1961), pp. 2029—2036
1.2.18 SPITZER, L. JR.: Physics of Fully Ionized Gases. London, 1956
1.2.19 LEES, L.: Combustion and Propulsion. Third AGARD Colloquium, New York 1958, pp. 451—498
1.2.20 ZÁVIŠKA, F.: Kinetická teorie plynů (Kinetic Theory of Gases) Prague, Vědecké vydavatelství, 1951
1.2.21 HANSEN, F. C.: Approximations for the Thermodynamic and Transport Properties of High-Temperature Air, NASA Technical Report R-50 (1959)
1.2.22 TZY-CHENG PENG—PINDROH, A. L.: ARS Preprint, 1955—61
1.2.23 AMDUR, I.: *Planetary and Space Science* **3** February (1961), pp. 228—235
1.2.24 DAHLER, J. S.: Theories of Gas Transport Phenomena. (Thermodynamic and Transport Properties of Gases, Liquids and Solids), pp. 14—24, N. York, Mc. Graw Hill, 1959
1.2.25 AHLBORN, B.—WIENECKE, R.: *Z. f. Phys.* **165** (1961), pp. 491—501
1.2.26 BROKAW, R. S.: Alignment Charts for Transport Properties, Viscosity, Thermal Conductivity and Diffusion Coefficients for Nonpolar Gases and Gas Mixtures at Low Density. NASA Technical Report R-81, 1961
1.2.27 BROKAW, R. S.: *Planetary and Space Science* **3** Febr. (1961), pp. 238—252
1.2.28 SAMUILOV, E. V.: Effect of Internal Degrees of Freedom of Particles on the Transfer Coefficients of a Multicomponent Gas Mixture (in Russian) Fizicheskaya gazodinamika, pp. 59—69, Moscow, 1959
1.2.29 BUTLER, R.—BROKAW, R. S.: *J. Chem. Phys.* **26** (1957), pp. 1636—1643
1.2.30 CALCOTE, H. F.: Relaxation Processes in Plasma. ARS Preprint 895, 1959
1.2.31 BROWN, S. C.: Basic Data of Plasma Physics. N. York, John Wiley and Sons, 1959
1.2.32 KIHARA, T.: *J. Phys. Soc. Japan* **14** (1959), pp. 128—133
1.3.1 VOTRUBA, V.—MUZIKÁŘ, Č.: Teorie elektromagnetického pole (Theory of the Electromagnetic Field) Prague, 1955
1.3.2 ALFVÉN, H.: Cosmical Electrodynamics, London—New York 1950
1.3.3 FINKELNBURG, W.—MAECKER, H.: Elektrische Bögen und thermisches Plasma (Electric Arcs and Thermal Plasma) Handbuch d. Physik, Bd. XXII, Berlin, Springer Verlag, 1956, pp. 254—444
1.3.4 KULIKOVSKI, A. G.—LJUBIMOV, G. A.: Magnitnaya gidrodinamika (Magnetohydrodynamics) Moscow, 1962
1.3.5 PIKEL'NER, S. B.: Osnovi kozmicheskoi elektrodinamiki (Fundamentals of Cosmic Electrodynamics) Moscow, 1961
1.3.6 LEHNERT, B.: *Nuovo Cimento Suppl.* **13** (1959) No. 1, p. 59
1.3.7 ALLIS, W. P.—BUCHSBAUM, S. J.: The Conductivity of an Ionized Gas in

a Magnetic Field. Gas Dynamics Symposium at Northwestern University (24. 8. 1959)
1.3.8 Dungey, J. W.: Cosmic Electrodynamics, Cambridge University Press 1958
1.3.9 Cowling, T. G.: Magnetohydrodynamics, London 1957
1.3.10 Brdička, M.: Mechanika kontinua (Continuum Mechanics) Prague 1959
1.3.11 Shih-I Pai: Introduction to the Theory of Compressible Flow. Princeton N. J., Van Nostrand, 1958
1.3.12 Kihara, T.: *J. Phys. Soc. Japan* **13** (1958), p. 473
1.3.13 de Groot, S. R.: *Plasma Physics* **2** (1961), pp. 188—194
1.3.14 Granovski, V. L.: Elektricheski tok v gaze (Electric Current in Gas) Moscow-Leningrad, 1952
1.3.15 Langmuir, I,: *Phys. Rev.* **43** (1933), pp. 224—251
1.3.16 Fay, J.—Riddell, F.: *JAS* **25** (1958), pp. 73—85
1.3.17 Motulevich, V. P.: The System of Equations of Laminar Boundary-Layer Flow Considering Chemical Reactions and Various Kinds of Diffusion (in Russian) (*Fizicheskaya gazodinamika* etc, pp. 159—170) Moscow, 1962
1.4.1 Fabrikant, V. A.: Mechanizm izluchenia gazovogo razriada (Radiation Mechanism of Gaseous Discharges) Moscow, Gosenergoizdat, 1940, p. 236
1.4.2 Larenz, R. W.: *Z. f. Phys.* **129** (1951), p. 327
1.4.3 Maecker, H.—Finkelnburg, W.: Elektrische Bögen und thermisches Plasma (Electric Arcs and Thermal Plasma), Handbuch d. Physik, Bd. XXII, Berlin, Springer Verlag, 1956
1.4.4 Unsöld, A.: Physik der Sternatmosphären (Physics of Stellar Atmosphere) Berlin, 1938
1.4.5 Unsöld, A.—Maue, A. W.: *Z. f. Astrophysik* **5** (1932), No. 1
1.4.6 Maue, W.: *Ann. Phys.* **13** (1952), p. 161
1.4.7 Maecker, H.: *Z. f. Phys.* **114** (1939), **116** (1940) p. 257, **139** (1954) p. 448
1.4.8 Gaunt, J. A.: *Proc. Roy. Soc. London* **126**A (1930), p. 654
1.4.9 Menzel, D. H.—Pekeris, C. L.: *Monthly Notices* **96** (1935), p. 77
1.4.10 Bartels, H.: *Z. f. Phys.* **125** (1949), p. 597
1.4.11 Bartels, H.: *Z. f. Phys.* **126** (1949), p. 108
1.4.12 Bartels, H.: *Z. f. Phys.* **127** (1950), p. 243; **128** (1950), p. 546
1.4.13 Unsöld, A.: *Z. f. Astrophysik* **12** (1936), p. 56
1.4.14 Holtsmark, J.: *Ann. Phys.* **58** (1919), p. 577
1.4.15 Lindholm, E.: *Ark. mat. ast. och. fys.* **28**B (1941), p. 3
1.4.16 Gross, B.: Měření vysokých teplot (Measuring High Temperatures) Prague, SNTL, 1962
1.5.1 Cohen, S.—Spitzer, L.—Routly, M. P.: *Phys. Rev.* **80** (1950), p. 320
1.5.2 Burnhorn, F.—Maecker, H.: *Z. f. Phys.* **129** (1951), p. 369
1.5.3 Maecker, H.: *Z. f. Phys.* **136** (1953), p. 119
1.5.4 Fast, J. D.: *Philips Res. Rep.* **2** (1947), p. 382
1.5.5 Wienecke, R.: *Z. f. Phys.* **146** (1956), p. 39
1.5.6 Schlüter, A.: *Z. f. Naturforschg.* **5**a (1959), p. 72

1.5.7 Maecker, H.—Peters, T.: *Z. f. Phys.* **144** (1956), p. 596

1.5.8 Gvosdover, S. D.: *Phys. Z. Sov. U.* **12** (1937), p. 164

1.5.9 Landau, L.: *Phys. Z. Sov. U.* **10** (1935)

1.5.10 Davydov, B.: *Phys. Z. Sov. U.* **12** (1937), p. 269

1.5.11 Druyvestein, M.: *Physica* **5** (1938), p. 561

1.5.12 Schulz, P.: *Z. f. Phys.* **112** (1939), p. 435

1.5.13 Schulz, P.: *Ann. Phys.* **166** (1947), p. 318

1.5.14 Maecker, H.—Peters, T.: Z. *Phys. Chem.* 198 (1951), p. 318

1.5.15 Ornstein, L. S.—Brinkmann, H.: *Proc. Roy. Soc. Amsterdam* **34** (1931), pp. 33, 498

1.5.16 Ornstein, L. S.—Wijk, W. R.: *Z. f. Phys.* **49** (1938), p. 315

1.5.17 Knaus, H. P.—Mc. Cay, M. S.: *Phys. Rev.* **52** (1937), p. 1143

1.5.18 Janin, J.—Eyraud, Ch.: *Revue d'optique theoretique et instrumentale* **28** (1949), p. 612

1.5.19 Gross, B.: Měření vysokých teplot (Measuring High Temperatures) Prague SNTL 1962

1.5.20 Jürgens, G.: *Z. f. Phys.* **134** (1952), p. 21

1.5.21 Larenz, R. W.: *Z. f. Phys.* (1951), p. 327

1.5.22 Maecker, H.: *Z. f. Phys.* **129** (1951), p. 108

1.5.23 Schnautz, H.: *Spectrochimica Acta* **1** (1941), p. 173

1.5.24 Bartels, H.: *Z. f. Phys.* **127** (1950), p. 243; 128 (1950), p. 546

1.5.25 Bartels, H.: *Z. f. Phys.* **125** (1949), p. 597; 126 (1949), p. 108

1.5.26 Burnhorn, F.—Maecker, H.—Peters, T.: *Z. f. Phys.* **131** (1951), p. 28

Part 2

Technical Sources of Plasma

2.1 Historical Introduction

At the present time technology is unimaginable without control over high temperatures and their industrial utilization. The most intensive continuous source of high temperatures, the electric arc, has been known ever since the early nineteenth century. Technically, however, it could not be utilized until an efficient source of electric power – the electric generator – was invented in 1883. High-temperature sources based on the exploitation of exothermic chemical reactions have, however, by no means lost their industrial inportance and therefore we shall briefly review them.

2.1.1 Chemical Reactions

Ways and means of obtaining high temperatures by burning solid, liquid or gaseous fuels are common knowledge. The temperature of the combustion gases may be enhanced by preheating the air or enriching it with oxygen. The Bunsen burner and the open-hearth or the blast furnace with a Cowper stove are examples from laboratory and industrial practice, respectively. Such installations have played a paramount part throughout the period of vigorous industrial advance in the later half of last century, but their improved designs still find extensive industrial application.

The design of flameless burners, utilizing the catalytic effect of ceramics for the accelerated combustion of gas, dates from the beginning of our century, although it did not find any extensive industrial application until the fifties. Without any preheated air they can supply combustion gases heated to 1,700°C.

Temperatures of the order of 1,750°C are also attained by advanced methods of pulverized-coal combustion in slag tap furnaces or cyclone burners. The demand for even higher temperatures led to the introduction of pure oxygen as an oxidizer.

In the oxyhydrogen flame, where heat is generated according to the reaction $H_2 + 1/2\, O_2 \rightleftarrows H_2O + 56{\cdot}7$ kcal/mol, the obtainable temperature is comparatively low (~2,525°C). Its value is determined by the thermal dissociation of steam, an endothermic reaction which prevents temperature from rising any higher. In industrial practice, the oxyhydrogen flame was soon superseded by the oxyacetylene flame which can yield temperatures up to ~3,135°C.

The hottest flame can be achieved by burning gases whose combustion products have a high dissociation energy. Carbon monoxide and nitrogen (257 and 225 kcal/mol, respectively) are typical examples of such gases. Under industrial conditions, the highest flame temperature is obtained by burning cyanogen gas in oxygen, the reaction being

$$C_2N_2 + O_2 \quad = \quad 2\,CO + N_2 \qquad (2.1.1)$$

The temperature of this flame is 4,850°K.

The oxidation of various metals is another way of obtaining high temperatures in industrial practice. The best known and most widely used instance is the combustion of iron with oxygen. This procedure is exploited for cutting low-alloy steels with an oxyacetylene flame. The acetylene only serves to preheat the steel to ignition temperature, whereas the cutting process proper relies on the heat liberated in chemical reactions. The simultaneous reactions which take place in this act are tabulated in Table 2.1.1.

Table 2.1.1

Reacting substances	Reaction product	Heat liberated [kcal/mol]
$Fe + 1/2\, O_2$	FeO	64·3
$3\, Fe + 2\, O_2$	Fe_3O_4	266·9
$FeO + Fe_2O_3$	Fe_3O_4	4·1
$2\, Fe + 1\, 1/2\, O_2$	Fe_2O_3	198·5

Which of these reactions is decisive depends largely on the thickness of the material to be cut. On an average, the amount of heat liberated is 4,020 to 4,170 kcal per kilogram of oxygen [Ref. 2.1.1].

Another well-known and extensively utilized source of high temperature is the reaction of iron oxide with aluminium:

$$3\,Fe_3O_4 + 8\,Al \rightarrow 9\,Fe + 4\,Al_2O_3 + 721{\cdot}1\ \text{kcal}. \qquad (2.1.2)$$

Chemical reactions are also the source of the high temperatures required in rocket engines. Actually, such an engine is a high-temperature burner with maximum acceleration of the combustion gases. Most of the rocket fuels currently used have only been developed after the Second World War. This subject, however, lies beyond the intentions of our book and the reader will find it summarized in Ref. [2.1.3].

2.1.2 Solar Furnaces

The urgent need of a high-temperature source in which no combustion gases pollute the object under investigation has led to the construction of solar furnaces for research laboratories. The sun rays are concentrated in the focal plane of a concave mirror where a solar image is formed. Objects placed there are heated to a high temperature, its value depending on the size of the mirror and its precision. As a rule, the mirror – made

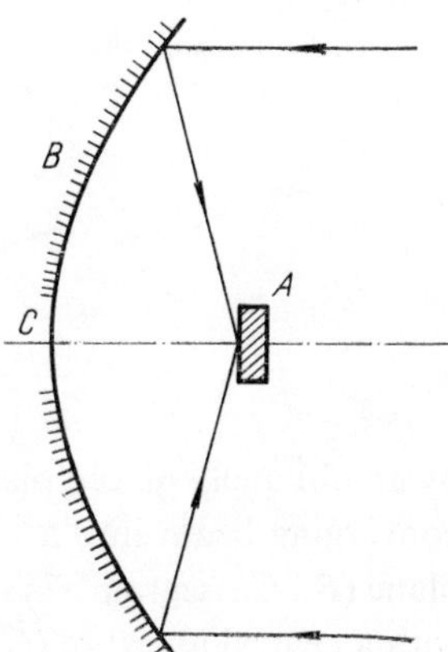

Fig. 28 Solar furnace with parabolic mirror

of glass or aluminium – is shaped as a paraboloid of revolution.

The layout of mirrors and specimens as used most commonly is schematically indicated in Figs. 28 to 31. In the simplest arrangement (Fig. 28) the specimen (*A*) is fixed in the focal plane of a parabolic

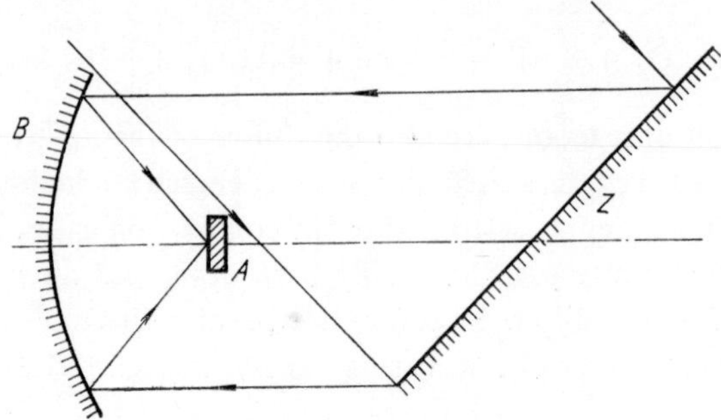

Fig. 29 Arrangement of heliostat

mirror (*B*). Usually, an aperture (*C*) in the centre of the mirror facilitates observation and measurements. In the "heliostat", the arrangement shown in Fig. 29, a plane mirror (*Z*) projects the sunbeam into a parabolic mirror (*B*). Newton's arrangement (Fig. 30) reduces the size of the device

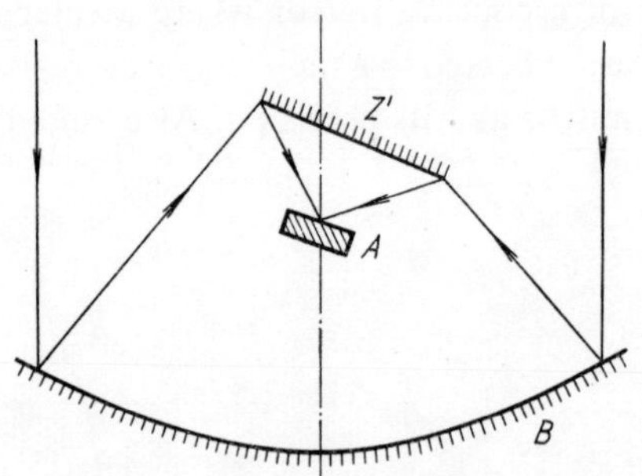

Fig. 30 Newton's arrangement

by an obliquely placed plane mirror (*Z*) which concentrates the conically converging beam into a point situated between the mirror and its focal plane (*B*). Cassegrain's system (Fig. 31) consists of a parabolic mirror (*B*) with a central aperture (*C*) and a hyperbolic mirror (*D*) which – through the aperture (*C*)—projects the rays concentrated by mirror (*B*) onto lens (*E*). The specimen (*A*) is located in the focus of the lens (*E*).

In installations for practical use, the diameter of the solar image is 6·1 to 8 mm; every increase in its size reduces the heat flux. The

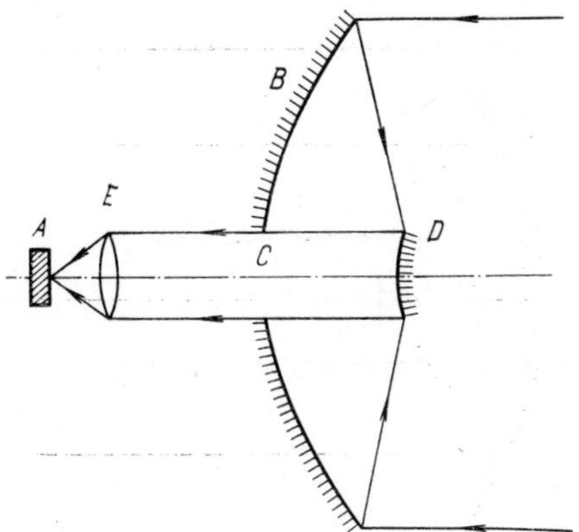

Fig. 31 Cassegrain's system

temperature gradient close to the solar image is very marked. The characteristic temperature curve reproduced in Fig. 32 shows that the slightest shift of the specimen from the centre of the solar image causes a very considerable temperature drop.

The distance plotted on the x-axis is measured from a suitable

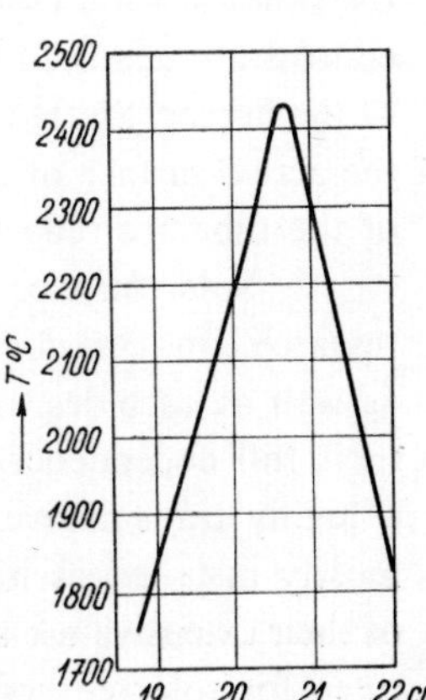

Fig. 32 Temperature curve of solar image

fixed point on the frame. In order to obtain a more uniform energy distribution over a larger area, special parabolic mirrors were made of three parts mutually somewhat inclined and thus producing three solar images next to each other. In current designs of solar furnaces the

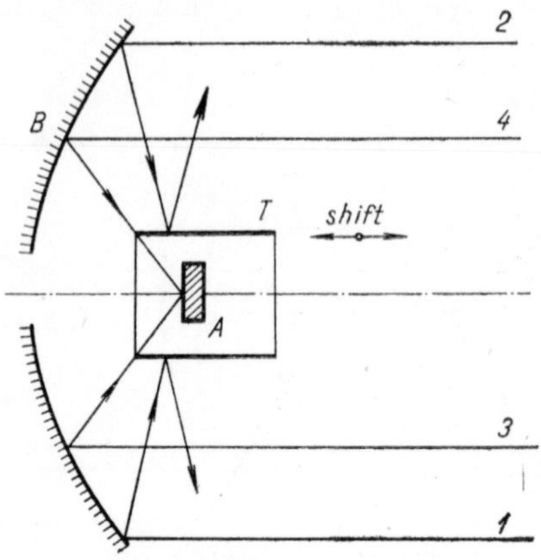

Fig. 33 Tubular screen governing amount of incident radiation

temperatures in the focus are about 3,000°C. The maximum attained so far is 4,000°K.

The temperature is controlled by screening the rays incident on the mirror. A few black cloth umbrellas fixed above the mirror are the simplest way. Good results and particularly continuous temperature control are achieved by a screening tube placed along the optical axis of the mirror. Shifted in the axial direction, the tube increases or reduces the active surface of the mirror. All rays incident on the outer surface of the tube are reflected. The principle is illustrated in Fig. 33.

Solar furnaces found their main application in high-temperature research on ceramic and refractory materials, heat-resisting alloys, as well as carbides, nitrides and borides. Their decisive drawback was their full dependence on meteorological conditions. As they called for a highly transmissive, calm and homogeneous atmosphere, they were mostly installed at high altitudes. These limitations stood in the way of their extensive use and later on a high-intensity electric arc was substituted for solar energy (cf. Sec. 2.1.5.3).

2.1.3 High Temperatures Obtained by Resistance Heating

Easy temperature control and cleanliness in operation account for the popularity of high-temperature resistance furnaces in laboratory work.

Bars of tungsten, molybdenum and other high-melting metals are at present produced from metal powders; compacted in suitable moulds, the material is pre-sintered in a protective atmosphere in an ordinary furnace. The bar is then re-sintered by high-temperature resistance heating in a vacuum or a protective atmosphere. The work-piece acts as a resistor and is heated by the current passing through it. Sintered carbides with very high melting points, such as tantalum carbide (4,148°K) and niobium carbide (3,763°K), are also prepared in this way. The most popular furnaces for obtaining high temperatures in the laboratory are graphite furnaces, and kryptol furnaces with carbon powder loosely heaped round a ceramic insulating tube acting as resistor element. Temperatures up to 3,000°C can be produced in the graphite furnace, while the temperature obtainable in the kryptol furnace is limited by the fusion point of the ceramic-tube material. If zirconium dioxide is used, temperatures up to 2,400°C may be obtained.

Furnaces with metallic heating elements are also built. In a vacuum or a protective atmosphere, temperatures up to 2,200°C may be reached with a molybdenum heating element, and up to 3,000°C with tungsten. Several screening tubes made of thin molybdenum sheet are mounted concentrically with the resistor element and act as thermal insulation. The furnace shell is externally cooled with water.

2.1.4 High-Frequency Heating

In metallurgical applications this method is usually called induction heating because of the inductive coupling between the coil fed by high-frequency current and the metal object introduced into the coil. A wide range of frequencies is used to this purpose. The metal to be heated acts as a short-circuited secondary coil of an air-core transformer or a transformer with incomplete core. The energy induced by the coil in the metal object is transformed into Joule's heat.

The high-frequency current heating the workpiece passes only through a thin layer on its surface (skin effect) and decreases exponentially as the depth increases.

2.1.5 The Electric Arc

Not until three quarters of a century after its discovery was the electric arc first used as a source of high temperature. Methods of its exploitation for welding were worked out almost simultaneously with its introduction in lighting engineering. The physical and technological properties of the arc are described in Div. 2.2; here we shall only briefly refer to some ways in which it is exploited.

2.1.5.1 The arc lamp

In the arc lamp, the arc burns between carbon electrodes which are subject to comparatively rapid consumption. The automatic control of the electrode position is achieved by a differential regulator which secures both a constant distance between the electrodes and the constant position of the source of radiation throughout the period of operation.

The carbon electrodes of modern arc lamps are "cored" with rare earths. The arc produced by such lamps – the Beck arc or "effect arc" – is distinguished by high temperatures in its anode region and the high luminosity of cerium vapours, cerium compounds being the main component of the core. Heavy-current Beck arcs (up to 1,200 A with positive carbons 20 by 30 mm) fitted with mirrors 300 mm in diameter have given luminous effects up to $9 \cdot 10^9$ Hefner candles. The positive carbons are mainly cored with a mixture of cerous fluoride and ceric oxide. Cerium oxifluoride CeOF is formed during the combustion, and investigation has shown that a cooled anode prevents the further oxidation of this compound to ceric oxide. This is a way to achieve steady combustion under higher current loads on the anode, while carbons only cored with ceric oxide burn unsteadily even with the current load 30% lower. Arc lamps with cooled anodes, developed in USA, have a luminance of up to 200,000 nit. [Ref. 2.1.7].

2.1.5.2 The welding arc

Two processes, developed in Russia at the end of last century, are used in arc welding: welding with consumable and nonconsumable electrodes (carbon or tungsten). In 1882 a method was found of joining, melting and parting metals by an electric arc burning between a carbon electrode

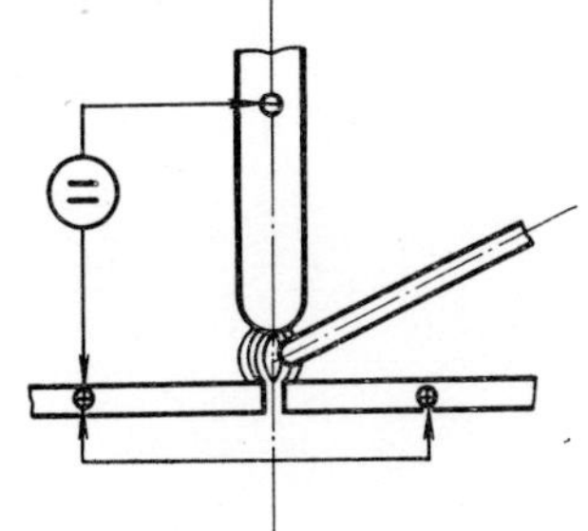

Fig. 34 Welding with carbon electrode

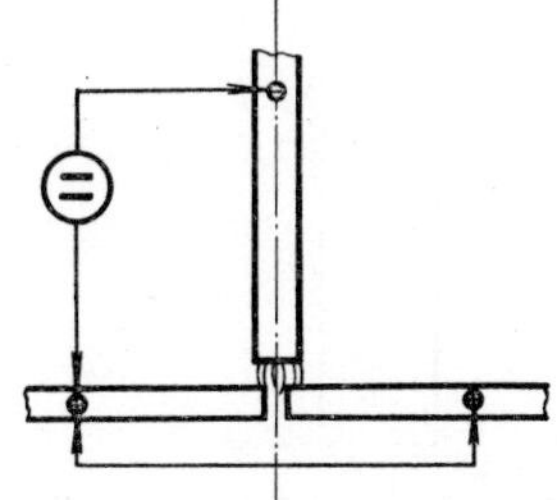

Fig. 35 Welding with consumable metal electrode

and the material; for welding, filler metal was introduced into the arc. This method is schematically shown in Fig. 34. Later, in 1888, the procedure of welding with a consumable metal electrode was worked out (cf. Fig. 35).

Many problems involving materials had to be solved before arc-welded joints became fully reliable, and thus 40 years elapsed before arc welding attained the popularity which it fully deserves. The original technique of welding with a bare wire was later improved by coating the wire with a flux that fuses together with the metal and protects the metal bath by a slag layer. Physically, the electrode covering stabilizes the arc, and some materials (such as fluor spar) permit the arc to be lengthened. As the arc fuses the metal more rapidly than the electrode covering, the latter extends beyond the metal core and directs and concentrates the flow of the molten filler metal.

When the welding process was automated, bare wires which can be coiled and continuously fed came to the fore again, but powdered flux is fed into the weld around the welding wire. This procedure (cf. Fig. 36) is far cheaper than the use of flux-covered electrodes. Originally, the heat generated by the resistance of the flux to the passage of the current was believed to supply much of the energy liberated in submerged-arc welding; later, resistance heating was found to account for only one per cent of the total heat liberated [Ref. 2.1.8].

Welding in a shielding atmosphere of carbon dioxide – with a bare electrode, either automatically or by hand – has found extensive

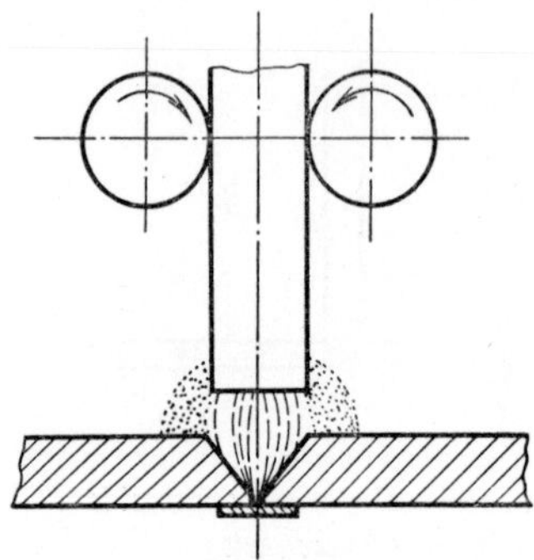

Fig. 36 Submerged-arc welding

application during the last decade. The slag-forming additions required for the production of high-quality welds, mainly manganese and silicon, are supplied as alloying additions to the filler metal.

The use of nonconsumable metal electrodes in combination with shielding atmospheres has also widely spread. Tungsten or thoriated tungsten, less frequently tantalum or molybdenum, are the materials used for this purpose. The current load must be selected at a value which will not cause the metal bath to be contaminated by the electrode material. The maximum permissible loss of electrode tungsten is supposed to be 50 μg per centimetre of weld joint on steel, and 25 μg/cm on aluminium [Ref. 2.1.2].

The high dissociation energy of hydrogen is utilized in atomic-hydrogen arc welding. The arc, entirely independent of the workpiece, is maintained between two tungsten electrodes including an acute angle; the shielding hydrogen is supplied through the current-carrying tube of the electrode holder. In the arc, the hydrogen molecules are dissociated according to the equation

$$H_2 \rightleftarrows 2\,H + 95\ \text{kcal.} \tag{2.1.3}$$

On recombination of the hydrogen atoms, as they strike the workpiece, the large amount of heat absorbed from the arc in the process of dissociation is liberated and transferred to the material. Atomic-hydrogen welding requires high ignition voltages (~300 V), while the welding voltage is 50 to 80 V, supplied by a.c. of 10 to 100 A. At a temperature of ~3,950°C the concentration of atomic hydrogen is about 76%. In the boundary zones where the hydrogen is in contact with atmospheric oxygen it is burned to water. Either manual or automatic techniques may be used for atomic-hydrogen arc welding.

2.1.5.3 **Electric arc furnaces**

Ever since the end of last century, the high temperatures and high heat concentration of the electric arc have been exploited in electric furnaces. The two main types of such installations are direct-arc furnaces (Fig. 37) with an arc burning between the electrode and the metal to be fused, and indirect-arc furnaces (Fig. 38) with an arc—independent of the furnace charge—burning between two electrodes. Direct-arc furnaces are more widely used; indirect-arc heating is mostly limited to

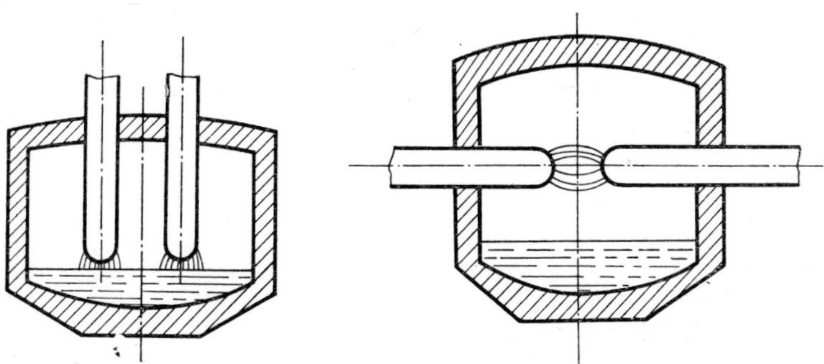

Fig. 37 Direct-arc furnace *Fig. 38 Indirect-arc furnace*

non-ferrous metallurgy and foundries, where the more advanced low-frequency furnaces have been superseding it in recent years. Electric-arc furnaces operate with very high currents and are fed by special regulating transformers with many switching steps. The efficiency of the furnace primarily depends on the correct proportion between the furnace size and the transformer output, moreover on the favourable layout of the feeder buses.

Apart from the differences in design, we distinguish – according to the way in which the electrical energy is transformed into thermal energy – between

- steel-melting furnaces of 220 – 380 V, with all the energy generated in the arc; and
- reducing furnaces (cf. Fig. 39) in which part of the thermal energy is liberated in the arc, and part is generated as the current passes through the charge which is highly conductive in the high-temperature zone next to the hearth. Depending on the conductivity of the charge, these furnaces operate at a voltage of 50 to 100 V.

Vacuum arc furnaces have been developed for high-melting metals, such as titanium, molybdenum, zirconium. In these furnaces the pulverized or sponge metal obtained by reduction is re-melted into water-cooled moulds. Either the metal powder is continuously compacted into a consumable electrode which is melted off by the arc, or else the metal powder is dropped into the arc burning between a cooled tungsten electrode and the metal bath. Especially in furnaces with a non-consumable electrode, a shielding atmosphere may be substituted for the vacuum [Ref. 2.1.9]. In this case the furnace is scavenged with argon. The furnace is fed with D.C. from several high-output welding

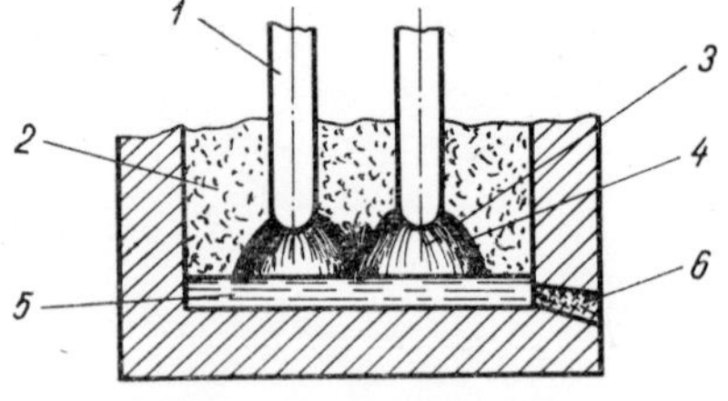

Fig. 39 Reducing furnace
(1) electrode (2) charge (3) arc zone (4) zone of resistance heating (5) bath (6) tap hole

sets connected in parallel. The working voltage is about 35 V and the current 2,000 to 5,000 A.

Solar laboratory furnaces, as used for research in the high-temperature region and studies of various radiation effects, were described in Chap. 2.1.2. The drawback of these furnaces, in particular their dependence on atmospheric conditions, has been overcome by substituting the anode crater of a Beck-type electric arc as the source of radiation. The optical systems employed in such furnaces are indicated in Fig. 40. In the ellipsoid mirror of the single-mirror system, the radiation source

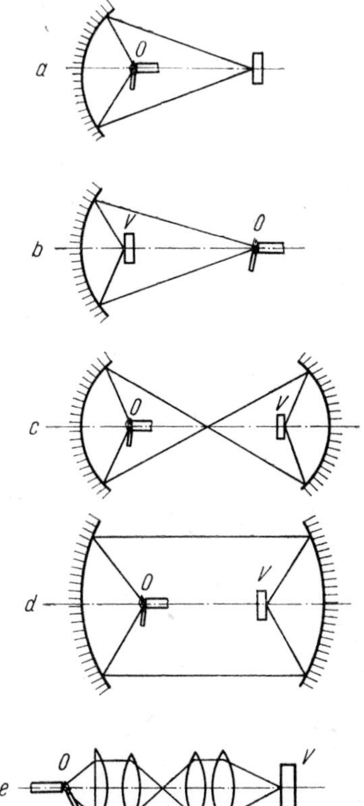

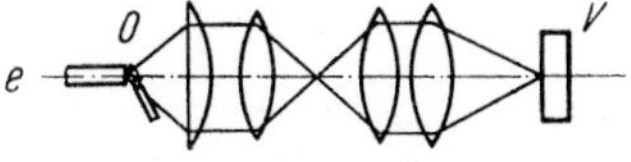

Fig. 40 Optical systems for radiation-type arc furnaces

is placed in either the near or the far focus (a or b); accordingly the image of the source is larger and the heat-flow density smaller, or vice versa. Most popular are two-mirror systems with ellipsoid (c) or paraboloid mirrors (d): Rays concentrated by condenser lenses, as shown in (e), yield the most uniform distribution of the heat flow on a comparatively large image of the radiation source.

Ever since the end of the Second World War, radiation furnaces have found extensive application in laboratories studying material properties at high temperatures; moreover, their convenient precise control and dosing of heat flux is used in research on skin burns.

Before discussing the development of plasma-generator designs, we have to mention the arc furnaces for combustion of atmospheric nitrogen in nitric-acid production. Although these furnaces, introduced at the beginning of this century, are hardly used any more, they were an important stage in the development of electric-arc installations [Ref. 2.1.10].

The reactions between atmospheric nitrogen and oxygen, described in Chap. 3.4.3, do not give sufficient yields except at high temperatures. Therefore methods capable of increasing the heat of the arc plasma have to be employed in furnaces for nitrogen combustion. The numerous designs of arc furnaces intended for this purpose may be classified in three main groups according to the manner in which the arc burns:

1. The motion of the arc is governed by the flowing reaction mixture.

2. The motion of the arc is governed by a magnetic field.

3. The arc does not move and its column is stabilized by an air vortex.

The main representative of the first group is Pauling's furnace, derived from the horn lightning arrester. The electrodes are arranged as indicated in Fig. 41. A.C. of 5,000 to 6,000 V is used. The arc is ignited

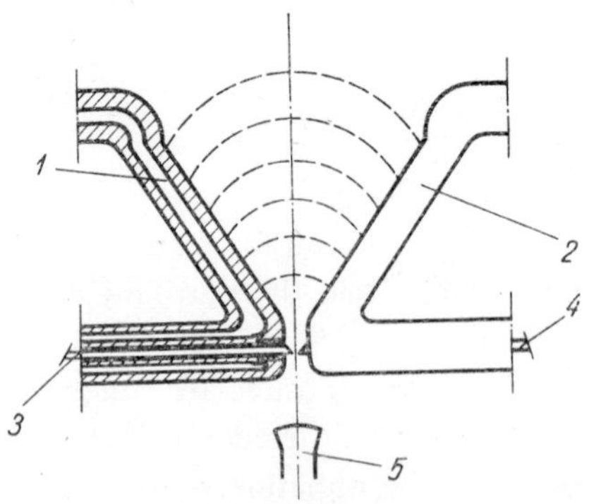

Fig. 41 Pauling's furnace

on the auxiliary electrodes (*3*) and (*4*), and the air streaming from nozzle (*5*) blows it into the extreme (dashed) position on the electrodes (*1*) and (*2*) where it tears off. The recovery voltage then causes the arc to be re-ignited in the gap between the auxiliary electrodes,

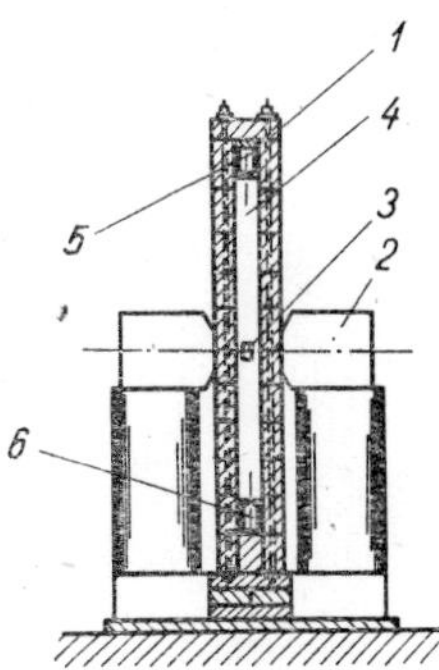

Fig. 42 Birkeland-Eyde's furnace

and the cycle is repeated. The rapid motion of the arc spots over the cooled electrodes accounts for a tolerable service life of the latter.

Birkeland-Eyde's method of widening the arc by a magnetic field belongs to the second group. In the diagram Fig. 42, (*1*) represents the furnace lining, (*2*) is a strong electromagnet whose poles are situated in the electrode plane perpendicularly to the centre line of the water-cooled electrodes (*3*). Apertures (*5*) and (*6*) are the air inlet and outlet respectively. In the flat furnace chamber (*4*), the magnetic field widens the arc to a plate shape of about 1·8 m diameter. Currents of 30 to 20 A at 2,000 to 3,000 V gave the best results.

The Schönherr-Hessberger furnace with a long and stable arc is the most important representative of the third group. The arc is stabilized by air rotating at high speed. This design is considered to be a forerunner of the present-day plasma torches with the arc stabilized by a gas vortex. In Fig. 43, illustrating the principle of this furnace, (*1*) is the electrode whose bottom part is centred in tube (*2*), the actual working chamber. Tube (*2*) simultaneously acts as the other electrode and is electrically insulated from the rod-type electrode (*1*). Tube (*7*) concentrically surrounds tube (*2*); in the upper part they are welded up to form a water reservoir. The space between them is divided by tube (*4*) and forms an air labyrinth acting as heat exchanger. The air is fed through aperture (*6*), rises between tubes (*7*) and (*4*), passes between tubes (*4*) and (*2*) into the bottom part of the furnace, and flows through tangential openings (*5*) past electrode (*1*) into the working

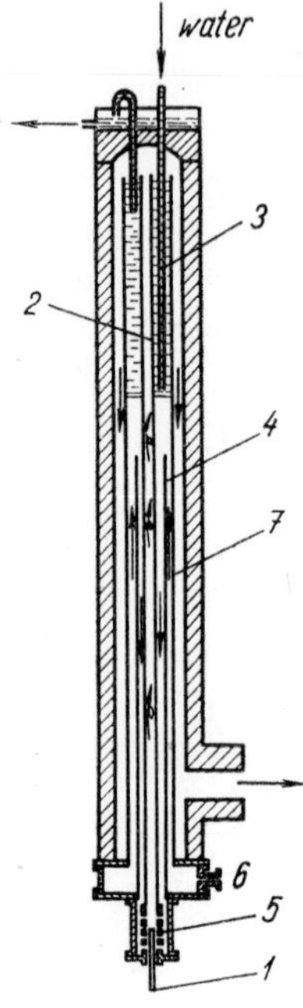

Fig. 43 Schönherr-Hessberger furnace

chamber. The arc is initiated by means of a thin wire which produces a short-circuit between electrode (*1*) and tube (*2*) in the bottom part of the furnace. Once ignited, the arc is blown into the water-cooled zone (*3*). An arc 8 m long and burning very steadily is thus produced. The arc output is about 500 kW.

This concept was later used in a methane-splitting arc furnace started at Hülls (Germany) in 1938. The furnace works at 7,500 V and 900 A. Unlike in the nitrogen-combustion furnaces, the arc is only 1 m long, but the current is 20 or 30 times stronger which results in a higher arc temperature and higher yields.

2.1.6 Short-Lived Processes

As a rule, the production of high temperatures by short-lived processes is accompanied by destructive effects which are not easily controlled. The highest temperatures produced so far on the Earth are associated with thermonuclear explosions. For the time being, however, thermonuclear and atomic explosions are not controllable. Violent chemical reactions can be controlled, up to a point, but they are still predominantly employed in destructive techniques. When Wartenberg photographed shock waves produced by explosions, 30 years ago, he found that in places where shock waves of two different explosions collide temperatures are far higher than in their centres; he measured up to 17,000°K. If the spatial arrangement of the explosive is such that the shock waves collide in one point, highly ionized high-temperature plasma is generated there. The anti-tank weapons known as bazooka and Panzerfaust in the Second World War were based on this principle.

In physical high-temperature research, metal wires are exploded by a current derived from the discharge of a large condenser (1 μF, 20–50 kV). Rough spectroscopical estimations put the resultant temperature of such experiments at 15–20,000°K, but it is too short-lived—in the range of 10^{-6} s—to be exploited technically. The same, except for special cases, applies to high temperatures experimentally obtained from electric sparks. Such a special case is welding by condenser discharges. Condensers for 500 to 5000 V are used, and the discharge lasts 20 to 80 μs; hence the fusion is strictly limited to the surface of the material. This method therefore lends itself to the welding of otherwise unweldable thin wires, foils, etc.

2.2 Physical and Technical Principles of Plasma Torches

Plasma torches are devices for transforming electrical energy into the thermal energy of a directed plasma jet. They make use of a stabilized electric arc which burns in a constricted space and is blasted with gas or the vapours of a liquid surrounding the arc column. The plasma jet, continuously generated from the stabilizing medium in the arc column, issues from the discharge nozzle at a high velocity.

2.2.1 Historical Development of the Stabilized Arc

In Sec. 2.1.5.3 we described Schönherr's design of an electric furnace used for the oxidation of atmospheric nitrogen at the beginning of this century. In this furnace an arc burning steadily in a long tube was secured by a vortex, formed by air tangentially introduced and rotating in the tube at high velocity. This device is the forerunner of the vortex-stabilized plasma torch.

The first theoretical studies and results of observations made on a water-stabilized arc were published by Gerdien and Lotz in 1922 [Ref. 2.2.1]. The apparatus consisted of two carbon electrodes with a cylindrical nozzle placed coaxially between them. Water tangentially introduced into the middle of the nozzle formed a vortex that covered the channel surface with a thin water layer throughout its length. On leaving the nozzle, the water uniformly issuing on both ends of the channel produced a spray cone, and thus did not wet the electrodes. The important result derived from these experiments was the discovery of a high concentraion of thermal energy (due to the constriction of the arc) and of the axial acceleration of the hot plasma as it issues from the vortex. When the gauge pressure in the stabilizing channel was measured later on, it was found to increase with the intensity of the electric current; the electric-field intensity in the arc increased similarly [Ref. 2.2.2]. After the properties of the stabilized arc had been described, propositions for its technical utilization began to appear in patent literature. First among them was an Austrian application [Ref. 2.2.3] whose author proved that the efficiency at which electrical energy is transformed into thermal energy could be increased by 75%. To this end the stabilized arc was blasted with air or other gases. The carbon rod was enclosed by a concentrically fitted tube of conductive material acting as the second electrode. Instead of a conductive tube, another variant featured a tube made of insulating material and carbon rods uniformly spaced at its circumference so as to protrude into the operating zone. A collateral patent [Ref. 2.2.4] protected the application of a stabilizing medium consisting of vapours of suitable liquids evaporated in the vessel of the apparatus.

Another Austrian application [Ref. 2.2.5] describes the arrangement of a tubular electrode fitted with a nozzle concentric with the internal rod-type electrode. The narrow aperture of the nozzle must be smaller than three times the electrode diameter in order to ensure stability.

A British patent [Ref. 2.2.6] protects the stabilization of the cutting or welding arc by the combined effect of axial blasting and a magnetic field excited by a field coil placed concentrically. The internal electrode—the cathode—is made of graphite.

A German Siemens patent [Ref. 2.2.7] protects a device whose principle is rather similar to that of the present-day plasma torches. The difference is in the nozzle, which consists of a complicated grating of carbon electrodes radially pointing towards the middle; as they wear in operation, they are shifted in such a way that their tips form a tapering circular cross-section (Fig. 44). The inner electrode is also a carbon rod. The arc is stabilized by axial gas blasting and by the burner section mechanically narrowed down by means of the electrode grid. The device was designed for A.C.

American authors also suggested various designs utilizing

a stabilized arc. The patents Ref. [2.2.8 and 9] describe plasma generators fed by either D.C. or A.C., and there is already a tendency to use water-cooled electrodes made of tungsten or other metals. The arc is stabilized by gas, and the motion of the arc spots is controlled by a magnetic field.

There is no evidence, however, that any of these patents has been industrially utilized. The designs were not elaborated and no devices suitable for industrial exploitation were built. The technical and economic conditions for the industrial application of a stabilized arc had not yet matured, as aluminium and thick stainless steels were not yet widely used in industrial production.

After the Second World War, the rapidly growing chemical, foodstuffs and power-generating industries called for large-size apparatus made of stainless or high-alloy steel, and the greatly increasing consumption of aluminium likewise necessitated rapid and economical methods of parting materials. For some time this need was satisfied by oxy-acetylene cutting, with iron or quartz powder added into the flowing oxygen. The beginning of the fifties saw the renewal of theoretical studies in the physics of the high-temperature arc, mainly in the German

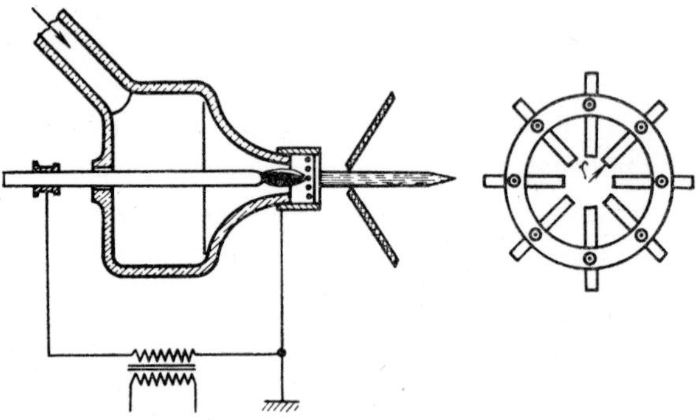

Fig. 44 Siemens torch (designed in 1935)

Federal Republic where several authors worked on this problem. Gerdien's method of stabilizing the arc by a water vortex, as published in 1922, was the starting point of the new research. All basic physical measurements were carried out with an apparatus designed by Maecker

[Ref. 2.2.10] (Fig. 45). The tubular part of the water turbine had a diameter several times as large as that of the water vortex, which in turn was determined by the diameter of the nozzles coaxially located at both ends of the tube. In this arrangement the water is introduced tangentially,

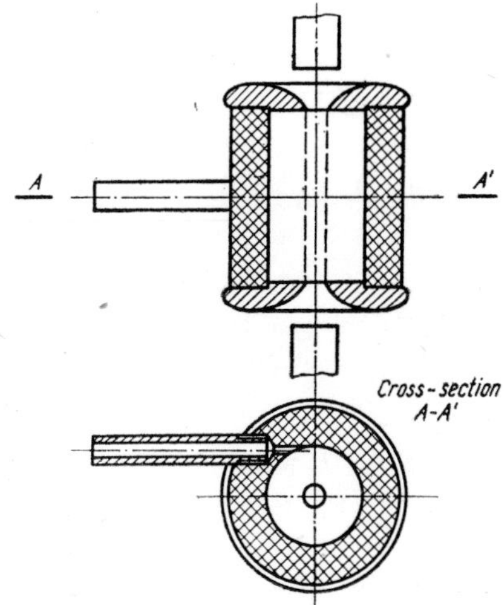

Fig. 45 Maecker's design for a water-stabilized arc

and the velocity of its rotation increases towards the middle of the water channel. The inner diameter of the water vortex is easily altered by exchanging the nozzles. The arc was observed by a sight glass in the stabilizer body. Both electrodes were placed outside the stabilizing channel. This arrangement facilitates convenient measurements of the physical and electrical arc parameters, but it is unsuitable for practical purposes as the energy liberated in the channel is carried away uniformly to both sides by the plasma jet.

Weiss [Ref. 2.2.11] shifted one electrode into the vortex channel, and adjusting simultaneously the design of the nozzles, he actually constructed the prototype plasma torch (Fig. 46). The front nozzle, made of graphite, acted as an electrode, and to prevent it from being covered with water, it had a smaller diameter than the water vortex. All the water fed into the vortex flowed off through the annular gap between the electrode and the edge of the rear nozzle. The rear nozzle was followed by the pressure chamber, which was sealed by the water preventing the plasma from issuing at the rear. Owing to the rapid wear of the front nozzle the device could not work longer than for a few seconds, which was sufficient for measuring the velocity of the flowing plasma. This design (Fig. 46) was also used in the USA for

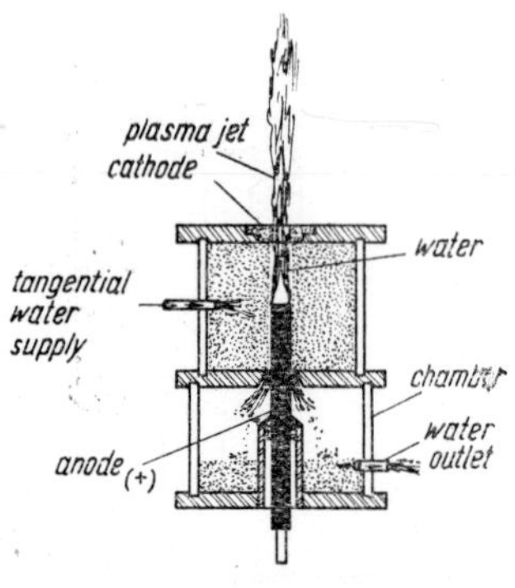

Fig. 46 Weiss' design for a water-stabilized arc

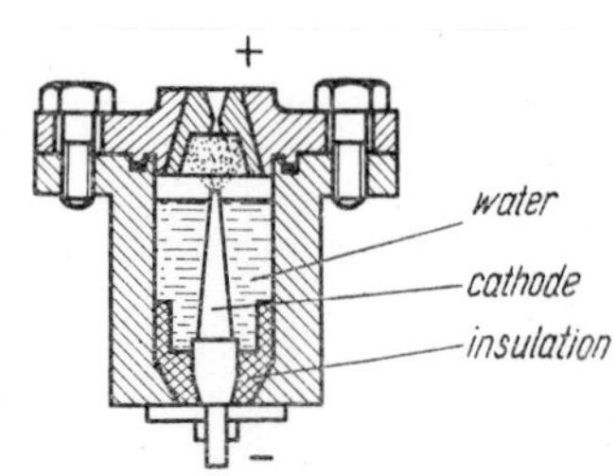

Fig. 47 Peters' design for supersonic plasma velocities

research into the re-entry conditions of ballistic missiles [Ref. 2.2.12 to 14] and other investigations of material behaviour at high temperatures and flow velocities.

High discharge velocities were obtained with an apparatus designed by Peters [Ref. 2.2.15]. It consisted of a pressure vessel with a graphite Laval nozzle (Fig. 47) serving for discharge. The vessel was filled with water, which was intensively vapourized by the arc burning between the nozzle and a graphite electrode; the vapours entered into the arc. This pressure device secured supersonic velocities of plasma flow.

The basic work quoted above was followed by devices for industrial utilization. Such was a plasma torch designed in the GFR, mainly for cutting metals (with the workpiece acting as the anode) [Ref. 2.2.16]. This design was derived from the concept presented by Weiss [Ref. 2.2.11] and was affected by all of its shortcomings; this was probably the reason why it was never put to any practical use. The same also refers to a prototype tried out in the GDR [Ref. 2.2.17].

A laboratory burner with combined arc stabilization was also designed [Ref. 2.2.18]. A stream of argon protected a tungsten cathode from the corrosive effects of steam, but the stabilizing channel proper was formed by a water vortex. The anode was a water-cooled carbon rod.

In the USA, liquid stabilization of the arc was also attempted; soon, however, the Americans developed devices stabilizing the arc by means of gases or gas mixtures [Ref. 2.2.20 and 21]. A number of patents granted in the later half of the fifties [Ref. 2.2.22–34] refers to various arrangements of details in the torch head. Simultaneously, several firms began the commercial production of plasma torches, mainly for metal cutting, application of protective coatings, special purposes and research.

Gas-stabilized plasma torches – manufactured under licence or based on the makers' own research – soon spread in all industrially advanced countries.

Plasma torches with water-stabilized arc, capable of industrial application, have been developed at the Czechoslovak Research Institute of Electrical Engineering (R.I.E.E. for short) in Prague-Běchovice. The design, which is based on various patented features [Ref. 2.2.35 to 38] is fully described in Div. 2.3.

2.2.2 The Electric Arc

The arc is an electric discharge in gas, capable of independent existence. Its characteristic features are

- slight cathode drop;
- slight potential difference on the electrodes;
- current ranging from a few to thousands of amperes;
- great current density in the cathode spot;
- intensive radiation of light both from the electrodes and from the discharge path.

Its position among electric discharges in gas is illustrated in Fig. 48.

According to the current supplied we distinguish between D.C. and A.C., arcs. Each of these kinds can be classified into freely burning and stabilized arcs, stabilization meaning in this case the limitation of the space in which the arc burns.

The static characteristic of a freely burning arc is largely influenced by the material of the electrodes. Hence the classification of arcs as burning between either carbon (graphite) or metal electrodes.

The cathode of the electric arc reaches a temperature of 3,200 to 3,600°K, the anode of 3,600 to 4,000°K. This difference is due to the

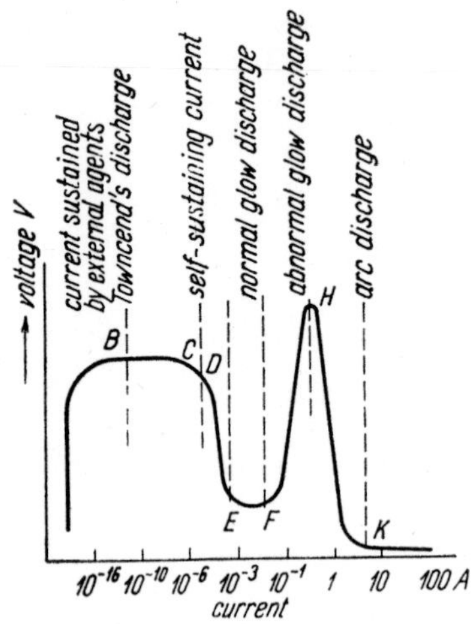

Fig. 48 Classification of discharges

fact that the anode is only cooled by heat conduction and radiation losses, while the cathode is, in addition, affected by heat losses corresponding to the work function of the electrons. The anode temperature depends on the current flowing through the arc. If the amperage rises to a value at which the anode material begins to evaporate, the anode phenomena alter the arc characteristic. A classification distinguishing between low- and high-current arcs has therefore been introduced [Ref. 2.1.4]. In a certain intermediate region between low and high currents, the arc burns unsteadily and emits a characteristic hissing sound.

The characteristic of a D.C. arc was first experimentally ascertained by Ayrton [Ref. 2.2.39]. In the range of steady burning, these hyperbolic characteristics are described by the formula

$$U_{\text{arc}} = A + Bl + \frac{C + Dl}{I}, \qquad (2.2.1)$$

where U_{arc} is the arc voltage; l the length of the arc; A, B, C, and D constants depending on the kind and pressure of the gas and the dimensions and cooling of the electrodes; and I is the current flowing through the arc.

Tables 2.2.1 and 2.2.2 [Ref. 2.2.40] show the dependence of constants A, B, C, D on the kind of the gas (and its circulation in the case of air) and on the air pressure, respectively.

Table 2.2.1

Constant	Air		Ar	N_2	CO_2
	circulatg.	non-circulatg.			
A	35·7	44·1	24·8	44·5	48·2
B	3·0	2·6	0·9	1·7	2·6
C	114·8	17·8	10·2	10·2	23·3
D	1·8	1·8	0·0	8·7	5·3

Table 2.2.2

Pressure [Torr]	A	B	C	D
740	38·5	2·15	54	6·1
200	25·5	1·84	38	8·3
50	33·7	1·51	26	10·8
5	23·7	1·20	0	15·7
10	27·0	1·35	19	13·4

Ayrton's characteristics (Fig. 49) are valid for an arc between carbon electrodes in air and an arc length of only a few millimetres. Arcs burning between metal or cored electrodes considerably deviate from Eq. (2.2.1). Nottingham [Ref. 2.2.41, 42] has later rewritten this relation for wider application:

$$U_{arc} = A + Bl + \frac{C + Dl}{I^n}, \tag{2.2.2}$$

where n, depending on the anode material, varies from 0·35 for zinc to 1·38 for tungsten. For an arc in air, exponent n was found to be linearly dependent on the boiling temperature T [°K] of the anode surface:

$$n = 2{\cdot}62 \times 10^{-4} T. \tag{2.2.3}$$

In accordance with Eq. (2.2.1), $n = 1$ for carbon electrodes, and $n = 0{\cdot}67$ for copper covered with an oxide casing.

The constants A, B, C, D differ for various metals and are greatly affected by external conditions. For arcs which are long enough,

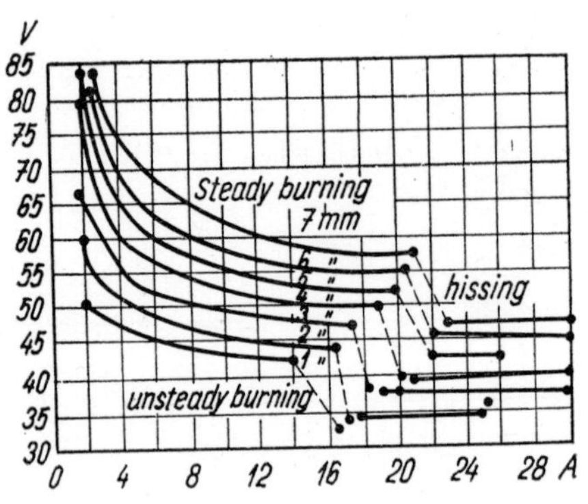

Fig. 49 Ayrton's volt-ampere characteristics

constants A and C may be neglected and Eq. (2.2.1) assumes the form

$$U_{\text{arc}} = l\left(B + \frac{D}{I}\right). \tag{2.2.4}$$

If high currents flow through the arc, the fraction $\frac{D}{I}$ is negligible and the voltage is only a function of the arc length:

$$U_{\text{arc}} = A + Bl. \tag{2.2.5}$$

Provided the arc is long enough, constant A may be neglected to a first approximation, and in this case the voltage is proportional to the arc length:

$$U_{\text{arc}} = B \,.\, l. \tag{2.2.6}$$

This approximation is for instance sufficient for the estimation of the voltage generated on the arc when a short-circuit is broken.

Conditions of a stable arc. Owing to the negative slope of its characteristic (Fig. 50) the arc is not stable in an electric circuit with constant e.m.f without series-connected stabilizing resistances. In the example shown in Fig. 50 the open-circuit voltage of the source is

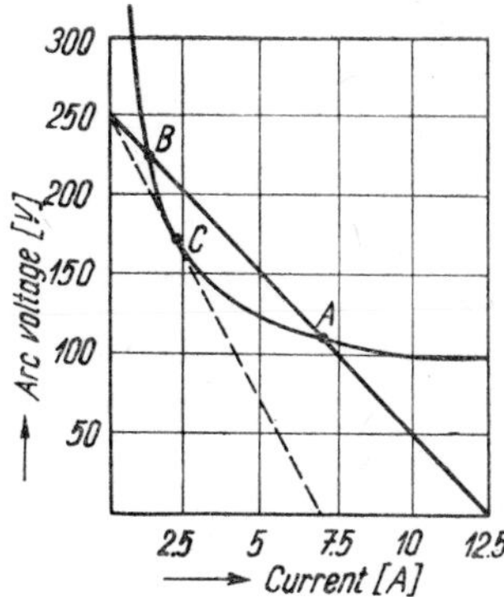

Fig. 50 Characteristics of arc and source

$U_{oc} = 250$ V, and the connection in series of a resistance $R = 20\ \Omega$ produces a falling characteristic of the source U_Z (full line):

$$U_Z = U_{oc} - RI = 250 - 20I \qquad [\mathrm{V}]. \qquad (2.2.7)$$

The arc can burn steadily in points A or B only, where the characteristics of the arc and the source intersect. In these points the arc voltage equals the voltage across the source terminals. In point B, however, the arc cannot hold out since any change in current unbalances the component voltage. If the current diminishes, insufficient voltage extinguishes the arc; if the current increases, the arc voltage drops and the increased source voltage raises the current flowing through the arc up to the equilibrium point A. At this point any change in current causes a change in voltage, which returns the current to position A.

The reduction of resistance R reduces the slope of the current-source characteristic, and point A is shifted into the region of higher currents. The tangent to the arc characteristic (dashed line) corresponds to the maximum value of resistance R at which the arc is still capable of burning, with the voltage and current corresponding to point C.

If the changes in current are rapid, the dynamic characteristic of the arc greatly differs from the static one. The effect of the cathode temperature on the arc discharge curve causes a hysteresis which appears as a deviation of the arc voltage from the curve of its static characteristic (cf. Fig. 51). A high-frequency sinusoidal current of lower amplitude than the D.C. voltage U_{arc} is superimposed onto the continuous arc

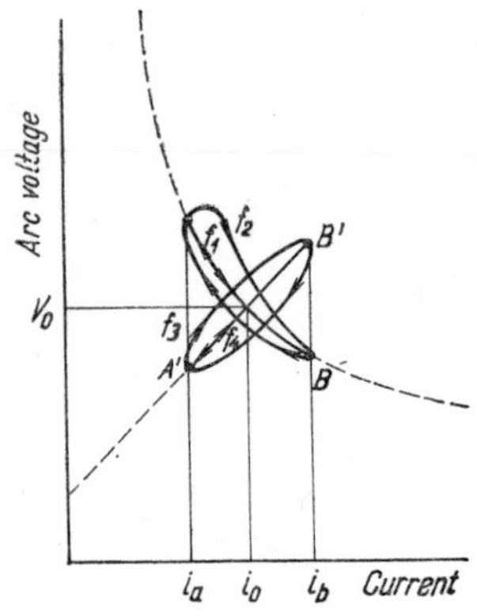

Fig. 51 Dynamic characteristic of D.C. arc

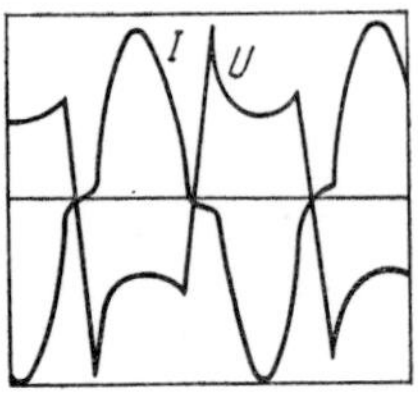

Fig. 52 Oscillogram of A.C. arc

current i_{arc} with the arc voltage U_{arc}. The current oscillates between the points i_a and i_h. At lower frequency, the voltage follows (in the sense indicated by arrows) the elliptic curve f_2 circumscribed around the static characteristic. As the frequency increases, the loop swivels round the operating point 0 until it reaches the position $A'B'$, and ultimately flattens out to form the straight line f_4.

The dynamic characteristic is very important in arcs fed by A.C. A theory of the A.C. arc has been published by Simon [Ref. 2.2.43] and elaborated by other authors. A simultaneous oscillogram of the current and voltage in an A.C. arc of 50 c.p.s. frequency is shown in Fig. 52, and the dynamic characteristic in Fig. 53. The alternating arc is extinguished at every transition of the current through zero and ignited again at the opposite polarity of the electrodes. At the industrial frequency of 50 c.p.s. this process is repeated 100 times a second. Heated from the preceding semiperiod of the discharge, the cathode emits electrons immediately at the beginning of the new semiperiod (which in the oscillogram appears as a slight rise of the current curve above the zero line). This thermal inertia of the electrodes permits the arc to be re-ignited as soon as the component voltage attains the flash-over value for the given medium. Between the ignition voltage (the sharp voltage peak) and the extinction point, the voltage curve follows the characteristic of the arc. However, the initiation is not an exclusively

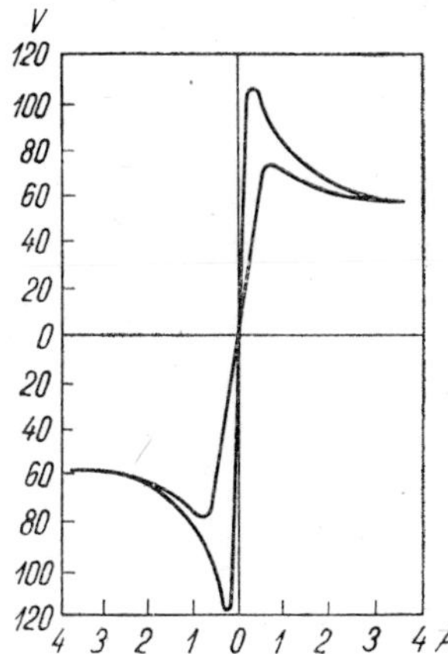

Fig. 53 Dynamic characteristic of A.C. arc

thermal process, and the electrode potential has to be higher than the starting voltage. The burning arc is stabilized by an effective or apparent resistance connected into the alternating current, similarly to the connection of the ohmic resistance R into the D.C. circuit.

An equation describing the static characteristic of an arc freely burning in the air between carbon electrodes has been derived from experimental values; it is similar to Eq. (2.2.1) with the member C omitted [Ref. 2.2.44] and reads

$$U_{\text{arc A.C.}} = 30 + 1.2l + \frac{12l}{I_{\text{ef}}} \quad [\text{V}] \tag{2.2.8}$$

where l is the length of the arc [mm] nad I_{ef} the current [A].

All the equations quoted yield only approximate values. Since the arc is very sensitive to external factors, not all of which have been considered in the formulae, the arc voltage may deviate from the calculated value.

2.2.3 Parts of the Arc

In the electric circuit every arc constitutes a certain resistance R_{arc} the magnitude of which depends on the parameters of the discharge. Hence a potential gradient is formed:

$$U_{arc} = I \cdot R_{arc}. \tag{2.2.9}$$

Since the resistance R_{arc} is not uniformly distributed throughout the gap – it is large next to the electrodes and small in the arc column – the

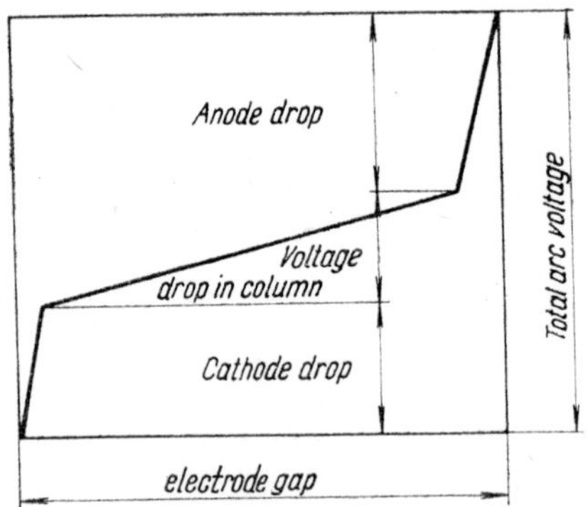

Fig. 54 Voltage drop along the arc

electric-field gradient is also non-uniform along the path of the arc. The potential distribution along the discharge path, indicated in Fig. 54, shows that the arc can be divided into three regions

– the cathode region with a considerable drop in voltage (the cathode fall);

– the arc column with comparatively very good electrical conductivity and therefore a slight voltage drop;

– and the anode region with a considerable drop (the anode fall).

All these regions are dependent on the magnitude of the current flowing through the arc and the medium in which the arc burns. The magnitude of the cathode and anode falls depends on the electrode material.

2.2.4 The Cathode

The flow of the current through the arc is effected by electrons liberated by thermal emission from the surface of the heated cathode. By acceleration in the region of the cathode fall, the electrons acquire a high measure of kinetic energy, which enables them to ionize neutral atoms in collisions. The positive ions thus formed are accelerated in the opposite direction, strike the cathode, and transfer their energy to it; thus they heat the cathode surface and keep up the thermal emission.

No unified theory explaining the cathode phenomena has been elaborated so far. The various theories are reviewed in Ref. [2.2.45]. The cathode phenomena may be classified in three groups: arcs with a constricted cathode region; arcs with an unconstricted cathode region; and arcs with a non-stationary constricted cathode region. The current density in the cathode spot ranges about 10^8 to 10^{10} A/m^2.

In the first case, the arc column is joined to the cathode by an annular neck, called the constricted cathode region, and the cathode spot does not move. The magnitude of the constriction depends on the cathode material, as well as the method by which the cathode tip is cooled and the degree to which this is done. The underlying theory has been worked out by Weitzel, Rompe and Schön [Ref. 2.2.46]. Intensified constriction of the cathode region causes a considerable rise of the current density in the cathode spot and consequently an increase in the heat supply to the cathode surface; heated to a higher temperature, this surface can then emit the requisite amount of electrons from a smaller area. Thermal overloading and evaporation of the surface layer of the cathode material may occur in the process. This phenomenon is especially marked in carbon or graphite cathodes, which form a crater at the tip after a certain period of operation.

In cathode materials with very high boiling or sublimation points (such as carbon, graphite, tungsten) and substantial emission of electrons at somewhat lower temperatures, the arc may—under certain circumstances—burn without constriction in the cathode region. The entire front face of the electrode is heated to a high temperature, and as soon as the electron current accounts for 97·5% of the total current value, the constricted region suddenly vanishes. Simultaneously the current density drops by a full order of magnitude and the temperature of the emitting cathode surface is lowered. The mode of the current transition from the arc plasma into the cathode and the form of the boundary layer depends on the heating of the cathode surface. This is borne out by the facts that the transition to a constricted cathode region takes place at a lower current when the currents drops than the opposite transition when the current rises (hysteresis); and also that for A.C. arcs the transition to an unconstricted cathode region requires higher currents. Moreover, in A.C. arcs with no constricted region, the re-ignition of the arc after the transition through the zero point passes without any marked ignition peak on the voltage curve.

In arcs with a non-stationary constricted region, the constriction is very marked and the focal spot is continuously moving on the cathode surface. This type occurs for instance where an arc with a low-vaporization-point cathode burns in a vacuum. The constriction is so substantial that it acts as a strong series-connected resistance, and thus several cathode spots may exist at the same time: up to 50 simultaneous cathode spots were found in vacuum circuit breakers under a current of 5 kA [Ref. 2.2.48].

The thermionic cathode emission is described by the equation

$$-i = AT^2 \, \mathrm{e}^{-\frac{\Phi_e - \sqrt{e^3 E}}{kT}} \tag{2.2.10}$$

where $-i$ is the current density, T the cathode temperature [°K], $A = \dfrac{2\pi emk^2}{h^3}$, which for most metals equals $6 \,.\, 10^5 \,\mathrm{A\,m^{-2}\,°K^{-2}}$; m and e are the mass and charge respectively of the electron, k the Boltzmann constant, h the Plack constant, E the intensity of the electric field in front of the cathode that reduces the work function Φ_e.

Eq. (2.2.10) makes it obvious that an emission can be greatly increased by the reduction of the work function Φ_e on the cathode. Since the work function is a property of the material, and elements with low work functions have mechanical properties that make them unsuitable for use as cathodes, such materials are merely added to well-proven cathode materials. As early as 1914 Langmuir observed the advantageous effect of thorium oxide added to tungsten. At temperatures above 2,500°K this oxide dissociates into thorium and oxygen. More recent measurements [Ref. 2.2.51] prove that the thickness of the thorium layer on the tungsten electrode has a marked effect on the work function. With a monomolecular layer the work function is 2·6 to 2·7 eV, with three times this thickness it increases to 3·55 eV; then it decreases again and reaches a constant value of 2·5 eV in the thickness range of a 5- to 40-fold monomolecular layer. Cesium and barium, likethorium, reduce the work function, and so do some metal oxides. Results of work-function measurements on activated surfaces together with the values of constant A for Eq. (2.2.10) are tabled below.

Table 2.2.3

Surface	φ/eV	$A \cdot 10^4$ [$A \cdot m^{-2}$ °K^{-2}]	Reference
Thorium on tungsten	2·63	3·0	[2.2.52]
Thorium on molybdenum	2·59	1·5	[2.2.52]
Barium on tungsten	1·56	1·5	[2.2.52]
Cesium on tungsten	1·56	3·2	[2.2.52]
Barium on oxidized tungsten	1·34	0·18	[2.2.52]
Cesium on oxidized tungsten	0·72	0·003	[2.2.52]
Aluminium oxide	3·77	1·4	[2.2.53]
Thorium oxide	0·016	2·94	[2.2.53]

The advantages derived from the activation of tungsten electrodes consist in the easier initiation of the arc, its steadier burning, and the sufficient emission achieved at far lower electrode temperatures, hence lesser electrode wear. The emission curve of a thorium-coated tungsten electrode is compared with that of a pure-tungsten one in Fig. 55 [Ref. 2.2.54].

Various authors differ as to the magnitude of the cathode drop. Older sources, such as Ref. [2.2.45], mostly quote 15–16 V as the value for copper, iron, zinc and tungsten cathodes, while more recent measurements seem to prove that the cathode fall is dependent on the current. In high-pressure discharges (30 atm) in xenon with tungsten electrodes, the cathode fall was found to diminish from 13 to less than

10 V as the current increased from 3 to 10 A [Ref. 2.2.55], and in an argon atmosphere 7 to 8 V were measured as the current increased from 60 to 200 A [Ref. 2.2.57]. The same paper reports a cathode fall

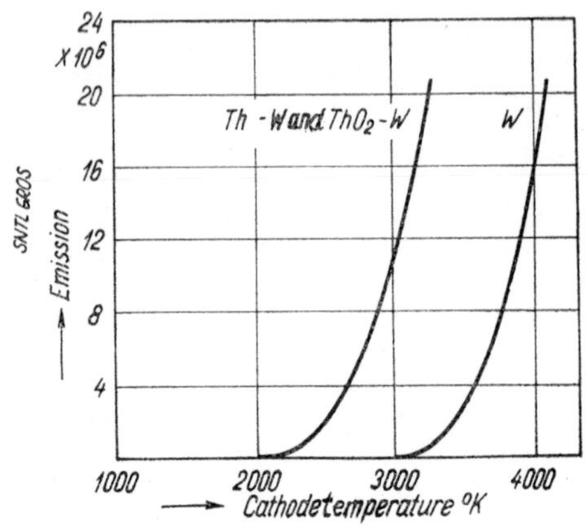

Fig. 55 Emission of thoriated and pure-tungsten electrode

of 7 ± 2 V for an arc burning in nitrogen, the current ranging from 50 to 200 A [Ref. 2.2.45].

2.2.5 The Anode

The anode is instrumental in the transition of the current into the hot plasma of the arc. It is intensively heated by electrons accelerated in the electric field. During operation, the anode temperature may rise to the boiling point of the material. The atoms evaporated from the anode enter the arc plasma and are ionized there. Unlike the cathode, however, the anode may be cold. The arc burns well even with a water-cooled anode, and no ions of the anode material have been found in the arc plasma in this case [Ref. 2.2.58]. The normal current density on a carbon anode at lower currents is about $4 \cdot 10^5$ A . m^{-2} (in the range of the steadily burning arc according to Fig. 49).

The hissing of the arc is explained by phenomena on the anode overloaded with current [Ref. 2.2.59]. As the current in the arc is increased, the current density (in the "microspots") of the anode rises to a value of 10^8 A . m^{-2} and a marked constriction appears in the anode region. The thermal load in the microspots rises to extreme values (10^6 kcal . m^{-2}). The eruptive escape of evaporated anode material (anode flame) is accompanied by a characteristic hissing sound, and the anode spot moves at a high velocity on the front face of the anode, during this process. The frequency of the hissing sound depends on the thermal conductivity of the carbon anode, and the sound effect is closely connected with oscillations observed in the voltage. The amplitude of these oscillations varies with the type of carbon and the current intensity in the range of 5 to 30 V. The changes in voltage are accompanied by changes in current amounting to 3 to 13 A [Ref. 2.2.45]. High-frequency oscillations of 3 V and 0·5 A are due to the existence of the "microspots" forming within the anode spot and reaching current densities as high as 3 . 10^8 A . m^{-2}. As the arc current continues to rise, the arc burns calmly again, the entire front face of the anode is uniformly heated, and the constriction of the arc column next to the anode disappears.

A marked constriction is also observed in oxidizing atmospheres on the transition of metallic anodes into the arc column In currents above 30 A and with the arc burning in inert gas or nitrogen, the transition between the arc and the anode is diffuse. Current densities measured on

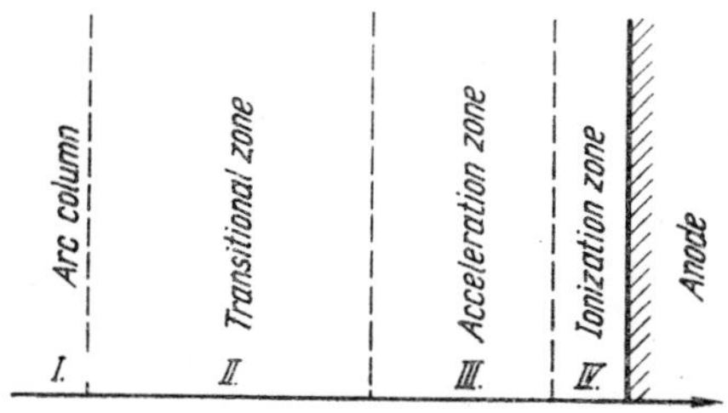

Fig. 56 Anode region according to Höcker

anodes (under currents of 100 to 500 A) were 2 . 10^6 to 2 . 10^7 A . m^{-2} in argon, and 5 . 10^6 to 6 . 10^6 A . m^{-2} in nitrogen [Ref. 2.2.62].

The author of Ref. [2.2.59] divides the anode region into four partial zones (Fig. 56): The arc column (*1*) is followed by the transition zone (*2*) where the directed motion of the electrons becomes uniformly accelerated towards the anode. In the acceleration zone (*3*) the electrons acquire the kinetic energy required for the ionization of neutral atoms, and in the ionization zone (*4*) the neutral atoms are ionized. The ion density is at its maximum at the transition from the 3rd to the 4th zone and then decreases towards the anode, whereas electron density rises in this direction. The validity of this theory is limited to low current densities and arcs with a high anode fall (20 V). At high currents, electric-field ionization gives way to thermal ionization, and the anode voltage may drop far below the value of the ionization potential.

The anode drop is also affected by the macroscopic flow of the plasma. In the Beck arc, for instance, high current loads on the anode cause the vapours of anode material to flow at velocities of 10 to 70 $m \cdot s^{-1}$ [Ref. 2.2.45]. At this rate the ions and part of the electrons are carried away from the anode surface. The electrons, which are very mobile, are only slightly affected. However, the directed (drift) velocity of the ions in the electric field is only of the order of metres per second, and the vapour flow causes more electrons to escape from the anode region into the arc column. A higher anode fall is therefore necessary to replace the ions removed by the flow from the anode region [Ref. 2.2.6].

The cathode emission in the axial direction produces an opposite effect on the cathode fall. If the velocity of this emission exceeds the drift velocity of the ions, these are prevented from migrating towards the cathode.

The hot region of the plasma is then in close proximity to the anode. The very thin lower-temperature layer ahead of the anode does not require a higher anode fall. Probe measurements have revealed an anode fall of 1 to 2 V in an arc between carbon electrodes under a current of 200 A [Ref. 2.2.60]. The anode fall of an arc burning in an argon atmosphere decreased with increasing current and was nearly zero when the current was sufficiently high [Ref. 2.2.61].

2.2.6 The Arc Column (Thermal Plasma)

In the column of the arc, the gradient of the electric field is relatively low and its magnitude is affected (apart from the current flowing through the arc) by various external factors—the kind of the gas, its pressure, the electrode material, the cooling of the arc, and external mechanical and magnetic forces.

As in the cathode and anode regions, the increase of the current flowing through a freely burning arc reduces the electric gradient in the column. In the arcs dealt with in this book, the density of charged particles in the plasma is dependent on its temperature, which in turn is determined by the total energy balance of the arc. The energy supply by Joule's heat has to equal the energy losses due to thermal conductivity, radiation and convection. As the current gains in intensity, the temperature of the plasma rises, hence its electric conductivity increases, and the electric gradient therefore diminishes. The effect which the higher current exerts on the arc voltage is apparent from measurements taken on an arc burning in argon [Ref. 2.2.62], where the arc voltage fell from the original 30 V at 6 A to 7·5 V at 150 A, the gap remaining unchanged.

The electric gradient of a freely burning arc is a function of the electric conductivity of the plasma, and consequently of its temperature. The heat losses are mainly due to conduction, and the quantity of the energy thus removed is proportional to the radius of the plasma column. As the current increases, so does the radius. At a certain magnitude of the current two extreme states can occur: the cross-section of the arc is either too large or too small. At excessive cross-sections, the temperature and hence the electric conductivity of the plasma is relatively low and the current transfer therefore requires a high gradient. Small cross-sections produce high plasma temperatures and consequently high electric conductivity; yet because of the small cross-section, the current transfer again requires a high gradient. The arc with the minimum gradient must be somewhere between these two extremes. Steenbeck [Refs. 2.2.63 and 64] calls this fact the principle of minimum voltage.

In a stable arc,

$$\frac{\mathrm{d}E}{\mathrm{d}T} = 0 \quad \text{or} \quad \frac{\mathrm{d}E}{\mathrm{d}r} = 0 \tag{2.2.11}$$

Steenbeck's principle of minimum voltage has many times been tested in practice, and the experimental results were found in good agreement with theory [Refs. 2.2.65 and 66].

For a symmetrical arc, the principle of minimum voltage can be formulated as follows: At given current and boundary conditions, the current-carrying region of a stationary arc has such a radius (or temperature) that the electric gradient of the arc has a minimum value.

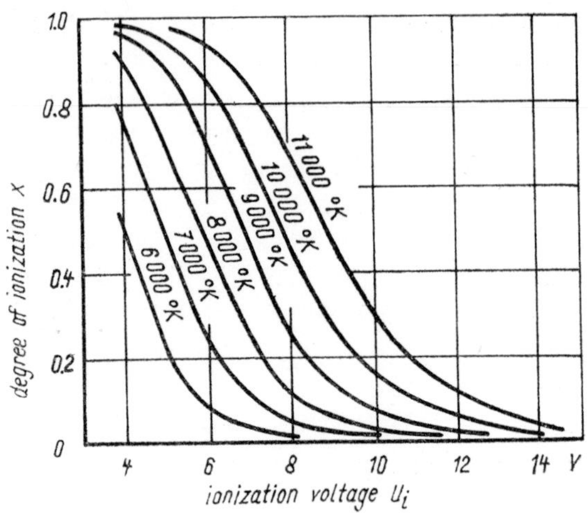

Fig. 57 Degree of thermal ionization vs. ionization voltage

Peters [Ref. 2.2.67] has proved that the principle of minimum voltage is identical with the principle of minimum entropy.

The kind of the gas greatly influences the arc voltage. Inert gases such as argon or helium form no molecules and require far less energy for their thermal ionization than do polyatomic gases, such as hydrogen, nitrogen, oxygen or steam, which require to be previously dissociated.

The degree of thermal ionization as a function of the ionization voltage of gases at temperatures up to 11,000°K and a pressure of 1 atm. is shown in Fig. 57. The enthalpies of some gases are plotted against temperature in Fig. 58, and their dissociation curves in Fig. 1. The curves of the diatomic gases show that regardless of relatively high energy supply, their temperature increases only slightly in the dissociation region.

The arc voltage is also affected by the losses which occur in the column, mainly owing to the thermal conductivity of the plasma; losses due to radiation and convection are often negligible in a freely burning arc.

The calculation of the thermal conductivity of plasma is carried out in Div. 1.2 and Chap. 1.5.4: here we only recall the results. The

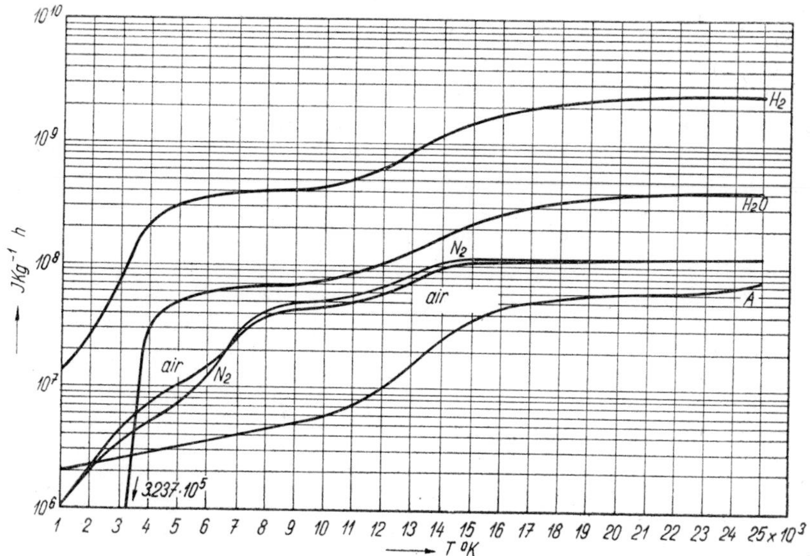

Fig. 58 Enthalpies of some gases vs. temperature

temperature dependence of the thermal conductivity of nitrogen [Ref. 2.2.68] in the interval between 400 and 35,000°K is plotted in Fig. 59. The resultant curve $-\varkappa-$ is mainly affected by thermal conductivity due to the diffusion of dissociation and ionization energy $\varkappa_d$, $\varkappa_i^+$, $\varkappa_i^{++}$. The curves of classical conductivity due to molecules and atoms or to free electrons are designated $\varkappa_c$ and $\varkappa_e$, respectively. The thermal-conductivity curve for other diatomic gases is similar. For monatomic gases, thermal conductivity rises with temperature up to 10,000°K, and there is no increase in conductivity due to dissociation [Ref. 2.2.69].

The kind of the gas and its thermal properties exert a greater influence on the arc plasma than does the ionization potential (cf. Fig. 60). The highest electric gradient in the arc column occurs where the plasma is formed by hydrogen (ionization voltage 13·59 eV), although the ionization potentials of helium, nitrogen and argon are higher. This is due to the superior thermal conductivity of hydrogen.

Experiments have proved that increased gas pressure in the medium surrounding the arc causes the arc voltage to rise. Theoretically,

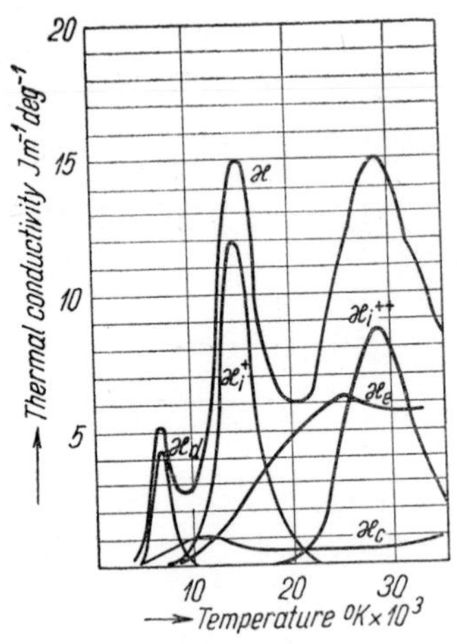

Fig. 59 King's temperature-dependence curves of thermal conductivity for nitrogen

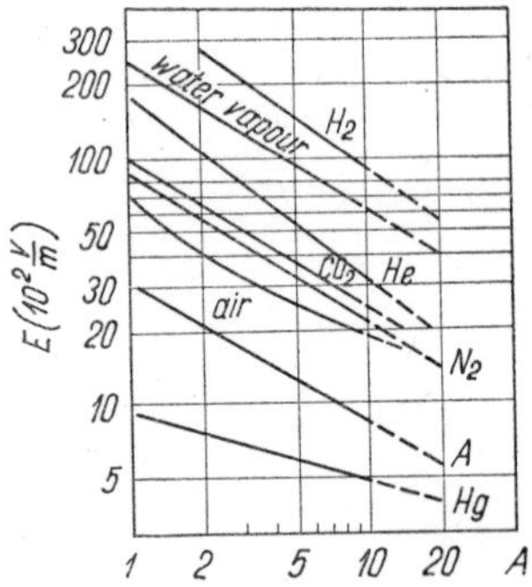

Fig. 60 Arc characteristics in various gases

the dependence of the arc-column gradient on the pressure has not yet been fully elaborated, because the gradient depends on many elementary processes which themselves are dependent on the gas pressure. The electric gradient can rise with pressure in the limits between $p^{1/2}$ and $p^{1/6}$ [Ref. 2.2.45].

The influence of the electric material is proportional to the amount of this material evaporating into the plasma. However, even a small amount of material evaporated from the electrode may produce a marked effect on the degree of ionization and the electric conductivity of the plasma, provided its ionization potential is low.

The influence which external materials can exert on the temperature in the boundary layers of the arc is utilized in stabilized arcs where the plasma is subjected to artificial cooling. Cooling reduces the conductivity of the outer layers of the column, and consequently the cross-section of the conductive arc core; the quantity of charged particles required for the flow of the current has to be secured by increasing either the electric gradient or the conductivity (by raising the temperature). If the voltage across the electrodes is increased, we also increase the gradient in the arc column, hence the current density in the reduced cross-section of the conductive core, and the core is heated to a higher temperature by Joule's heat. This increases its conductivity, and (if the source is soft) the arc voltage somewhat drops. An equilibrium is thus established, in which the higher gradient and plasma temperature (determining the conductivity) secure the same electric current as the uncooled arc. The higher output supplied to the unit length of the arc covers the energy losses from the arc column. These considerations explain the fact that in a cooled arc the plasma has a higher temperature than without cooling.

Outside mechanical and magnetic forces affect the arc column in the same way in which they act upon a conductor capable of deformation – they alter its shape and position. Even air streaming around a freely burning arc affects the shape of its column. With the electrodes placed horizontally, the rise of the hot air layers causes the column to assume the characteristic arched shape. With the electrodes placed vertically, the streaming air cools the bottom electrode and heats the top one; consequently the arc widens towards the latter. External blowing can not only modify the shape of the arc column, but can shift it altogether (Pauling's furnace, Sec. 2.1.5.3). In stabilizing the arc by means of gas, designers act upon the arc either by an axial gas stream or by a gas vortex rotating round the arc. By either method they cool the arc column in addition to the mechanical orientation they provide for the path of the discharge.

According to the relative direction of the field intensity, a magnetic field acts in two different ways upon an arc. It may produce an analogous effect as it does on a conductor under current, causing its macroscopic acceleration; the action of blowing magnets in circuit breakers is based on this principle. If the magnetic field is parallel with the axis of the arc, it stabilizes the position of the arc column against spontaneous transverse deviations; this is sometimes exploited in welding.

Moreover, a strong magnetic field may affect transport phenomena (electrical and thermal conductivity, diffusion) in a direction perpendicular to the force lines of the magnetic field.

2.2.7 The Stabilized Arc

The stabilization directs and limits the discharge path of the arc and efficiently cools the outer plasma layers. Moreover, in the plasma torches used in industry the plasma is macroscopically accelerated along the centre line of the arc. Apparatus manufactured at present is mostly fed with D.C., the rod-type internal electrode acting as cathode, and the flow of the plasma is accelerated in the direction towards the anode.

According to the stabilizing medium we distiguish gas- and liquid-stabilized plasma torches. Gas-stabilized torches are designed with various methods of arc stabilization.

The simplest method is wall stabilization, schematically represented in Fig. 61a. The arc column fills the nozzle up to its wall which cools the boundary layers of the gas. To secure the transport of the thermal energy by the flowing gas, the latter is supplied into the electrode chamber in small quantities. This method is used where a low jet velocity of the plasma is wanted for technical reasons. The hot gases of the column sheath are in direct contact with the nozzle walls; consequently, the maximum output obtained in such a design is relatively low. The transformation of electrical energy into the thermal energy of the beam is less effective than in methods employing an intensively blown arc.

Fig. 61b represents the vortex stabilization of the arc, with the gas – under pressure – introduced into the electrode chamber through a tangential opening. On discharging from the orifice, the gas is accelerated; forced to follow the curvature of the wall, it rotates around the electrode at a high velocity. The rotation of the gas also continues in the nozzle, and as the arc burns, the lighter hot gases – less affected by the centrifugal force – travel nearer to the centre line of the discharge, whereas the cooler gas approaches the walls. The nearer to the centre line, the larger the axial component of the velocity; it outweighs the radial component, and finally suppresses it, so that the plasma beam issues in laminar flow in the axial direction.

The sheath-stabilized arc according to Fig. 61c differs from the wall-stabilized arc by the quantity of the gas flowing axially: it is

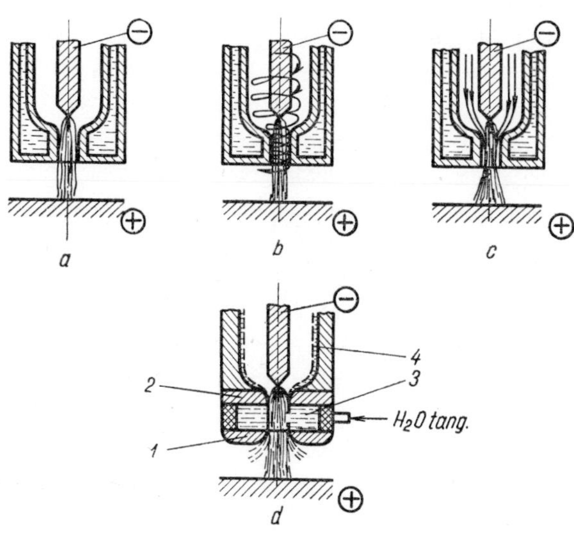

Fig. 61 Methods of stabilizing the transferred arc

so large that a relatively cold gas layer forms between the hot core of the discharge and the nozzle wall.

The principle of liquid stabilization of the arc, as industrially used for cutting metal, consists in forming water vortices with two diameters (Fig. 61d): a small diameter in the stabilization chamber (*3*), and a larger one in the electrode chamber (*4*). The vortex in the stabilization chamber is formed by the rapid rotation of tangentially introduced water, and the copper nozzles (*1*) and (*2*) determine its open-end diameter. In operation, the nozzles are protected by a layer of water which issues from the stabilization chamber, uniformly on both sides if the diameters of both nozzles are equal. In the electrode chamber, the rotation keeps the water at the walls where it forms a film. The problems of liquid stabilization are discussed in detail in Div. 2.3.

All the methods of stabilization listed so far (Fig. 61a–d)

feature what is called a transferred arc: it burns between an internal electrode—usually a rod—acting as cathode, and the material to be processed, which is situated out ide the plasma torch, acting as anode. These methods are mainly used for metal cutting and processing.

Other plasma torches employ what is called two-pole designs or non-transferred arc, i.e. the arc burns without depending on an external anode (Fig. 62). The anode, in this case, is the nozzle itself, suitably lengthened to this end, and amply cooled with high-pressure water. This design can only be used for an arc stabilized by argon or pure nitrogen; the amount of hydrogen added must not exceed 10% of the total gas quantity, because it increases the heat transfer to the anode, and the cooled anode material cannot stand the heat stress if more hydrogen is present. Plasma torches with a non-transferred arc are used for cutting and processing ceramic materials, and also for applying protective coatings. The powdered coating material—a metal or metal oxide—is fed to the plasma torch through an orifice opening into the nozzle, and the powder is entrained by the carrier gas.

In plasma torches with stabilization by inert or reducing gases, the cathodes are made of tungsten or thoriated tungsten, i.e. materials distinguished by very low wear.

In addition to the two basic forms of arcs referred to above, a third method, using two electric circuits, is being worked out. It is

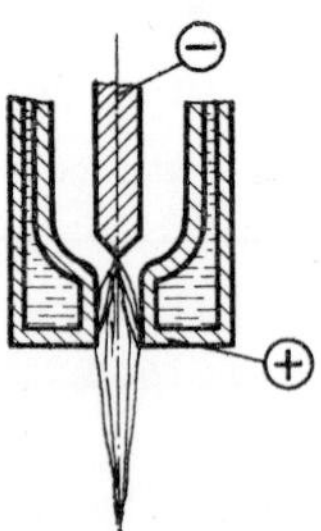

Fig. 62 Non-transferred arc

used for surfacing carbon steels with metal carbides [Ref. 2.2.77]. A plasma torch of this type together with the supply circuits is schematically shown in Fig. 63. The current source G_1 feeds the arc between the electrode and the nozzle, source G_2 is connected in parallel and supplies the current flowing from the electrode to the material. The current supplied by the G_2 source controls the depth of penetration. A protective gas, shielding the built-up surface from oxidation, is supplied through

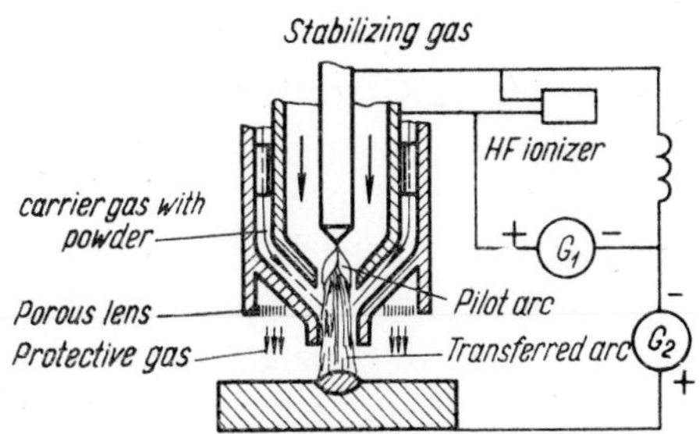

Fig. 63 Weld-surfacing plasma torch—powder-fed

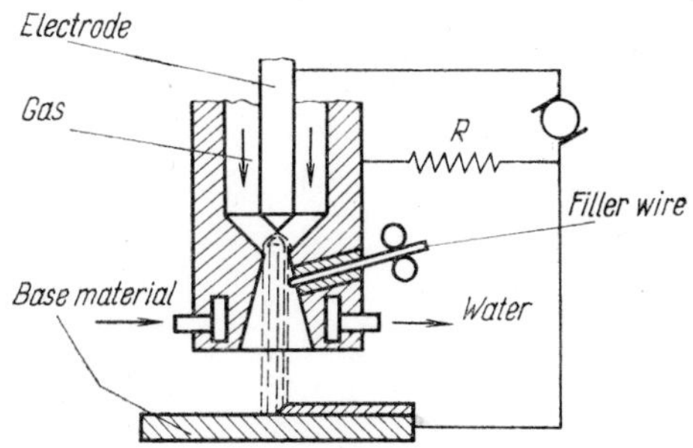

Fig. 64 Weld-surfacing plasma torch—wire-fed

a porous lens placed round the nozzle. The powdered facing material is fed to the torch by means of the carrier gas.

Fig. 64 schematically shows a plasma torch with a non-transferred arc, designed for surfacing with metal supplied in wire form, chiefly chromium-nickel steel.

In plasma torches stabilized with aggressive gases or water, the nozzle cannot act as anode, since it would be destroyed in a matter of seconds. In these cases a rotating water-cooled disk, slightly protruding into the plasma flow immediately ahead of the nozzle, is a satisfactory anode (cf. Fig. 79). The power circuit is closed by the disk. The heat supplied by the anode voltage drop is distributed over a large surface on the rapidly rotating disk, hence the service life of the electrode is relatively long. A more detailed description of this anode is given in Sec. 2.3.3.3.

2.2.8 High-Frequency Plasma Generators

In addition to apparatus in which plasma is obtained by an electric arc that heats a gas, plasma generators based on radio-frequency heating of ionized gas have also been designed and are at present undergoing laboratory tests [Ref. 2.2.71]. In plasma generators with arc discharges, the electrode material contaminates the plasma; such generators are

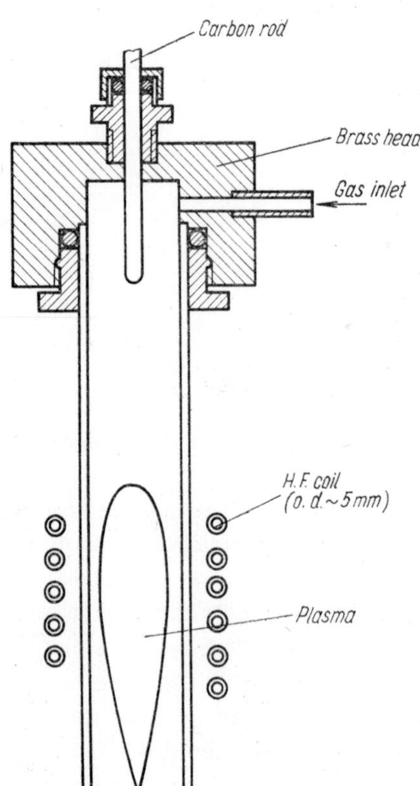

Fig. 65 H.F. plasma generator

therefore unsuitable for applications requiring high-purity plasma. The high-frequency plasma generator operates without electrodes and thus yields uncontaminated plasma. Another advantage of the h.f. generator, utilized for instance in growing single crystals, is the low axial velocity of the plasma (only a few metres per second).

The apparatus, schematically illustrated in Fig. 65, is very simple. The gas is conducted through a vertical silica tube, closed at its top by a cap that admits the gas; the bottom is open. For about one third of its length, the silica tube is surrounded by a high-frequency coil – five or six turns of copper tubing about 5 mm in diameter, internally cooled by running water.

The ionized gas required for starting the plasma generator is produced by the introduction of a carbon or tungsten rod (placed in the cap) into the working space inside the coil. The rod heated by a high-frequency current starts thermionic emission, and the ions accelerated in the high-frequency field ionize the gas. When the ionized region has sufficiently expanded, the electrode is withdrawn from the working space. The burner is started with pure argon and other gases are added later.

At low feed rates of the gas, the stability of the plasma is satisfactory. If, however, the gas velocity is such that the fresh gas cannot be heated to the ionization temperature (and no inductive coupling is possible below this point), then the plasma is extinguished. The total quantity of plasma issuing from the generator can be increased by tangential feed: the suction effect produced by the vortex starts an axial motion of the plasma in the reverse direction in the central i.e. the hottest region. The plasma of the central region then delivers the heat required for the ionization. Provided the frequency of the generator is correct, the optimum flow rate of the gas – ranging between 0·9 and 3·6 m^3 per hour – is primarily dependent on the amount of energy supplied to the plasma. For argon (pure or mixed with some other gas) the optimum feed rate is 0·9 to 1·2 m^3/hr. at an input of 3 – 4 kW, 1·8 to 2·1 m^3/hr. at 6 – 8 kW, and 3 to 3·6 m^3/hr. at 10 – 12 kW.

The correct frequency of the h.f. generator (1 – 40 Mc/s) is largely determined by the kind of the gas and the size of the silica tube. The best inductive coupling is obtained at the optimum frequency (f_{opt}) described by the following equation:

$$f_{opt} = \frac{\varrho \cdot 10^9}{5d^2} \qquad [\text{Mc/s}] \qquad (2.2.12)$$

where ϱ [ohm per cm] is the resistivity of the gas in the ionized state, and d the diameter [cm] of the ionized region in the silica tube.

The resistivity of this type of plasma and its specific heat are tabulated in Table 2.2.4 for some gases used most frequently [Ref. 2.2.73].

Table 2.2.4

	argon	helium	air	nitrogen	oxygen	hydrogen
resistivity [μΩ/cm]	$2 \cdot 10^4$	$5 \cdot 10^4$	10^5	10^5	10^5	$5 \cdot 10^5$
specific heat [cal/g/°C]	0·13	1·25	0·24	0·25	0·22	3·4

The efficiency at which the electrical energy is transformed into thermal energy is fairly low in these generators. Only about 52 to 57% of the input, measured by a wattmeter at the primary of the high voltage transformer, is delivered into the plasma. From the total energy input into an argon plasma, 20 to 35% is lost by ultraviolet radiation, and 38 to 45% by thermal conductivity; only 20 to 40% remains in the plasma. The highest temperature obtainable in the core of the plasma (without heating any object) is about 18,000°K with pure nitrogen, and 15,000 to 16,000°K with argon. In Fig. 66 the isotherms of an argon plasma are plotted according to Ref. [2.2.71]; the points indicate the spots where measurements were taken.

H.f. plasma generators have not been used in industry so far, but their applicability is being explored in various laboratory experiments. These are aimed at growing single crystals [Ref. 2.2.74] and at various chemical reactions in plasma. The output achieved so far in a plasma of this type is insufficient for successful industrial applications in chemical syntheses. However, this type of source holds out very real prospects

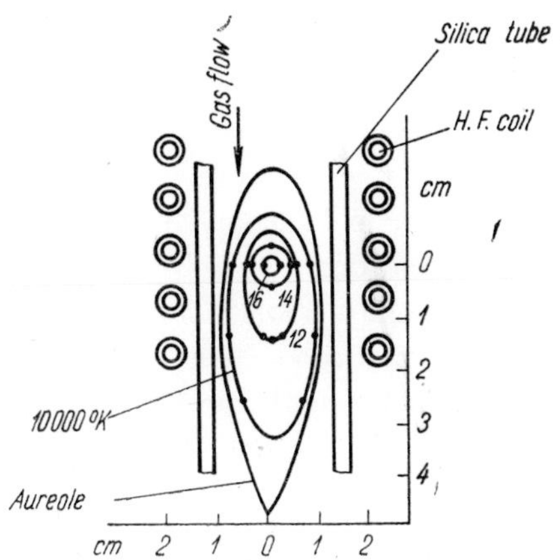

Fig. 66 Isotherms of argon plasma in H.F. plasma generator

in the metallurgy of pure metals with high melting points and in sintering high-melting oxides and carbides.

2.2.9 Radial Temperature Distribution of Plasma and Its Relation to Thermal Conductivity

In Chap. 2.2.6 we referred to the thermal conductivity of the gases used for stabilizing the arc. The difference between the thermal conductivities of monatomic and molecular gases is also evidenced by the differing temperature distributions along the radius of the arc.

By the end of last century it was already known that under higher currents the central part of the arc column tends to contract, i.e. to form a core. This phenomenon applies to arcs burning between either carbon or metal electrodes; it has not been observed in rare gases. Theoretical explanations have been given by King [Ref. 2.2.75] and Maecker [Ref. 2.2.76]. In the plasma of monatomic gases there is no core, and the radial temperature distribution is parabolic, the shape of the curve being described by the equation

$$T(r) = T_{\max}\left(1 - \frac{r^2}{R^2}\right), \tag{2.2.13}$$

where R is the radius of the boundary zone.

In plasma generated from polyatomic molecules, a similar radial temperature distribution only occurs at temperatures below the dissociation point. The thermal-conductivity curve of nitrogen has its break at about 7,000°K and drops sharply below this temperature. The same applies to hydrogen and steam, the break being at about 4,500°K. The rapid increase in thermal conductivity produces a thermal gradient which is sufficient for conducting the heat away from the axial portion of the plasma. If however the temperature in the centre line of an arc burning in nitrogen rises to say 11,000°K, then in the region above 7,000°K the temperature gradient $\frac{dT}{dr}$ next to the centre line must rise sufficiently to compensate for the diminishing thermal conductivity

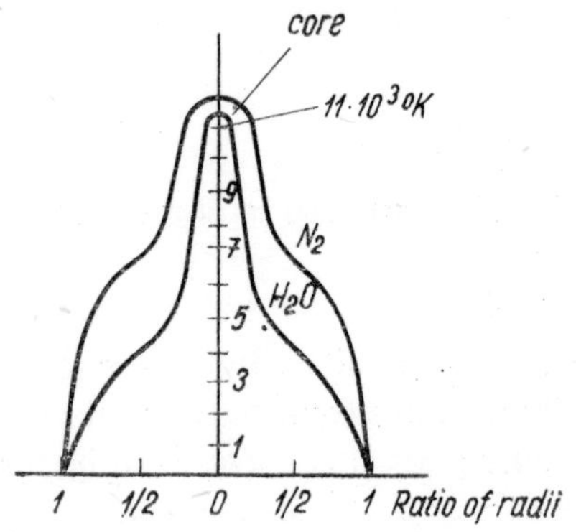

Fig. 67a Radial temperature distribution in arcs freely burning in steam and nitrogen, respectively

and insure a sufficient heat removal. In an arc burning in steam, the temperature gradient begins to rise earlier (at ~4,500°K) because the bond energy of the molecules is lower. However, owing to the high hydrogen content of the steam, the absolute value of thermal conductivity is fairly high ($14{\cdot}10^5$ erg/s cm°K). Hence the thermal peak is lower than in nitrogen. Fig. 67a shows the radial heat distribution in arcs freely burning in nitrogen and steam, respectively. The formation of a core in the arc plasma is dependent on the existence of a thermal peak in the centre line of the arc, caused by the relatively low thermal

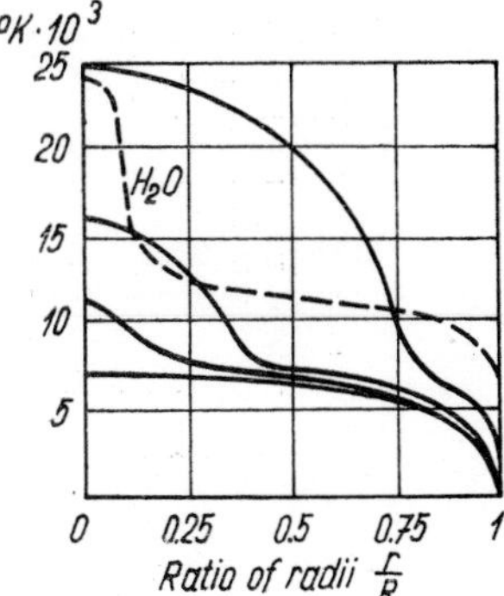

Fig. 67b Radial temperature distribution of plasma burning in nitrogen

conductivity of the gas in the temperature range above the dissociation temperature.

In the plasma torch the diameter of the arc is determined by the wall of the nozzle. An increase in the current flowing through the arc both raises the temperature in the axis and extends the ionized region, i.e. the core of the arc plasma. In Fig. 67b [Ref. 2.2.71] the radial temperature distribution in the plasma of an arc burning in nitrogen is plotted at various currents (and hence temperatures). The figure is supplemented by the analogous (dashed) curve for the plasma of a water-stabilized torch at a current of 280 A. The quantity important in the industrial utilization of a plasma torch is not the maximum temperature in the centre line of the arc, but the average enthalpy of the issuing gas. For gas torches the total specific energy content is easily computed according to the following equation:

$$M_c = \eta \frac{P}{M} \qquad [J \, . \, s^{-1}], \tag{2.2.14}$$

where η is the efficienty in the transformation of electrical into thermal energy, P the input [j . s^{-1}] of the arc, and M the amount of gas flowing through the torch [kg . s^{-1}]. The transformation efficiency is obtained by deducting the thermal losses in the plasma torch—ascertained from the heat gain of the cooling water—from the total input.

The approximate temperature can then be ascertained from the average enthalpy according to enthalpy curves (Fig. 58) or tabulated

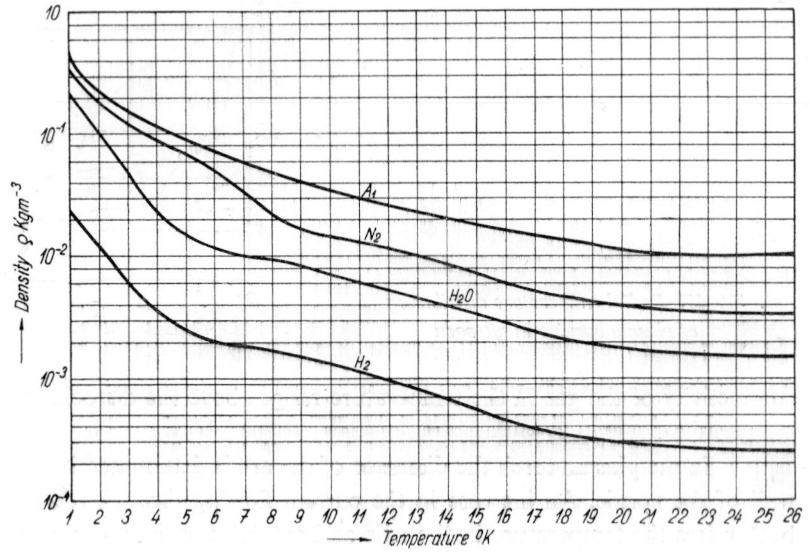

Fig. 68 Density of some gases vs. temperature

values. The average density ϱ belonging to this approximate temperature is found from the curves reproduced in Fig. 68. Moreover,

$$M = \frac{\pi}{4} d^2 \varrho \bar{v}, \tag{2.2.15}$$

where d is the diamter of the discharge nozzle, ϱ the average density, and $\bar{v}$ the average flow rate of the issuing gas. Hence

$$\bar{v} = \frac{M}{\frac{\pi}{4} d^2 \varrho}. \tag{2.2.16}$$

The total potential energy content h_c can be broken down into the specific enthalpy and the kinetic energy per unit weight:

$$h_c = h_0 + \frac{v^2}{2Jg}, \tag{2.2.17}$$

where h_0 is the energy content of the gas in the outlet, J Joule's constant, and g acceleration of gravity. From the value of h_0, the corrected temperature can be obtained with the aid of the graphs in Figs. 58

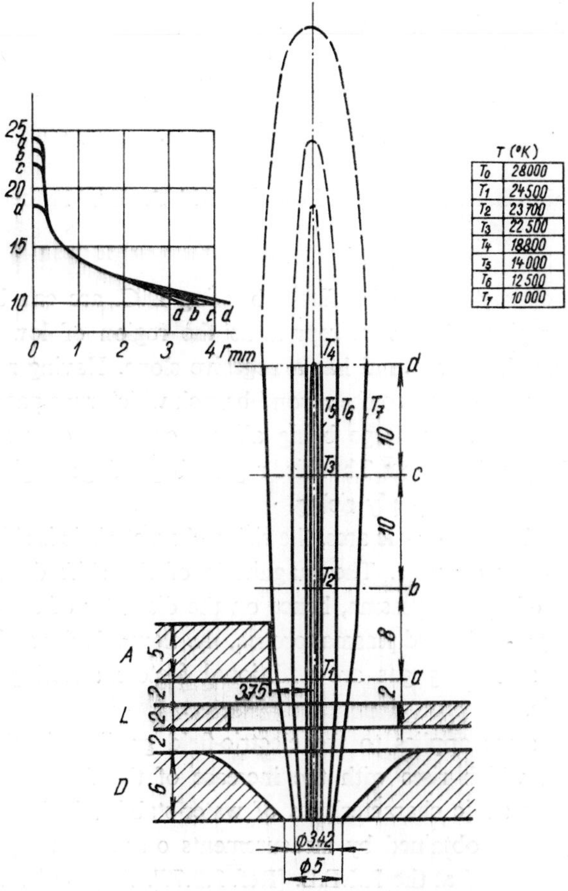

Fig. 69 Plasma temperatures of water-stabilized arc (R.I.E.E.)

and 68, the corrected velocity calculated according to Eq. (2.2.16), and the corrected value h_c can be derived. Thus we can estimate the average temperature T and the average velocity $\bar{v}$ with the required accuracy.

Spectroscopic temperature measurements and the radial temperature distribution in the plasma of a water-stabilized torch with independent arc are presented in Fig. 69. The nozzle diameter was 5 mm, and the channel of the water vortex 4 mm in diameter. D is the nozzle, L the trap catching the water that runs from the nozzle, and A is the disk-type anode. The values measured in the individual points are tabulated in the right-hand corner. Radial temperature distribution is also indicated. The current flowing through the arc was 280 A, and the voltage accross the arc 240 V.

2.2.10 Static Characteristic and Electric-Field Gradient in the Stabilized-Arc Column

The static characteristic of a freely burning D.C. arc can be divided into three regions. The first represents the region of lower currents flowing through the arc and has a negative slope. Having reached the minimum, it goes over into the second branch which runs parallel to the current axis, the arc voltage being almost constant over an interval of several hundred amperes. The third region, beginning at high currents – 1,800 to 2,000 A – is steeply rising.

In stabilized electric arcs, the rising characteristic is shifted to the region of lower currents. The magnitude of the shift depends upon the cooling of the arc plasma, hence on the diameter of the stabilizing channel. In gas-stabilized plasma torches, the flow velocity and composition of the stabilizing gas are additional factors affecting the static characteristic.

The same applies to the electric-field gradient in the positive arc column, which rises with the increase of the factors mentioned. Fig. 70 shows static characteristics of water-stabilized plasma torches. The curves were obtained by measurements on various types of such torches constructed at the R.I.E.E. [Ref. 2.2.77]. All nozzles were 5 mm in diameter except for the instance illustrated by curve (b), where the

VS-100 torch with a stabilization channel 15 mm long was fitted with nozzles of 4 mm diameter. Curves (*e*) and (*c*) are the characteristics of plasma torches VS 45 and VS 60 with stabilization channels 6 and 9 mm

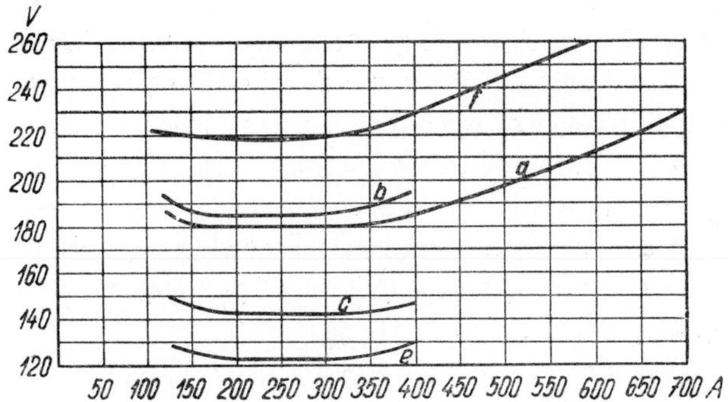

Fig. 70 Static characteristics of plasma torches with water-stabilized arc (R.I.E.E.)

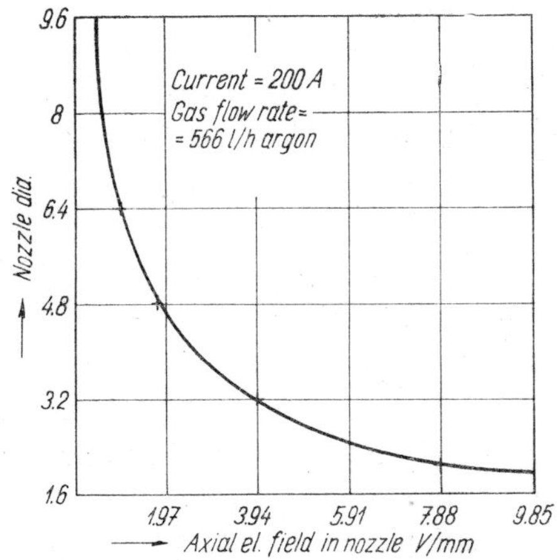

Fig. 71 Electric-field gradient in argon

long, respectively. Curves (*a*) and (*f*) are arc characteristics of the VS 100 plasma torch with a stabilization channel 15 and 20 mm long, respectively.

The electric-field gradient in the stabilization channel can be measured by a method based on the use of two channels of different lengths. In the interval from 200 to 500 A, the gradient rises from

67·5 V per centimetre to 80 V/cm. In the range 300 to 400 A, which is the current used in water-stabilized plasma torches for cutting and schooping, the gradient is 70 – 75 V/cm.

The static characteristic of gas-stabilized plasma torches shows a pattern similar to that of liquid-stabilized torches. The dependence on the kind and amount of the gas flowing through the nozzle is here added to the dependence upon the diameter of the stabilization channel. The effects of the amount (i.e. the intensity of cooling) and the kind of the gas have been discussed in Chap. 2.2.7, and the effect of the nozzle diameter on the electric-field gradient in the channel at constant current (200 A) and constant gas flow (566 litres per ampere-hour) is evident from Fig. 71 [Ref. 2.2.22].

2.2.11 Efficiency of Plasma Torches in Converting Electrical into Thermal Energy

The total losses in a plasma torch are measured by the heat gain of the cooling water. If Q_s signifies the losses in watts, then

$$\eta = \frac{U \,.\, I - Q_s}{U \,.\, I}, \tag{2.2.18}$$

and

$$Q_s = 4{\cdot}18 \,.\, C \,.\, V(t_2 - t_1) \qquad [\mathrm{W}], \tag{2.2.19}$$

where 4·18 is the conversion factor for converting cal/s into watts. C is the specific heat of water (1 cal cm^{-3}), V the amount of cooling water supplied [cm^3 . s^{-1}], and t_1 and t_2 the water temperature on the entry and outflow respectively.

In gas-stabilized plasma torches with an axially blown independent arc of minor output, the efficiency of the conversion of electrical into thermal energy is

$$\eta_n = 0{\cdot}6 \text{ to } 0{\cdot}65; \tag{2.2.20}$$

if hydrogen is used, it rises to 80% (input 62 kW, voltage 120 V) [Ref. 2.2.79]. In arcs stabilized by gas vortices, η_n is 0·8 for outputs above 100 kW, and below this limit efficiency drops; for 70 kW, η_n is 0·68 [Ref. 2.2.80].

The efficiency is far higher in plasma torches with transferred arc, where the anode losses are utilized for heating; argon- and hydrogen-stabilized plasma torches have an efficiency of

$$\eta_z = 0{\cdot}8 \text{ to } 0{\cdot}85. \tag{2.2.21}$$

In torches where the arc voltage is high (about 200 V) and the cathode loss relatively small [Ref. 2.2.81], the efficiency is up to

$$\eta_z = 0{\cdot}95. \tag{2.2.22}$$

In smaller water-stabilized torches the conversion efficiency is higher, but it drops as the output rises (provided the torch design is the same). At an output of about 45 kW the efficiency

$$\eta_z = 0{\cdot}75 \text{ to } 0{\cdot}78. \tag{2.2.23}$$

With rising output the efficiency becomes even smaller, as shown in Fig. 72 for a VS 100 plasma torch with a channel 20 mm long and

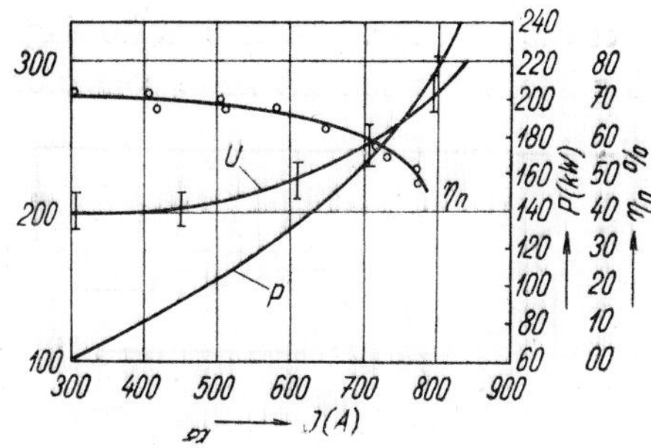

Fig. 72 Efficiency curve η_n for VS-100 plasma torch

a nozzle diameter of 4·8 mm at the front and 5·75 mm at the back. With an input of more than 130 kW the operation of this design is disadvantageous.

In connection with a non-transferred arc, anode losses must also be deducted, and with inputs up to 100 kW efficiency ranges between 0·70 and 0·60. The value of η_z must be known if the amount of water flowing through the plasma torch is to be calculated. This amount is then given by

$$V = \frac{P \cdot (1 - \eta_z) \cdot 0{\cdot}24}{C \cdot \Delta t}, \qquad [\mathrm{cm}^3 \cdot \mathrm{s}^{-1}] \qquad (2.2.24)$$

where P is the total imput required [W], C the specific heat of water [cal . cm^{-3}], and Δt the permissible heating of the water. If the water is heated above 50 to 55°C, the amount of water vapour generated is apt to cause trouble.

2.2.12 **Igniting the Arc**

In plasma torches stabilized by inert or inactive gases, i.e. in designs where the stabilization nozzle can act as the anode, the arc is ignited by an h.f. current which strikes a pilot arc with a very low current, as a rule 5 to 25 A. The output is sufficient to ionize the arc path, even in the case of a transferred arc. The entire ignition process passes in three stages: When the auxiliary circuit is switched in, a high-frequency spark discharge between the cathode and the anode nozzle ionizes the path and strikes the pilot arc. The current flowing through the arc attains the working intensity when the main contactor is switched in;

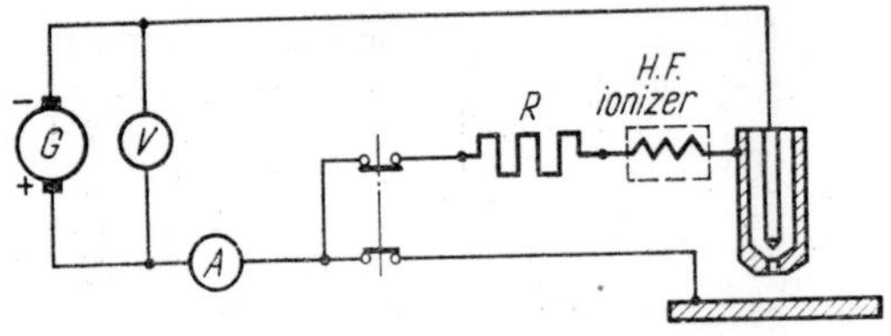

Fig. 73 Pilot circuit fed by main source

in the case of a transferred arc this contactor by-passes the current-limiting resistor, and in the case of a non-transferred arc it connects the conductive material to be processed into the direct circuit.

There are two basic ways of connecting the ionizer: either the main current source feeds an auxiliary circuit into which the nozzle is connected, or the auxiliary circuit has a current source of its own.

In the former instance, a current-limiting resistor (R in Fig. 73) is connected into the pilot circuit. It follows from the static characteristic of the arc that the arc voltage is high if the current is low. The magnitude of the resistance to be included in the circuit is determined by the voltage of the current source; it must be such that the pilot arc burns at the

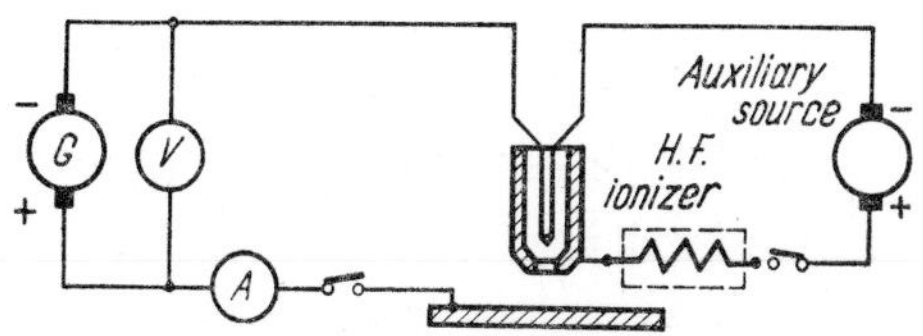

Fig. 74 Pilot circuit fed by separate source

lowest current at which it can exist at the given voltage. For the reasons stated in Chap. 2.2.2, the arc is ignited in pure argon, and the valve admitting the added molecular gas (hydrogen) is only opened when the arc has jumped over onto the material.

If the source has a rather flat characteristic curve, it is more advantageous and sometimes even unavoidable to use a separate source for the pilot arc (Fig. 74). The current of the auxiliary source can then be limited to 5 or 7 A at a higher voltage which depends on the kind of stabilizing gas used. In some types of torches (Plasmadyne, von-Ardenne Institute) an h.v. spark ignites a pilot arc between the nozzle and a separate auxiliary (starting) electrode (Fig. 94) which is disconnected as soon as the arc jumps over onto the main electrode.

The arc may also be initiated by mechanical methods instead of being started by h.f. currents; this applies especially to Soviet-designed torches. The connection is similar to that shown in Fig. 73, but there is no ionizer and the tungsten electrode together with its mounting is movable. The insulated top part of the electrode is depressed by hand, which establishes a contact with the nozzle. When the electrode is released, a spring returns it into the operating position (IMET design). Even simpler is a design in which the tungsten cathode is fixed, and a thin carbon pencil introduced through the nozzle establishes the contact between the latter and the cathode. As the carbon pencil is withdrawn from the nozzle opening, it ignites the pilot arc.

In liquid-stabilized plasma torches the ignition is more difficult; h.f. ignition cannot be employed, since the water connects the individual parts of the apparatus and short-circuits the h.f. current. The stabilizing channel formed by the surface of the rotating liquid has a strong de-ionizing effect, and a comparatively long auxiliary low-current arc cannot exist unless the auxiliary source has a high voltage.

For initiation proper these torches therefore rely on mechanical means, such as an igniter needle of 2 mm diameter, very accurately guided because its end must touch the rotating water without affecting the channel. With its end well insulated, the needle can also be withdrawn by hand or by means of an electromagnet. This method, however, is only used in special cases. As a rule, the arc of water-stabilized plasma torches is ignited by means of a 1·5-mm aluminium wire. An axial hole, 8 to 10 mm deep is drilled into the flushed end of the electrode to introduce the igniting wire. When the wire has been inserted into the hole, the protruding end is adjusted to the length required for bridging the gap between the anode and the cathode.

2.3 Design Principles and Construction of Czechoslovak Plasma Torches

2.3.1 Introduction

Czechoslovak plasma torches developed and designed for industrial purposes have their arcs stabilized by either a gas or a liquid. A type with liquid-and-air-stabilized arc developed for heating purposes combines the advantages of both systems.

2.3.2 General Design Principles for Plasma Torches

Up to the present, torches with inputs between 15 and 100 kW have been designed in Czechoslovakia. Applications in metallurgy, chemistry (facilitating chemical reactions at elevated temperatures) and other industries that require very high temperatures call for large-size units capable of long-term continuous operation. Installations of this kind are in the research and development stages.

2.3.2.1 Gas-stabilized plasma torches

In principle, Czechoslovak units of this description do not differ from foreign torches described in Div. 2.4. The use of inert gases as stabilizing medium permits comparatively simple designs of torches. Tungsten cathodes used in such gases eliminate the problem of electrode feeding. The designer has to secure sufficient cooling of the functional parts by flowing water and reliable sealing of joints; no cooling water must be allowed to penetrate into the working chamber since this would destroy the tungsten cathode. Good electrical insulation is required between the stabilizing nozzle and the rear part of the torch, which is at the electrode potential. Cooling the nozzle poses no particular difficulties in the case of transferred arcs, where the material to be processed acts as anode and the nozzle is not included in the current circuit. The non-transferred arc, where the nozzle acts as anode, poses the difficult problem of the optimalization of the gas and plasma flow, and the nozzle has to be intensively cooled in order to carry off the output (ranging in kW per cm^2) supplied to the anode. One of the most essential tasks facing the designer of gas-stabilized torches is the perfect centring of the cathode with respect to the nozzle aperture. Unless this condition is fulfilled the one-sided thermal load on the inner surface of the nozzle produces local heat flow of such high values that it can no longer be carried off by cooling; hence the nozzle surface melts.

The simple design of gas-stabilized plasma torches is partly offset by the relatively high price of the inert gases, which have to be very pure if excessive wear of the expensive tungsten electrodes is to be avoided. Argon with an addition of hydrogen is used most frequently. For metal cutting, i.e. work with a transferred arc, the optimum mixture is 65% argon and 35% hydrogen; with a non-transferred arc, however, the proportion of hydrogen must not exceed 10%. A worse drawback than the gas price—particularly in plants remote from supply centres—is the supply, storage and handling of the cylinders. This applies especially where the installation is not continuously and regularly used. Nevertheless, in some cases gas-stabilized plasma torches are more advantageous or even the only type applicable to the given job.

2.3.2.2 Water-stabilized plasma torches

Because of the reasons set forth in the preceding Section, water-stabilized plasma torches are often found preferable in industrial practice, notwithstanding their comparatively more complex and bulky design—necessitated by the carbon or graphite electrodes employed and the appurtenant electrode feeding device. Another disadvantage, the greater complexity of the arc initiation, is virtually negligible if viewed against the background of the total preparation time (including the positioning of the workpiece on the bench, the setting of the traversing device to the starting point of the cut, etc.). On the other hand, the advantages of such torches include the cheap and easily accessible stabilizing medium and the simple control of its quantity: the gauge pressure is set to the minimum of 2·5 atm required for the formation of the water vortex, and the arc then produces the amount of steam required for the generation of the plasma. Hence the cost of operation is low. The drawbacks inherent in the original concept of water-stabilized plasma torches (cf. Chap. 2.2.1) limited their practical application to certain instances of high-temperature material testing [Refs. 2.2.12, 2.2.13]. For unless the water vortex has a larger diameter than the electrode, the end of the arc erodes the electrode tip. The water flowing around the electrode is decelerated by friction, it penetrates easily into the crater and extinguishes the arc. Many applications require a channel of no more than 4 to 5 mm diameter; an electrode of the corresponding thickness, however, could not carry the current of 300 to 400 A which is necessary for generating a sufficient quantity of plasma in a channel of the dimensions indicated.

It follows from these considerations, verified by numerous experiments, that a liquid-stabilized torch must fulfil the following conditions in order to be suitable for industrial practice:

- The design must secure the formation of two water vortices with different diameters – a small diameter in the stabilization channel and a larger one around the electrode.
- The parts exposed to the heat of the plasma must be shielded by a water film.
- The tube draining the electrode chamber must be wide enough for a quantity of water which sufficiently cools the inner electrode and prevents the escape of plasma.

These conditions were already satisfied by the earliest plasma torch with a non-transferred arc designed at the Czecholovak Research Institute for Electrical Engineering (R.I.E.E.). In the schematic cross-section shown in Fig. 75, the diameter of the water vortex in the stabilization channel is a dashed line. Long stabilization channels (more than 15 mm) absolutely require the partition of the channel by an auxiliary nozzle (*1*) to reduce the number of ignition failures. Separate tangential openings introduce the water into the individual chambers, and it flows off uniformly to both sides provided the nozzles have equal diameters. A continous film of water is also required to protect the walls of the widening electrode chamber against the thermal effects of the arc. To this end, additional water is tangentially introduced at (*2*); it accelerates the retarded rotation of the liquid issuing from the stabilization chamber and causes the entire electrode chamber to be covered by a coherent water film. The valve at (*3*) is in its rear position when the arc is being ignited; the water then flows off over seat (*4*), and the initiation of the arc is thus facilitated. When the arc burns, the valve is in the position shown in the drawing, and the water is forced to run off through the central opening close to the electrode which is thus cooled.

In trial operation [Ref. 2.2.77], the plasma torch constructed according to this design gave a satisfactory performance, but certain insufficiencies emerged. Optimum results were only obtained with perfectly straight and well-burned graphite electrodes; in other words, the torch was highly sensitive to any eccentricity of the electrode position

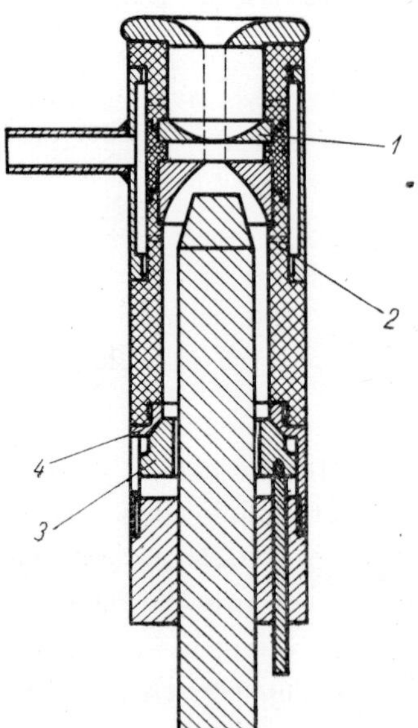

Fig. 75 Water-stabilized plasma torch (Czechoslovak patent 100 498)

in the channel. Carbon electrodes could not be used: the cooling effect of the water, which in the long part of the electrode chamber was not forced onto the electrode surface, was uncertain, non-uniform, and caused the electrodes to crack. The electrode wear under such conditions was 20 to 25 mm per minute.

After a thorough investigation into the hydraulic conditions, and experience gathered in plate-cutting on a pilot-plant scale, the

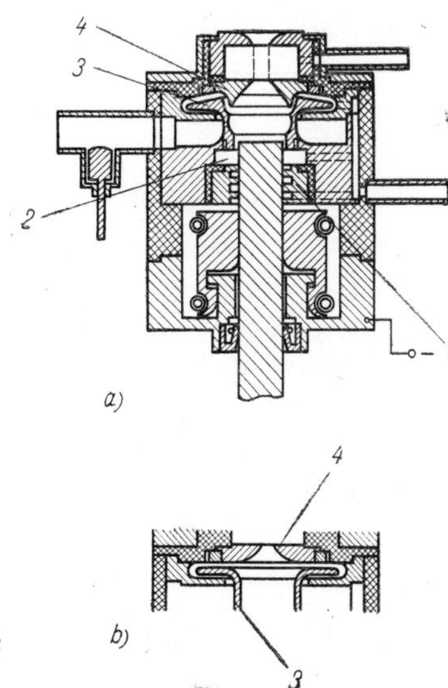

Fig. 76 Water-stabilized torch redesigned according to Czechoslovak patent application 4627-61

electrode chamber was re-designed, with the flow of the cooling water directed against the plasma pressure (cf. Fig. 76) [Ref. 2.2.36]. With the water running in a helical cooler along the entire circumference of the electrode, the cooling effect is more uniform, no excessive internal stresses are set up, and not even carbon electrodes crack. The water is introduced tangentially into the flat cooler helix (pos. *1* in Fig. 76a). The lead of the spiral is directed into the electrode chamber, hence the water flow counters the effect of the pressure due to the rapid steam generation in this chamber. From the spiral, the water enters the wider chamber (*2*), where its rotation is intensified by an additional tangential water jet; a vortex is formed, its diameter being determined by that of the tubular nozzle (*3*). The inner diameter of this nozzle changes either by steps or continuously, and its widened discharge edge encircles the lengthened socket of the rear nozzle (*4*). The whirling water adheres to the nozzle wall, exactly follows all changes in the nozzle diameter, and runs off—together with the water coming from the stabilization channel—through an annular siphon trap and the outlet pipe (fitted with a throttle valve) into a reservoir. This design, illustrated in Fig. 76a, has given very satisfactory results. With the throttle valve permitting the operator to control the counterpressure in the outlet pipe depending on the gauge pressure in the chamber, the siphon trap sufficiently seals the collector chamber up to an arc input of about 150 kW [Ref. 2.3.2].

The electrode chamber can also be designed as shown in Fig. 76b, where higher operational reliability is achieved at the price of reducing the efficiency of energy conversion by one or two per cent. Small electrode fragments or remainders of the ignition wire are more easily flushed out from chambers of this design. In torches for lower inputs, up to about 70 kW, both the comparatively involved siphon trap and the throttle valve can be dispensed with. This version is shown in Fig. 85. The electrode chamber is here sealed by a rotating water layer of annular shape which permanently covers the tangentially situated water outlet and thus prevents the steam from escaping at such minor overpressures as are generated at low inputs.

The design principles indicated above haven been embodied in several types of water-stabilized plasma torches designed at the R.I.E E.

The VS-45 unit with 45–80 kW input is intended for metal cutting, a Praga-500 welding set supplying the current.

The VS-60 unit with 60–100 kW input is also designed for metal cutting; however, fitted with a rotary electrode, it is used to advantage for applying high-melting protective coatings. The front nozzle has to be exchanged to increase the input to the peak of 100 kW.

The VS-100 unit with 100 kW optimum and 120 kW maximum input has so far mainly been used as a heat source for heating in plasma. An experimental unit has been tried out at a maximum input, so far, of 1000 kW (1800 A – 550 V).

2.3.2.3 **Plasma torch with combined water-and-air stabilization**

This torch with combined arc stabilization has been designed as a heat source for applications where the oxidizing effect of air does not matter (glass and ceramics industries). A torch with an arc stabilized by air only is not capable of operation, because the carbon electrode is too rapidly consumed for normal feed mechanisms. With the electrode indirectly cooled by water (i.e. a water-cooled copper sleeve enclosing the electrode), the poor heat transfer from carbon into copper causes a considerable length of the electrode to be overheated and the incandescent electrode reacts with atmospheric oxygen. Under these conditions the electrode wear is up to 80 or 100 mm/min (with an electrode of 13 mm diameter and a current of 350 A).

The plasma torch with combined arc stabilization, based on the design principle of the water-stabilized torch, has been found unsatisfactory for higher outputs. The water draining from the electrode chamber permits the simultaneous escape of the air supplied into the stabilizing nozzle; the escaping air carries off a certain amount of energy which increases with higher inputs. At low input, however, the electrode chamber of a torch with water-stabilized arc can be sufficiently sealed. This type of torch with combined arc stabilization by water and air has been described in Ref. [2.3.4].

In a more recent design (cf. Fig. 87), the electrode chamber is fully sealed, hence gases cannot escape except by way of the stabilizing nozzle. Graphite or carbon rods are used as electrodes. In order to keep the electrode wear low, a small amount of water is injected between the copper sleeve and the electrode; owing to the heat generated by the cathode losses, virtually all of this water is evaporated. On being discharged from the cooler, the part which has not been evaporated on the cathode is atomized by the whirling air and evaporated by the radiant heat of the arc before it can enter the stabilizing nozzle. Depending on the output, the water supply is 30 to 100 cm^3/min. The steam generated by the evaporation of the coolant serves for the formation of the plasma. The electrode wear is virtually the same as in water-stabilized plasma torches; with currents ranging between 200 and 450 A, the corresponding wear is between 2 and 8 mm/min. As part of the energy corresponding to the cathode losses is gained by the steam, very little energy is lost.

2.3.3 Design Principles for Plasma-Torch Components

2.3.3.1 Electrodes

The electrodes are the most important and thermally most stressed part of the torch. They conduct the current to the arc column and constitute a thermally-insulating transition between the hot plasma and the remaining parts of the arc circuit. Their design thus deserves close attention.

Except in water-stabilized plasma torches constructed according to Fig. 46 and designed for special tests of materials, the cathode is usually a rod-type electrode. Provided the stabilizing medium in inert gas or nitrogen with no more than 10% hydrogen added, the stabilizing nozzle can be employed as anode. If active stabilizing gases – air, steam, etc. – are used, the anode has to be placed outside the torch proper, and either appropriate feeding arrangements must continuously compensate for the wear, or the electrode – intensively cooled – must be moving at such a rate that the energy supplied is distributed over a sufficiently large area; this latter condition is fulfilled by water-cooled rotary disk anodes.

2.3.3.2 The cathode and its heat balance

Since plasma torches require a thermal arc with an incandescent cathode, high-melting metals or alloys and carbon or graphite are the appropriate cathode materials. The metal used most frequently is pure tungsten. Its current-carrying capacity can be enhanced by alloying with 1·5 to 2% of thorium oxide, and a further increase can be obtained by also adding zirconium. Metal cathodes can only be employed in combination with gases which remain inactive at the high temperatures occurring in the cathode-spot region. Such designs are usually referred to as torches with non-consumable electrodes, although strictly speaking this designation is not quite correct. It merely indicates that the wear is very slight and the shape and position of the electrode are not seriously affected even by working cycles continuing over several hours.

The best choice for torches with arcs stabilized by active gas or water are carbon or graphite rods. They are comparatively cheap and readily available; their more rapid wear enforces the use of a feeding device working at a controllable rate, but it does not really affect the operating cost. Appropriate cooling can greatly reduce the electrode consumption; however, considering the need to keep up the thermionic emission of electrons, cooling must be kept within limits and a compromise has to be struck between the two requirements.

In calculating the cooling of the electrode, we start from its overall thermal balance which is briefly indicated below.

Heat supply. The cathode is heated by the following processes:

(a) Supply of energy by the cathode fall

$$q_1 = i(U_k - \Phi - U_e), \qquad [\mathrm{W} \,.\, \mathrm{m}^{-2} \text{ per unit time}] \quad (2.3.1)\ *$$

where i is the average current density [A . m^{-2}], U_k the cathode fall [V], Φ the potential equivalent to the work function [V], U_e the potential equivalent to the mean thermal energy of the electron in the plasma [V].

By substituting the values measured for tungsten electrodes, $U_k = 8$ V (for electrodes under high loads).

$$\Phi = 4{\cdot}5 \text{ V}, \qquad U_e = 1 \text{ V},$$

we obtain

$$q_1 = i \,.\, 2{\cdot}5 \,.\, 10^4 \qquad [\mathrm{W} \,.\, \mathrm{m}^{-2}] \qquad (2.3.2)$$

Similarly, for carbon electrodes $U_k = 10$ V and $\Phi = 4{\cdot}36$, we obtain

$$q_1 = i \,.\, 4{\cdot}64 \,.\, 10^4 \qquad [\mathrm{W} \,.\, \mathrm{m}^{-2}]. \qquad (2.3.3)$$

(b) Energy supplied by convection and radiation from the positive arc column. Its amount per unit time is

$$q_2 = -\mathrm{div}\,(\varkappa \text{ grad } T) + S(T). \qquad (2.3.4)$$

The energy supplied to the cathode by convection is of considerable importance in water-stabilized plasma torches. The factors affecting

---

* According to Ref. [2.3.8] the energy supplied by the cathode fall can also be expressed by the following equation:

$$W_z = [xE_k - (1 - x)\,\varphi_{ef}]\,J_{(t)},$$

where x is the ionic component of the current, φ_{ef} the work function of the electrons, eV_i ioniziation energy, $E_k = aU_k + eV_i - \varphi$ is the energy transferred by the ions, and a is the coefficient of accommodation (0.98—1).

the amount of this energy will be described in Sec. 2.3.3.4 discussing the electrode chambers. As the size of the electrode chamber is determined by the diameter of the electrode, this dimension must not be exaggerated.

(c) Part of the electrical energy flowing through the electrode tip heats the electrode and generates the following amount of heat per unit time:

$$q_3 = R_1 \,.\, i^2. \tag{2.3.5}$$

The resistivity of cathode materials cannot be precisely specified: it varies in fairly wide limits with both their processing and the additions. For pure tungsten (ϱ = 18·7 to 19·1 g . cm^{-3}) resistivity at 20°C is 5·48 . 10^{-6} Ω cm the temperature-dependence of the coefficient of resistivity 4·6 . 10^{-7} Ω cm deg^{-1} and thermal conductivity $\varkappa$ is 125 kcal . m^{-1} . h^{-1} . deg^{-1}.

The properties of carbon and graphite materials depend upon the manufacturing procedures, and the differences between products of different manufacturers are very considerable. The Czechoslovak Elektrokarbon Works specify the following values:

carbon rods: maximum resistivity 5,000 microohms . cm

specific gravity 1·45 to 1·65 g . cm^{-3}

electrographite rods: maximum resistivity 2500 microohm . cm

specific gravity 1·5 to 1·65 g . cm^{-3}

The coefficients of thermal conductivity $\varkappa$ are

20 − 0·030 Δt [kcal m^{-1}h^{-1}deg^{-1}] for carbon rods, and
140 − 0·035 Δt [kcal m^{-1}h^{-1}deg^{-1}] for graphite rods.

(d) The resistance to the flow of the current on the contact between the supply lead and the electrode contributes considerably to the heat balance; this refers in particular to carbon electrodes where the contact resistance amounts to 6 mΩ. The energy liberated per unit time is

$$q_4 = R_2 i^2 \tag{2.3.6}$$

Heat removal. The heat supplied to the electrode is removed by conduction, evaporation of cathode material, endothermic chemical reactions, and radiation:

$$q_1 + q_2 + q_3 + q_4 = q_{cd} + q_{ev} + q_{ch} + q_{rad} \qquad (2.3.7)$$

Unless the electrode is cooled, evaporation and chemical reactions are the predominating modes of heat removal; intensive cooling entirely suppresses the chemical reactions and reduces evaporation to about one tenth of the former value.

In Fig. 77 the wear of a graphite electrode 13 mm in diameter is plotted against the current at various modes of cooling. Curve (a)

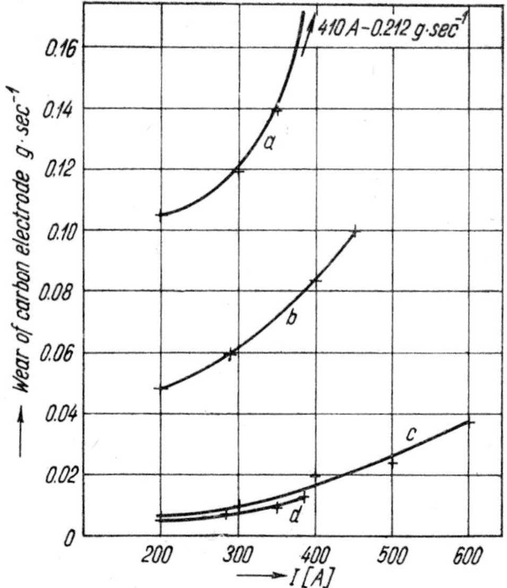

Fig. 77 Wear of graphite electrodes

refers to a cathode with minimum cooling, curve (b) to average values measured in torches with electrode chambers of the design schematically shown in Fig. 75, and curve (c) to a plasma torch with improved cooling (cf. Fig. 76). Curve (d) shows the loss in weight [$g \cdot s^{-1}$] measured on a 9-mm carbon electrode. It is slightly lower, in this case, although the reduction in length is nearly twice as much as in the 13-mm electrode.

Fig. 78 schematically represents worn electrode tips at different intensities of cooling: (*a*) belongs to an insufficiently cooled cathode,

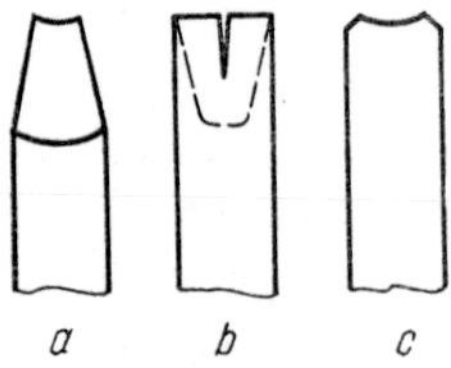

Fig. 78 Worn electrode tips at different cooling intensities

(*b*) to a cathode cooled excessively (considering the given current), and (*c*) is the shape to which the tip wears on a correctly cooled cathode.

In designs used in industrial practice, the following relation exists between the diameter d_e of carbon electrodes and the diameter d_{ch} of stabilization channels:

$$d_e = 2{\cdot}5d_{ch} \text{ to } 3d_{ch} \tag{2.3.8}$$

The proportionality factor in this relation is the result of a compromise between economic considerations and the postulate of the smallest possible electrode diameter for conducting the required current. The efficient conversion of electrical into thermal energy is promoted by $d_e = d_{ch}$; in other words: equal cross-sections of the electrode and the stabilization channel would be most advantageous.

The carbon electrodes of water-stabilized torches are very intensively cooled even in smaller units. For gas-stabilized torches indirect water cooling is sufficient. The tungsten electrode is, moreover, cooled by the stabilizing gas which flows around it on its way to the arc. Some designs (such as Plasmadyne and METCO) feature internal cathode cooling (cf. Chap. 2.4.1).

In torches with lower outputs, the electrode tip is cone-shaped, whereas for higher outputs the cone is truncated. The electrode must by carefully located to be exactly coaxial with the stabilizing nozzle; in operation it must not change its shape which is chosen to make the arc burn in the centre line of the nozzle. Changes in the shape of the electrode always result in damage to the nozzle. The electrodes used in the Czechoslovak plasma torches R6 and R6a (65% argon and 35% hydrogen) have conical tips.

Current loads, electrode diameters and nozzle diameters are summarized in Tables 2.3.1 [Ref. 2.3.5] and 2.3.2.

Table 2.3.1 GAS-STABILIZED PLASMA TORCHES

current [A]	nozzle dia. [mm]	electrode $W + 2\% ThO_2$ dia. [mm]	nozzle length [mm]
up to 200	3	3	3
up to 400	4	3	3
above 400	4·5—5	4	3

Table 2.3.2 WATER-STABILIZED PLASMA TORCHES COOLED ACCORDING TO FIG. 76

Current up to [A]	vortex channel diameter [mm]	electrode diameter [mm]	electrode material
360	4	9	carbon
450	4	9	graphite
400	4	13	carbon
600	4	13	graphite
1,200	8	20	graphite
1,200	8	25	carbon

2.3.3.3 The anode and its heat balance

The heat balance of the anode is significant in plasma torches with independently burning arcs, such as are used to shape electrically non-conductive materials or to face base materials with protective layers of high-melting metals, carbides or oxides. If the arc is stabilized by inert or inactive gas, it is as a rule the stabilizing nozzle which serves for an anode (cf. Div. 2.4). Its thermal load approaches the critical value and it must be cooled by the removal of heat, the maximum specific heat flow—q_{max}—can be expressed by the following relation:

$$q_{max} = \varkappa \frac{T_m - T_0}{s} \quad [\text{kcal} \,.\, \text{m}^{-2}]; \qquad (2.3.9)$$

$\varkappa$ is the coefficient of heat conduction of the nozzle material [kcal . . $m^{-1} h^{-1} deg^{-1}$], T_m the melting point of the material [°K], T_0 the temperature of the cooled nozzle wall [°K], and s its thickness [m]. By substituting the material constants for copper and $s = 2 \cdot 5 . 10^{-3}$ for the wall thickness, we obtain $q_{max} = 16 . 10^4$ kW . m^{-2}.

In addition to q_a, the heat energy generated in the anode fall, the anode receives by conduction and convection a heat flow q_v from the rapidly flowing central region of the plasma. The heat flow q_a directly affects only a certain limited area of the anode. Let us designate the surface of the active anode region – i.e. the region through which the current flows – by F_a [m^2]; the amount of heat energy supplied to the unit of this surface is then

$$q_a = i_a \left(\Phi + V_a + \frac{3kT}{2e} \right) \quad [\mathrm{kW} . \mathrm{m}^{-2}], \qquad (2.3.10)$$

where i_a is the current density in the active region $\left(i_a = \frac{I}{F_a} \right)$, Φ the potential equivalent to the work function for the given material, V_a the anode fall, k Boltzmann's constant, and $\frac{3kT}{2e}$ the potential corresponding to the thermal energy of the electrons entering into the anode.

Let us designate the active area of the anode in relation to conduction and convection (member q_v) with F_k; the density of the heat flow is then

$$q_v = \frac{St . m . h}{F_k} = \frac{St . \eta . P}{F_k}, \quad [\mathrm{kW} . m^{-2}] \qquad (2.3.11)$$

where St is Stanton's number, m the mass flow rate, η the efficiency of energy conversion, and P the input of the torch [kg . s^{-1}]. The transition of the arc plasma into the anode, and hence the value of q_a, is dependent on the medium in which the arc burns. In argon, the conditions are most favourable, the active surface is largest (the arc is cone-shaped, widening from the cathode towards the anode), moreover the anode fall is very low. In the range above 200 A, Busz-Peukert and Finkelnburg [Ref. 2.2.61] have measured a value of only 1 – 2 V for the anode fall in argon. In other media (for instance on a copper anode in air or steam), a spot of very high current density appears on the anode (the column contracts intensively) [Ref. 2.2.45]. In such cases the heat flow q_a is too great to be removed by cooling.

Theoretically, q_{max} can be as high as $16 . 10^4$ kW . m^{-2} (with the wall $2{\cdot}5 . 10^{-3}$ m thick); for safely controlled operation, however, the

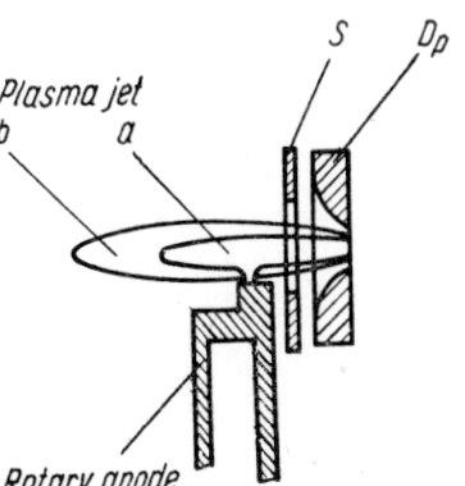

Fig. 79 Closure of circuit on rotary-disk anode

maximum load on the anode surface should be below $10 . 10^4$ kW . m^{-2}. The lower q_{max}, the longer the service life of the nozzle-anode. In active gases (air, carbon dioxide or steam), a nozzle serving as anode does not live longer than a few seconds. Therefore plasma torches with an independent arc burning in active gas have water-cooled disk anodes; rotating at a high circumferential speed, they ensure the distribution of the heat load over a large surface.

Fig. 79 schematically represents the way in which the current circuit is closed on the anode. The electrically conductive zone of the high-temperature plasma (*a*) flows past the edge of the rotating disk which is only hit by the relatively cool outer fringe layers (*b*). A secondary arc closes the current circuit. The distance between the edge of the disk and the centre line of the plasma must not be too large, for unless the fringe layers of the plasma sufficiently heat the surface of the edge, the anode spot "sticks" instead of readily sliding: the spot is entrained by the disk until the secondary arc is ruptured, and only then it jumps back into the neighbourhood of the plasma stream. This process is repeated in quick succession and causes the main plasma jet to oscillate and lose its axial symmetry.

Disk anodes have a relatively long service life (30 to 100 hours). To minimize the wear, their circumferential speed should be inverse proportional to the diameter (10 to 20 m/s for diameters of 220 to 60 mm).

Fig. 80 shows a disk electrode with interchangeable copper disk; this design has been tried in operation at currents up to 380 A.

The copper disk (*6*) is screwed to the carrier of a shaft (*2*) which is seated in ball bearings (*5*) and (*7*) in a duralumin sleeve (*1*). A recess

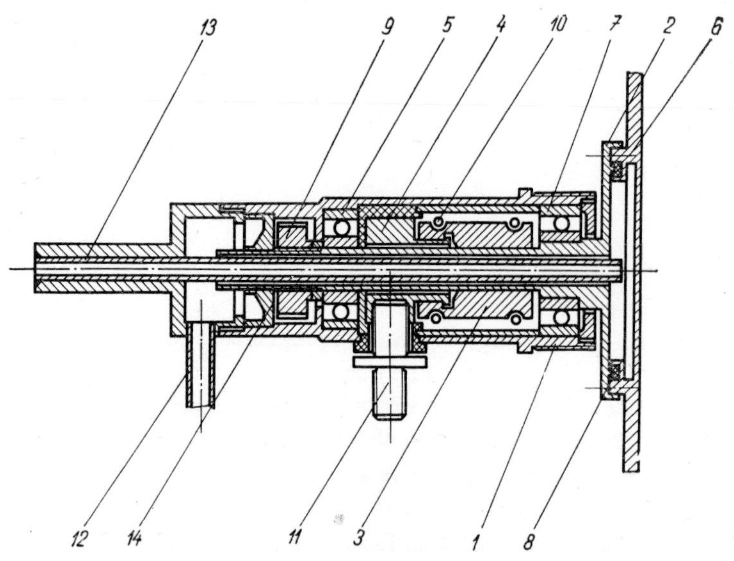

Fig. 80 Rotary anode with interchangeable copper disk

in the copper disk makes for good cooling, while a rubber gasket (*8*) prevents the water from escaping. The gasket is made of soft rubber to facilitate the tightening of the copper disk onto the carrier, and thus the satisfactory flow of the current. A connection rose (*3*) made of copper-graphite material leads the current to the shaft, via copper holder (*4*) and pin (*11*). A spring (*10*) uniformly presses the contact fingers of the rose onto the shaft. Tube (*13*) supplies the cooling water, which flows through the hollow shaft and drain tube (*12*) into a container. A rubber gasket (*14*) protects the bearings against the penetration of water. A flexible shaft (not indicated in Fig. 80) drives the disk through gearing (*9*).

At higher currents, disks of a larger diameter with the current directly supplied by brushes are preferable, for a longer service life.

As a rule, the disk is separable from the shaft, but welded together with a water chamber. This improves the cooling of the disk edge, which can be built up when the material–low-carbon steel–is worn. The shape of the edge is shown in Fig. 79. It is obvious from the schematic representation of the circuit-closing in this Figure that the employment of a rotary disk anode produces a current loop whose magnetic field deflects the plasma from the axial direction. This deflection becomes marked at higher currents (above 500 A), and in gas-stabilized arcs it exposes the stabilizing nozzle to unilateral heat stress which may damage it. This phenomenon has been countered by constructing an anode which consists of three disks spaced at 120° (Fig. 81); they are mutually offset to form a little aperture through which the plasma beam tangentially touches the three of them at a time. The disks are driven from an electric motor by a V-belt encircling the three shafts. The plasma torch is introduced from the rear between the sleeves of the disk

Fig. 81 Three-disk rotary anode

electrodes. The location of the current-supply brushes is also apparent from Fig. 81.

2.3.3.4 Stabilization channel and electrode chamber

In gas-stabilized plasma torches, the anode-nozzle is also the stabilizing member defining the maximum diameter of the arc column. In the case of non-transferred arcs, its cylindrical part is 10 to 12 mm long, depending on the gas employed. A transferred arc threatens with the hazard that it may jump onto the wall inside the nozzle, and another arc will then burn in series between the outer edge of the nozzle and the material. This phenomenon is countered by shortening the nozzle, so that the sum of the cathode and anode losses exceeds the voltage loss in the part of the arc column which is constricted by the nozzle. In the R-6_a plasma torch, the nozzle is 3 mm long [Ref. 2.3.5].

The relation between the nozzle diameter, the current flowing through the arc and the electrode diameter is apparent from Table 2.3.1. If the electrode diameter is denoted d_e and the diameter of the electrode chamber d_{ch}, their mutual relation in gas-stabilized torches can be expressed by the equation

$$d_{ch} = (2 \text{ to } 2{\cdot}5)\, d_e. \tag{2.3.12}$$

Although in gas-stabilized torches, too, the electrode chamber widens towards the stabilization channel, the flow of the plasma inside the apparatus is different: Introduced into the electrode chamber under a certain gauge pressure, the gas issues through the nozzle into the open, owing to the pressure gradient. In the water-stabilized torch, on the other hand, the radial heat losses of the arc evaporate the water surface, and the amount of vapour thus generated – and therefore the overpressure produced on the individual sections – is dependent on the specific heat load on the corresponding surface.

The amount of energy liberated on a given section of the water channel is dependent on the current and the electric gradient in that section. The wider channel in the electrode chamber reduces the electric gradient, and therefore also the amount of energy generated per unit length of the arc, while at the same time the water surface increases. Hence, a pressure gradient from the stabilization channel to the electrode chamber is set up, which accelerates part of the plasma in the axial direction from the stabilization channel to the cathode. This explains why in Sec. 2.3.3.2 we emphasized that electrodes of the smallest diameter compatible with technical and economic considerations (wear) should be used: all the energy flowing into the electrode chamber is carried away by the water and is thus lost. These considerations also explain the importance of a well-sealed electrode chamber. The minimum diameter of the stabilization chamber depends on its length. If the tangential inlet openings are laid out correctly, a good vortex surface is obtainable up to 12 mm length for a chamber diameter of 16 mm. If the stabilization chamber (i.e. the channel) is longer than 12 mm, it is preferable to use a third (internal) nozzle of equal diameter and have two chambers with independent tangential water inlets; this "stiffens" the stabilization channel [Ref. 2.3.2] and prevents it largely if not absolutely from widening when the arc burns (cf. the VS-100 torch).

The design of the nozzles is shown in Fig. 82 (a) to (d). Nozzles (a) and (b) according to Maecker [Ref. 2.2.10]—further on called M-type nozzles—are suitable for torches with non-transferred arcs

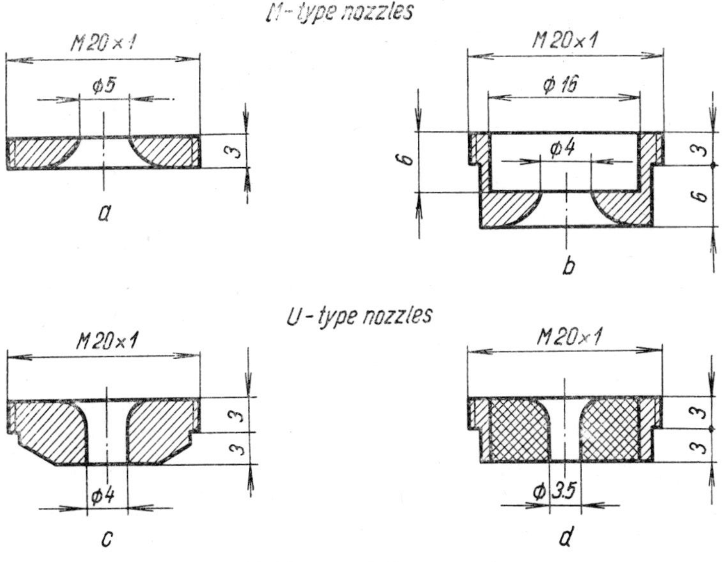

Fig. 82 M-type and V-type nozzles

(and disk anodes). Nozzle (d) with a cylindrical part in a ceramic insert—further on called V-type—has been developed at the R.I.E.E. and is employed with a rear nozzle 5·5 mm in diameter (cf. description of VS-45 plasma torch in Sec. 2.3.4.3); it is suitable for cutting material. The effect of the nozzle shape will be discussed in Chap. 3.1.2.

2.3.4 Designs of Plasma Torches

2.3.4.1 Gas-stabilized plasma torches

Gas-stabilized plasma torches types R-6 and R-6a, with arcs axially stabilized by a mixture of argon and hydrogen, have been developed at the Bratislava Welding Research Institute. The R-6 torch (Fig. 83)

Fig. 83 VS plasma torch, hand-operated version

is intended for work with either a transferred or a non-transferred arc at outputs of about 12 to 15 kW; it is produced with a handle or a holder, for manual or machine cutting respectively. The R-6a type, designed with a transferred arc, operates in the ORS-arg mechanical cutting set. It differs from the R-6 torch by a nozzle adapted for more convenient manufacture and fitting, as well as a simplified head, which can be screwed off the casing with the holder.

The electrode, gripped in a collet whose position is adjustable by a ring nut, is in both types screwed into the torch casing. In the variant shown in Fig. 83, a handle with two switches is fixed to this casing: one controls the ionizer, the other one the coil of the solenoid valve for the admission of hydrogen. The interchangeable nozzle, directly cooled by water, is integrated with the bottom part of the electrode chamber and tightened to the head by a screw cap. Rubber rings are used to seal the electrode chamber and prevent the penetration of water. A tube feeds the stabilizing gas into the centre part of the torch; from there it flows into the electrode chamber proper where it flushes and cools the electrode. The tubes supplying and draining the water also act as electric supply leads. An insulating piece connecting the torch casing with the head is necessary because these two parts differ in their potential when the transferred arc is being started or the non-transferred arc is burning. In a hand-operated cutting torch it is a distinct advantage to have the arc burning independently at least while the torch is at a distance from the material; this obviates the necessity of frequent re-starting. However, the hydrogen supply must be cut off by the switch controlling the solenoid valve.

The electrodes are made of tungsten alloyed with 2% thorium oxide. The diameters of the nozzle and the taper-tip electrode are dependent on the operating current (cf. Table 2.3.1).

The maximum rated output of the machine-cutting R-6a torch is 70 kW, and of the hand-cutting R-6 type 35 kW. Both values refer to operation with transferred arc. The maximum current load on the electrodes is 500 A. Depending on the output, the gas consumption is 1·8 to 6 m^3 per hour, the mixture being 65% argon and 35% hydrogen.

In addition to these metal-cutting torches with an axially stabilized arc, the Welding Research Institute has also developed a set for hard-facing with high-melting materials. Its main part is a plasma torch operating with a non-transferred arc stabilished by a gas vortex (the nozzle acting as anode). The cathode – tungsten alloyed with 2% thorium oxide – is 5 mm in diameter, indirectly water-cooled. A mixture of nitrogen and hydrogen is employed as principal stabilizing medium, but the cathode is especially protected by a stream of argon, since commercial nitrogen as supplied by Czechoslovak makers contains a higher percentage of water vapour. Two nozzles in series are employed, the front one being 4·6 and the rear one 3·5 mm in diameter. Between the cathode and the rear nozzle, argon flows at the rate of 0·12 m^3/hr., and between the rear and the front nozzles a mixture of nitrogen and hydrogen at the rate of 1·68 m^3 (1·2 m^3 nitrogen and 0·48 m^3 hydrogen) per hour. An additional amount of 0·15 m^3 of nitrogen per hour is consumed for the transport of the pulverized material into the plasma.

Cooling water at 4·5 atm gauge pressure is employed at the rate of 6 to 8 dm^3 per minute. The conversion of electric into thermal energy is performed with an efficiency of 75%. The service life of the rear nozzle is stated to be 80 hours of operation, and that of the front one 20 hours.

The arc is started by an r.f. current. The maximum permissible current load on the anode is 320 A, which permits continuous operation with about 25 kW output. The spraying performance under these conditions is approximately 1·5 kg of aluminium oxide per hour. The torch is 70 mm in diameter and 130 mm long, and the weight (without supply hoses and cables) is 2·3 kg.

2.3.4.2 The VS-60 plasma torch

This basic development type of a plasma torch with water stabilization (type denomination "VS") has been designed in two versions – for metal cutting (i.e. with a transferred arc) and for various other applications (i.e. with a non-transferred arc). In the latter design, for heating and metallizing purposes, a water trap slipped over the torch-head casing catches the water issuing from the front nozzle, and a separate discharge tube drains it away, as in the VS-100 torch (Fig. 86). This modification with an independent arc is supplemented by a rotary-disk anode (Fig. 80). Metal-cutting is now usually done with the VS-45 torch, especially designed for this purpose, and the VS-60 is mostly employed in the non-transferred-arc version. It differs from the VS-100 head by the shorter stabilization channel; consequently there is no divided stabilization chamber with an additional nozzle (*3*) as in Fig. 86 [Ref. 2.3.7].

2.3.4.3 The VS-45 plasma torch

The industrial modification of this plasma torch is shown in Fig. 84.

Especially designed for metal cutting, this torch has been fitted with a V-type cylindrical nozzle, which is particularly well suited to this end. The original all-metal copper nozzle 82 (c) was superseded by a nozzle with a porous ceramic insert (d).

The newly developed ceramic front nozzle has a long service life and is resistant to the plasma jet which tends to erode and widen the orifice. Moreover, this arrangement permits a considerable increase in the proportion of the front- to rear-nozzle diameters; hence all the water issues through the rear nozzle, and there is no spraying of water from the front nozzle, which otherwise troubles the operator. The front nozzle is 3·5 mm and the rear nozzle 5·5 mm in diameter.

The ceramic material is readily machined with ordinary tools and fitted with a copper ring threaded for screwing into the torch head. Capillary sorption fills the pores of the ceramic material with water from the vortex chamber, and unlike a metallic nozzle, it does not need

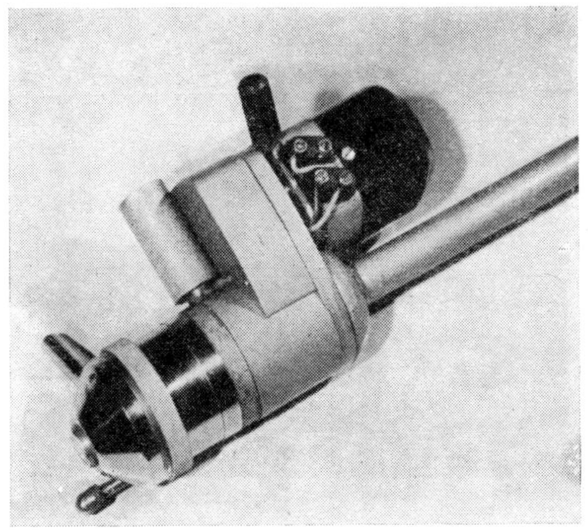

Fig. 84 VS-45 plasma torch—view

a water film to protect its surface. The ceramic insert is hot-pressed into the copper ring at about 250°C. The head is cone-shaped so as to facilitate the cutting of bevels. All parts of the feeding mechanism which operate at the potential of the cathode are thoroughly insulated. The main contactor is blocked and cannot be switched in unless the PVC protective tube of the electrode, fitted with a brass ring at its lower end, is screwed into a threaded split bushing mounted on an insulating cover.

The superior cutting properties of torches fitted with ceramic nozzles are due to the higher specific current load of the channel, and hence the increase of the region with an ionized and dissociated medium at the expense of the steam-carrying region. Water droplets entrained by the stream are fully evaporated as soon as they enter the cylindrical part of the ceramic nozzle. This results in a higher amount of steam and plasma from a shorter channel and – in addition – a more advantageous proportion between steam and its dissociation products.

The metal-cutting VS-45 plasma torch is shown in cross-section in Fig. 85. The conical head (*1*) permits the torch axis to be inclined at 45° to the material, for cutting bevels. The discharge nozzle (*2*) screwed into the head, is separated from the metal ring (*4*b) by insulating

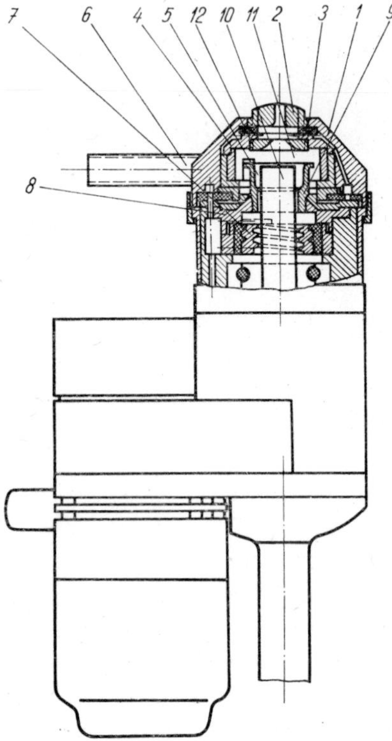

Fig. 85 VS-45 plasma torch—cross-section

washers and an annular rubber packing (*3*). The ring, which forms the wall of the stabilization chamber, has two tangential water inlets. Nozzle (*5*), screwed into it, opens into the water collecting chamber (*11*) with tangential drainage into tube (*6*). The vortex channel is formed in the electrode chamber (*12*) which is confined by tube (*9*). The electrode (*10*), intensively cooled by water flowing through flat helical grooves (*7*) on its circumference, reaches into the widened portion of tube (*9*) at one end, and is held at the other by the contact claws (not shown). Tube (*9*) is screwed into insert (*8*) which forms an acceleration chamber.

The electrode feeding device, placed behind the power supply terminals, secures both the axial feed (2 to 6 mm/min.) and the rotation of the electrode. It is driven by a geared fractional-horsepower D.C. motor whose infinitely variable speed is controlled by the voltage input (up to 24 V). The electrode need not be accurately centred, since any eccentric erosion on the tip is countered by its rotation. Hence the electrodes need not be ground absolutely straight as in the previous design according to Fig. 75. Commercial-grade carbon or graphite rods with standard dimensional tolerances are good enough. The rods used in the VS-45 torch are 500 mm long and 13 mm in diameter.

A rubber hose with a coaxial conductor feeds water and current to the torch. Channels in the insulant casing distribute the water in approximately equal proportions to the head and to the helical cooler and electrode chamber. The entire electrode periphery is intensively and uniformly cooled thanks to the tangential entry of the water into the helical cooler through which it moves towards the electrode chamber (i.e. against the pressure generated by the burning arc). The electrode chamber is thus efficiently sealed. The water flowing from the cooler is accelerated by the tangential injection of more water into the acceleration chamber (*13*) and forms the vortex channel in electrode chamber (*12*). Any water not evaporated there runs off—together with water from the stabilization chamber—into collecting chamber (*11*). The water rotating in this chamber forms a ring which seals off the draining tube.

2.3.4.4 **The VS-100 plasma torch**

A further development stage, intended mainly for high-temperature heating, is the torch head for 100 kW input. The design concept is the same as in the VS-60 torch, but the 13-mm electrode is made of graphite instead of carbon.

The cooling system permits graphite electrodes to be used up to 700 A without any cracking. Equipped for an independent arc, the head shown in Fig. 86 is adapted as a heat source for an experimental melting furnace [Ref. 2.3.6]; an external rotary electrode is employed as anode.

The stabilizing head is fitted with three nozzles: the front nozzle (*1*) is 4 mm in diameter, the rear nozzle (*2*), as well as the

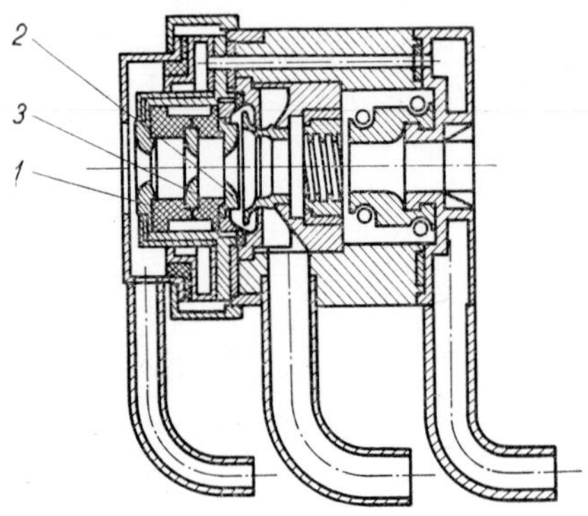

Fig. 86 VS-100 plasma torch

auxiliary intermediate nozzle (*3*) 5 mm. Other functional parts are similar to those of the torch types described earlier.

2.3.4.5 The VV-250 plasma torch for combined arc stabilization by water and air

This torch with an inexpensive stabilizing medium is designed for heating purposes with outputs above 100 kW. Particularly at higher outputs and high arc voltage (220–240 V) it converts electrical into thermal energy with a very satisfactory efficiency (η_z = 0·94 to 0·95 and η_n = 0·88 to 0·90] [Ref. 2.3.3]. The design of the head is apparent from Fig. 87. Unlike other gas-stabilized torches, this set operates with graphite electrodes. For more intensive cooling, water is tangentially injected at the rate of 80 cm^3 per minute into the flutes of the cooler, and all of it is evaporated. This keeps the electrode wear low, the rear parts are not affected by heat conduction from the electrode, and the latter is uniformly burned off throughout its cross-section. Air is supplied through tangential bores opening into the rear part of the electrode chamber.

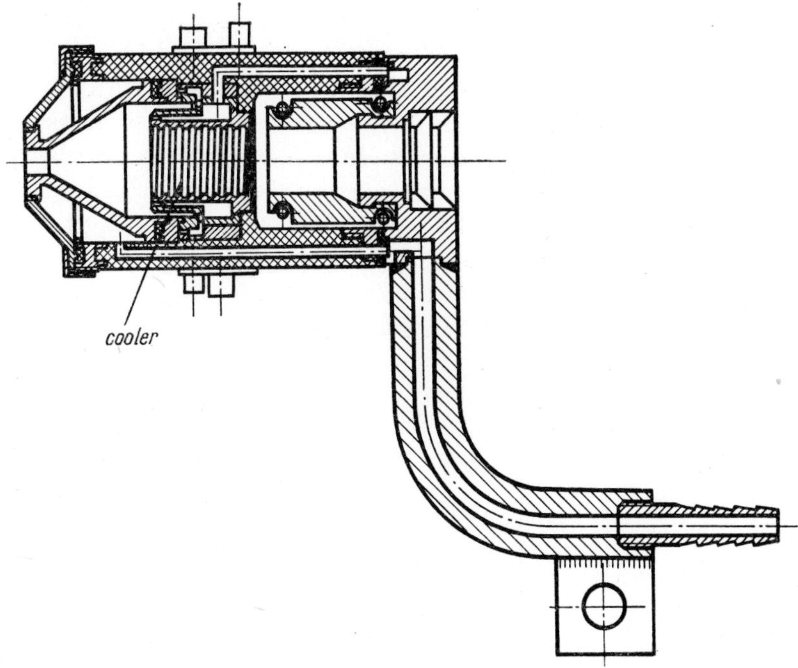

Fig. 87 VV-250 torch (gas-liquid stabilization)

2.3.5 Current Sources and their Characteristics

Both the gas- and liquid-stabilized torches developed in Czechoslovakia are fed with direct current.

Current sources for water-stabilized metal-cutting torches must satisfy the following conditions:

1. The ignition voltage (no-load voltage) must be higher than the voltage corresponding to the point where the extended descending branch of the arc characteristic intersects the voltage axis.

2. No excessive short-circuit current must flow through the electrode as the arc is being started; otherwise the igniting wire explodes and the arc is blown out of the channel.

3. For high-quality work, the active plasma zone in the kerf

(where the current passes from the plasma into the material) has to be blown into the bottom third of the cut.

4. Changes in the arc voltage (due to the altered length of the arc) must not substantially affect the current, i.e. the static characteristic in the working region of the current source must be steeply descending.

These conditions are most conveniently met by rotary convertors with cross-field dynamos or by transducer-controlled static converters.

Rotary converters, such as originally intended for welding purposes, mostly do not have a voltage sufficient for cutting thicker materials. It follows from the static characteristic of water-stabilized

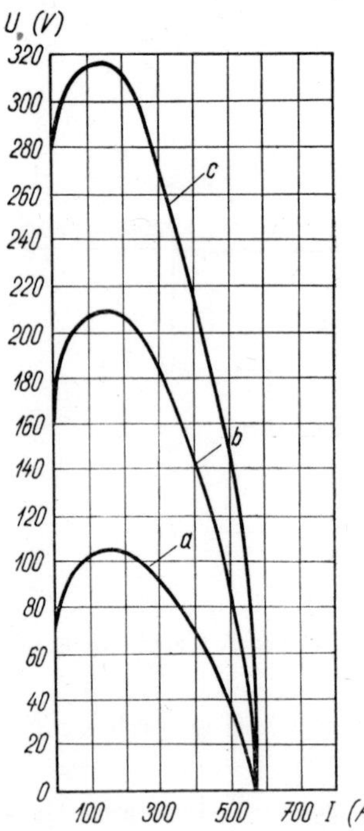

Fig. 88 Static characteristics of Praga-500 welding sets

arcs that at least two series-connected sets are required to fulfil conditions 1, 3 and 4; three sets must be employed for cutting thicker materials.

A single rotary welding set is sufficient for cutting material thinner than 25 mm with a gas-stabilized plasma torch; two sets connected in series are used to cut materials thicker than 25 mm with machine-operated gas-stabilized plasma torches. Safety regulations do not permit higher voltages than 100 V for cutting by hand.

Static characteristics of welding sets suitable for cutting metal with plasma torches are shown in Fig. 88.

2.4 Foreign Designs of Plasma Torches

In the west, the production of plasma torches is mainly concentrated in the USA. The West-European countries satisfy their needs partly by import from the USA, partly by production under USA licences (GFR). British Oxygen owns several patents, but the designs do not materially differ from those of American plasma torches. The USSR, Czechoslovakia, the GDR and Poland have developed their own designs some of which are produced in quantity. In this Division we review the principal designs of plasma torches produced abroad, at the present time.

2.4.1 Plasma Torches Manufactured in the USA

When – in 1955 – the Linde Air Company developed the first plasma torch for cutting aluminium [Ref. 2.4.1], this method of parting metals aroused much interest. The new apparatus found soon another field of application – the surfacing of various materials with overlays of high-melting-point metals and oxides. The gases used in the new installations attained temperatures several times as high as the combustion products of oxy-acetylene torches. This opened up new perspectives for some manufacturing procedures already well known and practised. The development of plasma torches in the USA was greatly accelerated by government grants for research within the framework of the extensive rocketry programme adopted in 1957. This support was due to the suitability of plasma torches for testing the high-temperature behaviour of materials and for modelling the re-entry conditions of ballistic-missile heads. Experience gained in laboratory work was utilized in the development of industrial installations for metal cutting and overlaying.

One by one, several USA firms began to produce plasma torches on a commercial scale – Linde Company, Thermal Dynamics Corp. Plasmadyne Corp.; METCO and AVCO manufacture almost exclusively overlaying torches.

Linde Co. (a Division of Union Carbide Corporation) manufactures plasma torches with an axially blown arc and markets them under the trade mark of "Plasmarc". The process of cutting with a plasma generated from a mixture of 65% argon and 35% hydrogen is protected by a patent [Ref. 2.2.22] and the mixture is denoted as Linde H-35.

The first torch types had a low plasma output (30 kW) and the gas mixture was fed in relatively small quantities. Cuts on aluminium were satisfactory, but large burrs were formed on stainless steel. The maximum depth of cut was about 25 mm. By increasing the output of the torch as well as the quantity of the gas supplied to it, the designers improved the quality of the cut and secured the possibility of cutting thicker materials. Simultaneously nitrogen was substituted for the expensive argon, mainly for cutting alloyed and structural steels [Ref. 2.4.2]. Structural steel is successfully cut with air, the tungsten electrode being protected by a stream of pure nitrogen directed into its close vicinity.

The Linde "Plasmarc" torch PT-2, a more recent universal metal-cutting type with transferred arc, can be operated on various gases; designed for mechanical cutting, it is mounted on a CM 37 or a CM45 carriage. A torch head on a CM37 carriage is shown in Fig. 89. The recommended gas mixtures, their throughput quantities, the nozzle diameters to be used for various metals and thicknesses, as well as the output and cutting velocities recommended are found in Div. 3.1.

The maximum output of the PT-2 plasma torch is 200 kW. The head is cooled with pressure water; the use of distilled water is advisable, both because of its higher dielectric strength and in order to prevent the deposition of boiler scale on the cooled parts. The pressure and throughflow quantity of the water depend on the kind of gas used and the output. For satisfactory cooling of the head when it is operated with the maximum output of 200 kW, the makers prescribe a flow of at least 15 l/min.; the water pressure should be 4·2 to 5·6 atm for cutting with inert gases, and 7 atm with air stabilizing the arc. Distilled water is cooled in a heat exchanger and circulated by a special pump.

The cutting set includes a type PCC-1 control panel containing solenoid-valve controls, flow meters for the gases, and remote controls for the cutting operation. Once the correct gas flow has been set and the pilot arc – a small-output arc burning between the tungsten electrode and the torch nozzle – has been initiated, all further operations are automated. Automatic controls open the valve which admits additional

Fig. 89 Linde "Plasmarc" torch type PT mounted on CM-37 carriage (By courtesy of Linde Company)

gas (hydrogen), start the carriage, stop it at the end of the cut, and extinguish the arc. The pilot arc is ignited by an r.f. generator. The carriage speed can also be controlled by hand.

A type PHC-250 transformer with a current rectifier of 400 V no-load voltage is used to supply the PT-2 plasma torch with electric current. Under continous load this source has an output of 50 kW, in peaks it may be overloaded up to 75 kW. Higher outputs are obtained by series connection of several power sets of this type. The static characteristic of the PHC-250 source is shown in Fig. 90. In addition

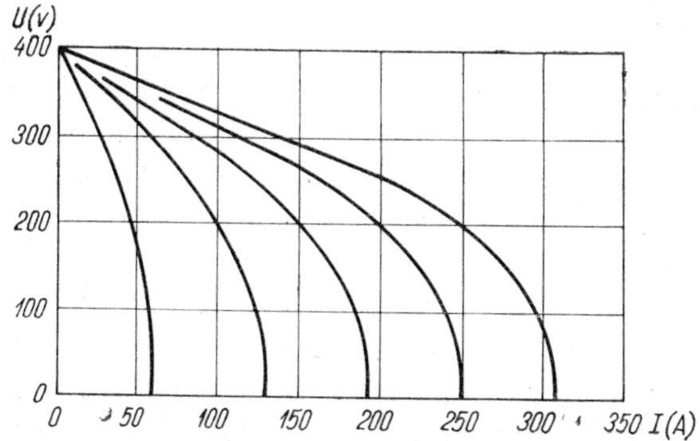

Fig. 90 Static characteristic of Linde PHC-250 power source

to the PT-2 set for mechanical cutting, Linde also manufacture a small lightweight plasma torch for cutting by hand with an arc stabilized by H-35 gas mixture. Moreover, they also produce on a commercial scale equipment for depositing hard-metal overlays from powdered surfacing materials. The operating principle of this set is shown in Fig. 63. The PT-1 fusion-surfacing torch, which is the main component of this set, has its arc axially stabilized with argon. The discharge path between the electrode and the workpiece is divided into two sections. A pilot arc, maintained by a separate current source, burns between the electrode and the small-diameter inner nozzle. A stream of argon, used as carrier gas, delivers the powdered overlay material into the jet of neutral plasma which issues from the inner nozzle and is directed by the outer one. Another current source feeds the transferred arc between the electrode and the workpiece material. In the second section of the discharge path, between the inner nozzle and the material, the current is adjustable within fairly wide limits; hence the heat supplied into this region and the depth of penetration into the base metal is well controlled. The heated metal particles deposited on the workpiece melt completely on contact with the fused surface of the base metal; a perfect metallurgical bond is thus produced between the overlay and the parent metal. A separate stream of shielding gas—argon—supplied through the annular porous lens on the face of the torch, shields both the overlay and base metal from harmful gases. From the metering hoppers—item (*1*) in Fig. 91—two plastic tubes convey the gas-carried metal powder into the torch head (*2*). An oscillator (*3*) secures weaving motion across the feed path of the torch; the frequency (up to 300 cycles per minute) and the amplitude (up to 2 inches) are adjustable. The welding process is fully mechanized, and the head with its ancillary equipment moves at a controlled speed along the side beam (*8*). The thickness of the overlay and its width per pass are controlled by the preset amount of powder feed, rate of advance and amplitude of torch oscillation.

The control panel (*7*) houses—in addition to the remote-control circuits for powder feed, advance and oscillation amplitude—three flow meters measuring the stabilizing gas for the non-transferred arc, the carrier gas conveying the powder to the head, and the shielding gas fed through the porous lens onto the molten-metal bath. The way of controlling the current sources is similar to the system used in the cutting torch, and the sources are mutually independent. Ordinary D.C. sets can also be used for the purpose. Source I supplies a controllable current of approximately 75 A, and a current of about 300 A is drained from source II. The makers also supply a special combined source, type designation C-75-300 (*5* in Fig. 91) for weld surfacing. The arc is initiated by means of the high-frequency generator (*6*). Pressure switch (*4*) ensures the minimum water pressure required in the cooling circuit and cuts out the whole set if the pressure in the piping falls below the permissible minimum level [Ref. 2.4.3].

Plasma equipment manufactured by Thermal Dynamics Corporation includes both cutting torches and apparatus for special research purposes.

The U-50 plasma torch, used in the USA as well as in western Europe, is primarily designed for metal cutting with a nitrogen-stabilized arc. The amount of hydrogen added to improve the cutting properties depends on the kind and thickness of the material to be cut (cf. Div. 3.1). However, the torch can also be operated on a mixture of argon and hydrogen. The arc is stabilized by a gas vortex. After exchanging the

Fig. 91 Linde fusion-surfacing set (By courtesy of Linde Company N.Y).

nozzle and the feeding hoses (cf. Fig. 92) which also conduct the current, the torch may also be used for metallizing work. In operation with a transferred arc (i.e. in cutting) the U-50 torch has a maximum output of 200 kW (1000 A, 200 V on the arc), and with a non-transferred arc the output is 100 kW [Ref. 2.4.4]. Fig. 92 shows the head while cutting stainless steel 25 mm thick.

The H-30 plasma torch for manual cutting works with an output of about 50 kW. The torch head is modified for arc stabilization with

Fig. 92 Thermal Dynamics plasma torch type U-50 (By courtesy of Thermal Dynamics Corp.)

pressure air. On structural steels, air speeds up cutting and yields a kerf with good quality edges (cf. Div. 3.1) [Ref. 2.4.4].

The F-80, another Thermal Dynamics plasma torch, operates with a non-transferred arc and its output is up to 500 kW (Fig. 93).

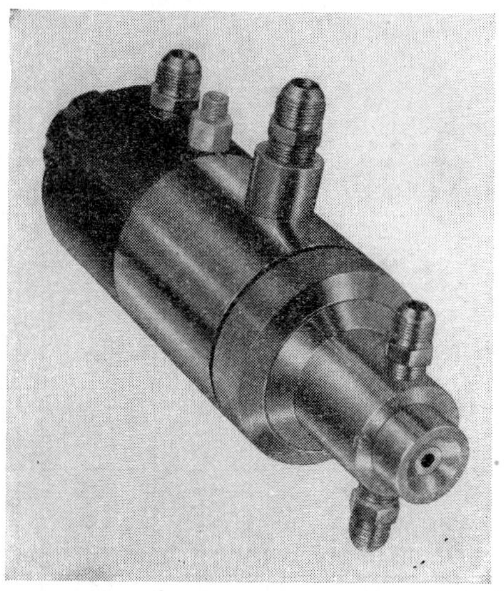

Fig. 93 Thermal Dynamics plasma torch type F-80 (By courtesy of Thermal Dynamics Corp.)

This apparatus is probably used for laboratory purposes only, for no information is available of such plasma torches being installed for industrial exploitation, so far.

The F-5,000 plasma torch with an output of approximately 5,000 kW is also intended for research applications. The arc is stabilized by gas under a pressure of 100 atm. This apparatus is the principal piece of equipment of special high-temperature supersonic air tunnels.

Giannini Scientific Corporation command a leading position in research and development of plasma torches with gas-vortex-stabilized arc. Their sister company, Plasmadyne Corporation, manufactures these torches which are intended for various applications.

Most popular are the plasma torches of the S-range, designed for coating practice. Various exchangeable nozzles supplied as accessories permit fuse-spraying with materials whose melting points lie in the range between 260 and 3600°C. The mean temperature of gas and plasma at the nozzle mouth can be varied between 550 and 20,000°C. The S-range consists of three types of torches whose characteristic data are tabulated below [Ref. 2.4.5].

Table 2.4.1

Type	SG 1	SG 2	SG 3
Output [kW]	4—40	5—80	2—25
Amount of stabilizing gas [m^3 $hr.^{-1}$]	0·425 to 2·265	0·850 to 2·831	0·425 to 1·700
Weight (without supply hoses) [kg]	1·6	2·0	0·454
Weight of deposit (tungsten surfacing) [kg $hr.^{-1}$]	2.7 to 3·6	9·0	1·8 to 2·2
Input required for this deposit [kW]	40	60	25

The design principles of the S torches are shown in the diagrammatic cross-section Fig. 94. The conical cathode (*1*), made of thoriated tungsten, is held in a copper housing (*2*). Intensive cooling makes for long electrode life: the makers guarantee 200 hours in operation with argon. The cathode reaches far into the hollow cone of the copper anode (*3*) whose taper ultimately passes into a cylindrical nozzle. The vortex swirl of the gas distributes the active anode region over the entire circumference of the nozzle. The powder supply duct (*4*) traverses the handle and is then inclined through 4° from the perpendicular to the torch axis; hence the powder is injected against the direction of the plasma stream. Cooling water and current is supplied through hoses with a coaxial lead [connection inlet and outlet at (*6*) and (*5*), respectively]. The cooling water passes through an aperture in the plastic insulator (*7*). The gas supply tube (*8*) opens tangentially into the working chamber. The arc is initited through a capacitive spark discharge from a pilot electrode (*9*). The plastic cover (*10*) tapers into a handle.

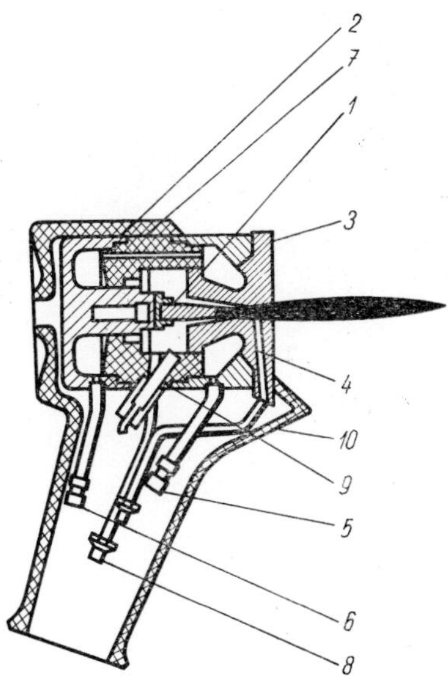

Fig. 94 Plasmadyne SG-1 torch head (By courtesy of Plasmadyne Corp., Santa Anna, Calif.)

In addition to the interchangeable nozzles for various materials, accessories also include nozzles with two openings for the supply of powders; they increase the efficiency of the coating process by up to 20% and permit two materials to be sprayed at a time with independent control of their amounts. Another accessory is a mixing chamber to be mounted on the front nozzle, with one feeding hole (SM 776) or three of them (SM 780) for additional gas or powder materials.

Fig. 95 shows a schematic diagram of a torch modified for fusion surfacing. Unlike in the Linde overlaying torch, the powder is injected into the plasma beam through a single opening below the nozzle mouth, which includes an angle of about 45° with the centre line of the nozzle.—For the S-type plasma torches, the makers also supply an attachment for feeding the overlay metal in form of rods 1/4 or 3/16 in.

in diameter; the surfacing process is carried out by the method schematically indicated in Fig. 64.

S-type plasma torches for manual and mechanical cutting are also used in a modification with non-transferred arc. Both the cutting

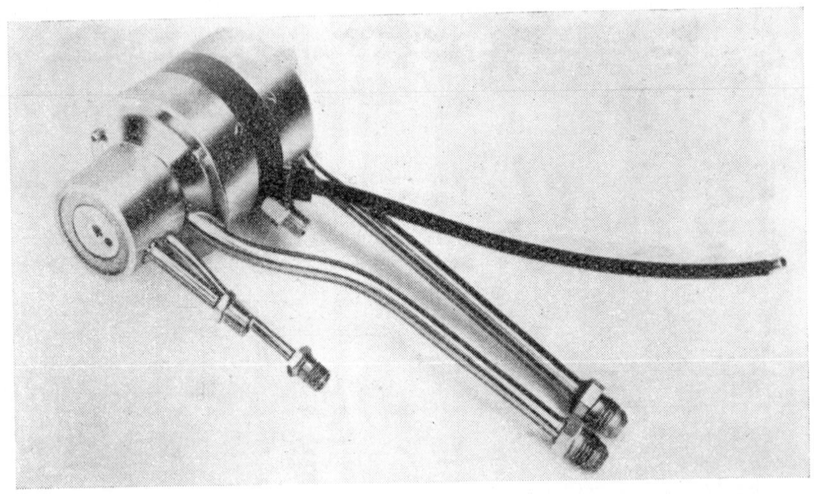

Fig. 95 Plasmadyne torch modified for fusion surfacing. (By courtesy of Plasmadyne Corp.)

speed and efficiency are far lower in operation with a non-transferred arc. Such torches are therefore better suited for grooving or chamfering of edges before welding, and in particular for cutting high-melting non-conductive materials.

Plasmadyne Corporation are also makers of current sources for plasma torches. Their PS-42 and PS-62 types are three-phase transformers in conjunction with three-phase fullwave bridge rectifiers. The former has a continuous power of 14 kW with a peak output of 20 kW, the latter a continuous rating of 40 kW, and 60 kW peak output. Control panels are supplied in either a stationary or a mobile type. They do not materially differ from other makes. Fig. 96 shows a complete SG-1 equipment (40 kW) consisting of a powder-fed hand-operated plasma torch (*1*), a mobile control box (*2*) (with independent regulation of both power sources), powder feeder (*3*) and two PS 42 power sources (*4*) and (*5*).

The P range of Plasmadyne torches, equipped with a variety of accessories, is mainly intended for research work. The M-4 type has

Fig. 96 Plasmadyne SG-1 set (40 kW) (By courtesy of Plasmadyne Corp.)

a maximum output of 120 kW, and the M-8 type of 300 kW. Plasmadyne Corporation are also makers of complete equipments for hyperthermal tunnels with an arc output of 1,000 kW [Ref. 2.4.6].

The METCO firm, producing all kinds of metallizers, also manufacture single-purpose plasma torches intended for coating. The first plasma torches marketed by METCO were of a design similar to that of the F-80 torch (Fig. 93), though with a smaller output

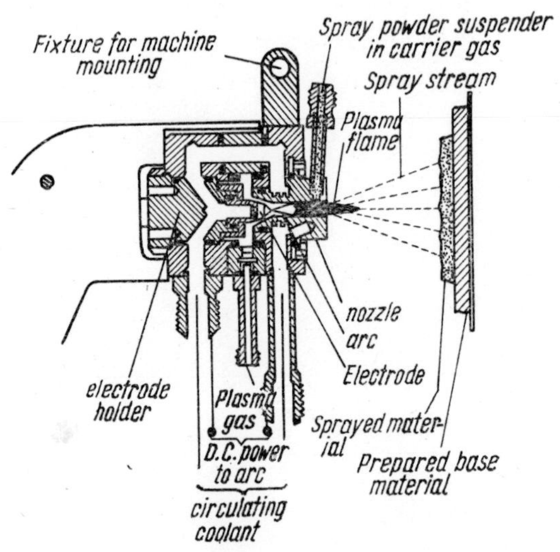

Fig. 97 METCO MB-2 spray-coating plasma torch

(25 – 40 kW). The nozzle, somewhat shorter, had no independent supply of cooling water. The torch was operated on nitrogen mixed with a maximum of 10 % hydrogen. The 2 MB type (cf. Fig. 97), a more recent design, is lighter in weight and resembles the Plasmadyne concept. As in other types described above, other gases or their combinations can also be used for stabilization [Ref. 2.4.7].

The AVCO Corporation (Research and Advanced Development Division) participate in the R & D programme of the US Air Force and have developed methods for the thermal protection of ballistic projectiles (Titan and Minuteman) on their re-entry into the atmosphere. Work on such problems required the construction of thermal sources with high-intensity heat flux and high-velocity flow of hot gases. Some of these sources were also modified for industrial utilization and are being produced on a commercial scale. The PG-030-1 plasma torch is used to apply coatings to test materials and serve as a high-intensity source for spectroscopic purposes. The arc, axially stabilized with argon, works at a low voltage (20 to 35 V) and a current ranging from 75 to 800 A, yielding an output of 2 to 30 kW.

The new PG-100 plasma torch, modified for spray coating, is notable for its neat lightweight design. The variable gap between the cathode and the nozzle-anode facilitates the adjustment for various stabilizing gases without exchanging nozzles. In the same manner, the thermal conditions in the plasma can be modified to suit various coating materials.

2.4.2 Plasma Torches Made in the USSR

Several Soviet research institutes have studied and developed plasma torches. The A.A. Baikov Institute of Metallurgy (IMET) has developed low-output torches with an argon-stabilized arc; the Soviet National Research Institute for Autogenous Metal Treatment (VNIIAVTOGEN) has studied metal cutting with nitrogen and argon plus hydrogen; and the Soviet National Research Institute for Welding Equipment (VNIIESO) has investigated metal cutting with both pure hydrogen plasma and argon-hydrogen mixtures.

IMET has developed three types of torches: IMET 104 and 105 were designed for research purposes, and IMET 106 was also produced commercially in modifications for hand- and machine cutting. This axially stabilized torch can be operated with either a non-transferred or a transferred arc, and the maximum output is 15 kW. A special device is used to initiate the arc. The cathode – an internal thoriated tungsten electrode – is movable. When an insulated button on the top of the head is pushed, the cathode tip touches the anode. The push-button is released, a spring returns the electrode into its previous position, and an arc is struck between its tip and the anode. The nozzle is screwed into the head and indirectly cooled with water [Ref. 2.4.8].

VNIIAVTOGEN has developed the UDR-58 cutting torch which works with a transferred arc; it is manufactured in separate modifications for operation by hand and by machine. The control box for automatic operation is small enough to be mounted on the carriage for mechanical cutting, together with the cutting head [Ref. 2.4.9]. A more recent type, the EDR-1-60, can also be mounted on the shape-cutting sets SGU-1-58 or SGU-1-60 manufactured by the Avtogenmash Works of Odessa [Ref. 2.4.10]. The arc is axially stabilized by nitrogen or argon, both either pure or mixed with hydrogen. The thoriated tungsten electrode is 3 mm in diameter, and the maximum operating current is 400 A at 70 to 80 V. Argon mixed with 20 to 35% hydrogen is recommended for cutting aluminium, and nitrogen—pure or with up to 50% hydrogen—for stainless steel up to 25 mm or thicker, respectively.

Gas consumption is 24 to 30 litres of argon, 8 to 13 litres of hydrogen, or 30 to 150 litres of nitrogen per minute. The nozzle diameter is 3 mm for currents up to 250 A, and 4 mm up to 400 A. The gauge pressure should not be higher than 3 atm for argon and 5 atm for nitrogen. The control panel has no flow meter, and the amount of gas is set to the gauge on the cylinder pressure regulator. Recent modifications for manual cutting are designated UDR-61 and RDM-1-60.

The arc is initiated either by means of an oscillator or by a contact method: a thin pointed starting electrode made of thoriated tungsten is introduced through the nozzle until it touches the inner electrode; as the pilot electrode is withdrawn, an arc is struck between the inner electrode and the nozzle. A PS-500 welding set, or two of them connected in series, is used as current source. Cooling water should be supplied at the flow rate of 1·5 to 2 l/min, and no particular demands are made on its quality.

The UPR-1 cutting torch developed at the VNIIESO operates on argon mixed with a higher percentage of hydrogen resulting in a higher arc voltage—250 to 300 V. The rated output is 75 kW and gas consumption 0·6 m^3 of argon and 2·4 m^3 of hydrogen per hour. Since the relatively high arc voltage cannot be supplied by a welding set, a special D.C. source, using ignitrons, has been developed for this torch [Ref. 2.4.11]. The tungsten electrode is reported to be consumed at the rate of 0·05 g/A hr. The torch is designed for cutting thick materials.

2.4.3 The State of Development Work in the GDR

Industrial plasma apparatus has been developed at the Forschungsinstitut Manfred von Ardenne in Dresden and the Central Institute for Welding (ZIS) in Halle. The physical problems involved are being studied at the Institute of Physical Technology of the German Academy of Sciences (DAW), at its Berlin department for radiation sources, as well as in the Meiningen branch for arc research.

A cutting torch with a transferred arc of 100 kW maximum output has been developed at the Ardenne Institute, and since 1964 Kjellberg of Finsterwalde manufactures it on a commercial scale. A special feature of the clever design (cf. Fig. 98) is the axial combination of hoses and cables connected to the top of the head and supplying the control and work currents, cooling water and gas. A minute magnetic gas valve is located inside the torch.

The rod-type thoriated tungsten cathode is movable in the axial direction. The nozzle – 4·5 mm in diameter – opens out into a conical electrode chamber; it is directly cooled with water. A well-designed low-input ignition, the good aerodynamic properties of the electrode chamber and the nozzle, as well as the intensive direct cooling of the nozzle walls result in a long service life of the nozzle. The arc is stabilized by a gas vortex. Fed by a special rotary convertor manufactured by Kjellberg, the plasma torch has an output of 100 kW. The gas is supplied into the cathode space and flows on into the arc chamber through a ceramic insert with a spiral groove, which sets it into a whirling motion. The ceramic insert simultaneously protects the plastic insulation of the cathode from heat radiated by the arc. For higher outputs and thicker material to be cut, the arc is stabilized by a mixture of argon with 35% of hydrogen, the feed rate depending on the arc output; a flow rate of 4·6 to 6·2 $m^3hr.^{-1}$ is recommended for outputs ranging between 75 and 110 kW. Remarkable results achieved in cutting metal with this head, as well as high reliability in operation and clean cuts are reported in Ref. [2.4.13].

An improved design, derived from the 100 kW torch, is a smaller head with 18 kW maximum output. It is notable for its very narrow beam, obtained from a nozzle with an orifice as small as 1·1 mm in diameter. The cylindrical portion of the nozzle is only 2·5 mm long and opens out conically (as in the design shown in Fig. 98) with an apex angle of 52°. A conical screwed cap with a 90° apex angle presses upon the tip of the conical nozzle, when drawn in. This design greatly

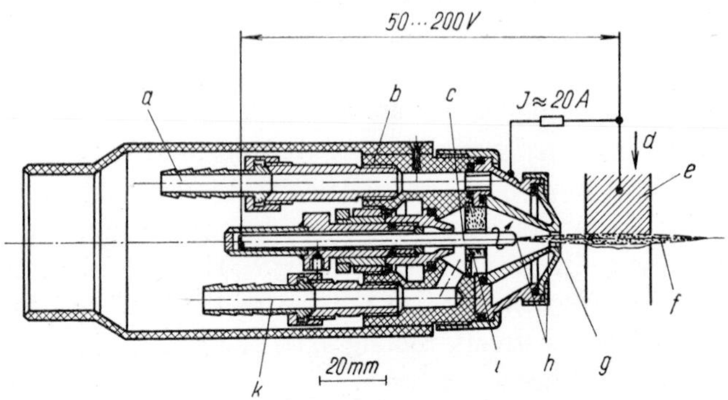

Fig. 98 Plasma cutting torch for 100 kW maximum output developed at Forschungsinstitut Manfred von Ardenne, Dresden (By courtesy of v. Ardenne Institute)

simplifies the nozzle, and direct cooling with water results in reliable operation and long service life.

The current is supplied from a transductor-controlled selenium rectifier. The no-load voltage is 350 V, and in operation 170 to 180 V at a current intensity of 100 to 110 A. The maximum width of the kerf is 3 mm, and the quality of the cut edges and surfaces is excellent. Owing to the narrow kerf, i.e. the small amount of fused material, the cutting speed is fairly high; up to 15 mm thickness, the maximum speed is only about 10% smaller than that of medium-output plasma torches. As the material becomes thicker, the difference in speed increases; at the maximum thickness—which is 30 to 35 mm—the difference in cutting speed is already 20 to 25%.

This is a very suitable tool for engineering plants where no thick materials have to be cut: the initial cost is low, the current source inexpensive, and the operation very economical; the consumption of argon and hydrogen is only 0·7 and 0·5 m^3 per hour, respectively.

The design is similar to that of the preceding type, with hoses and cables entering axially into the head. Fig. 99a shows the assembled head; in Fig. 99b (with the cover removed) the miniature magnetic valve (*1*) and the mixing nozzle (*2*), working on the injector principle, can clearly be seen.

The Ardenne Institute also works on high-pressure plasma torches with 20 kW maximum output and an extremely narrow kerf. Tungsten sheet 0·1 mm thick is cut with a kerf 100 μm wide, the non-transferred arc having an output of 500 W at 10 A current intensity,

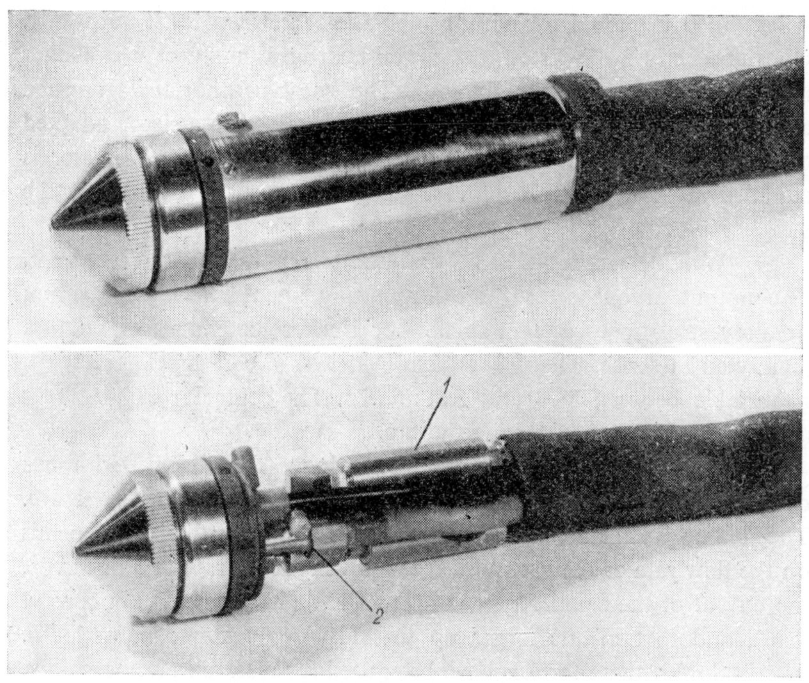

Fig. 99 Narrow-beam plasma torch (By courtesy of M. v. Ardenne Forschungsinstitut)

$U = 50$ V. The argon pressure in the head is 40 atm, and the nozzle diameter 200 μm.

A torch of about 400 kW output is in the development stage; no data referring to applications have been released, so far.

The Central Institute for Welding (Zentralinstitut für Schweisstechnik) in cooperation with the Leuna Works and Institut für Leichtbau of Dresden have developed a cutting set for minor outputs. The hand-operated version, the ZIS-224 torch, is suitable for cutting, grooving or argon-shielded welding.

The machine-cutting torch head, ZIS Leuna 247, is designed for transferred arc with up to 50 kW output. The arc is stabilized with an argon-hydrogen mixture (approximately 65 + 35%) for cutting aluminium or copper, and with an argon-nitrogen mixture (the proportion being 40/60 to 35/65) for cutting high-alloy steels up to 25 mm thick. On thicker high-alloy steel the argon-hydrogen mixture gives better results. The arc is axially blown.—The rod-type thoriated tungsten electrode, either 3 or 4 mm in diameter, is mounted in a sleeve adapted to the diameter used. The maximum current load is 500 A. Water is supplied at the rate of 3 l min.$^{-1}$ at a gauge pressure of 3 atm in the supply piping.

Welding sets KW 510 VC are used to supply the electric current. The output of one set suffices for cutting aluminium up to 30 mm or alloy steel up to 25 mm thick. For thicker materials two sets are connected in series. The unit is equipped with a carriage whose speed is variable between 0·08 and 5 m min.$^{-1}$. The control panel includes an H.F. generator for starting the arc.

The Institute of Physical Technology (DAW) in Berlin has developed two types of plasma torches with a non-transferred arc. They feature axial stabilization with argon. The electrical output depends on the flow rate of the gas. With argon flowing at the rate of 3 m^3 hr.$^{-1}$, the output of the smaller type is up to 6 kW (300 A maximum current load), and that of the larger up to 20 kW (600 A maximum load).

So far, these torches have been used for research purposes only, except for a single practical application: the production of ceramic globules, less than 0·5 mm in diameter. In this process, the end of a ceramic rod is heated to melting point, and the plasma stream blows the thin molten layer off the heated tip. As the droplets fly, surface tension forms them to a spherical shape.

From the types developed in Berlin, the Meiningen Branch of the Institute has derived the HBI-6 torch, adapted for spray coating with metals or oxides. The arc is stabilized by a gas vortex. As a rule, it operates on nitrogen mixed with hydrogen, though an argon-hydrogen mixture can also be used. The nitrogen consumption is reported to be 2·8 m^3 $hr.^{-1}$ at an output of 20 kW; the maximum current load is 350 A. The cathode, with indirect water cooling, has a diameter of 5 mm.

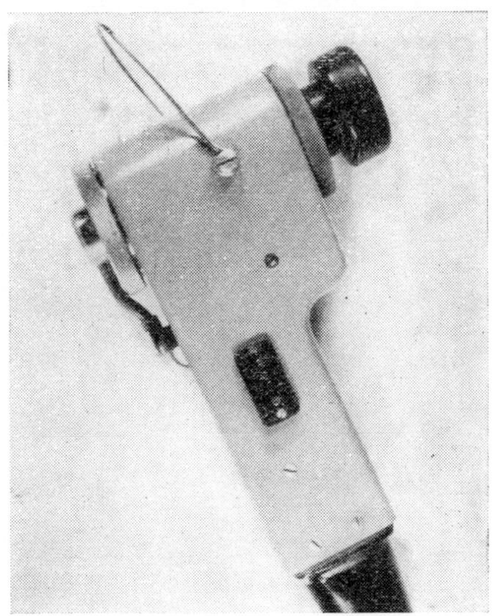

Fig. 100 Spray-coating torch HNI-6 (By courtesy of DAW, Meiningen)

Fig. 100 shows this small, lightweight, yet fairly efficient fuse-spray gun, weighing no more than 1·27 kg (excluding supplies). So far, this torch has mainly been used for studying the effect of various parameters on the quality of the coating; production on an industrial scale is being prepared [Ref. 2.4.14].

2.4.4 Development Work in Poland

In Poland, the Gdansk Institute for Welding Machines, Polish Academy of Sciences, were the first to study plasma torches. The output of their present PT-5 head ranges between 5 and 20 kW, with 350 A maximum current load. The arc is stabilized by a vortex of argon or nitrogen or of a mixture of both gases.

The Nuclear Research Institute (IBJ) of Swierk has developed the RC-4 torch with axial argon stabilization, operating with a maximum output of 22 kW at 600 A. The stabilizing nozzle is 6 mm in diameter and the argon consumption 3·5 $m^3hr.^{-1}$. This Institute has also developed a metal cutting torch operating with transferred arc and having a maximum output of 100 kW. The nozzle diameter is 4 mm, and the arc is stabilized by a vortex. Technical-grade nitrogen without any additions is used up to the maximum thickness (100 mm), the flow rate being up to 10 $m^3hr.^{-1}$. The quantity production of both these torches has been started.

References

2.1.1 GUZOV, S. G.: Materialni i teplovoy balans kyslorodnoy razdelitelnoy rezki (Material and Heat Balance of Oxygen Cutting). *Trudy VNIAVTOGEN*, V Mashgiz, 1959, p. 115

2.1.2 CONN, W. M.: Die technische Physik der Lichtbogenschweissung (Technological Physics of Arc Welding) Berlin, Springer Verlag, 1959

2.1.3 PENNER, S. S.—DUCARMA, J.: Chemistry of Propellants, Oxford, Pergamon Press 1960

2.1.4 Conn, W. M.: *Z. f. Angew. Physik* **6**, 1954, p. 284

2.1.5 FINKELNBURG, W.: Hochstromkohlebogen (High-Current Carbon Arc) Berlin, Springer Verlag, 1948

2.1.6 LOZINSKI, M. G.: Promyshlennoye primenenie induktsionnogo nagreva (Industrial Application of Induction Heating) AN SSSR, Moscow 1958

2.1.7 FINKELNBURG, W.: *Z. f. Angew. Physik* VI (1954) **5**, p. 239

2.1.8 POGODIN, N. N.—ALEXEYEV, A. V.: Theorie der Schweissprozesse (Theory of Welding Processes) Berlin, Verlag Technik, 1953

2.1.9 KUHN, W. E.: Arcs in Inert Atmospheres and Vacuum. New York, John Wiley and Sons, 1956

2.1.10 PAULING, H.: Elektrische Luftverbrennung (Electrical Air Combustion) Halle, Wilh. Knapp Verlag, 1929

2.2.1 GERDIEN, H.—LOTZ, A.: *Wiss. Veröff. a. d. Siemens Werken*, Bd. II 1922, p. 489—496

2.2.2 LOTZ, A.: *Wiss. Veröff a. d. Siemens Werken*, Bd. XIV, p. 25

2.2.3 Austrian patent specification 119, 328 (1926)

2.2.4 WIST, E.: Austrian patent specification 129, 236 (1930)

2.2.5 Austrian patent specification 126, 939

2.2.6 British patent specification 371, 814 (1929)

2.2.7 German patent specification 685, 455 (1934)

2.2.8 RAVA, A.: U.S. patent specification 2, 011, 873 (1935)

2.2.9 RAVA, A.: U.S. patent specification 2.052,796 (1936)

2.2.10 MAECKER, H.: *Z. f. Phys.* **129** (1951) p. 108

2.2.11 WEISS, R.: *Z. f. Phys.* **138** (1954) p. 170—182

2.2.12 Bonin, J. H.—Price, C. F.: Conference on extremely High Temperatures, Boston 1958, p. 237
2.2.13 Dickermann, P. J.: Conference on Extremely High Temperatures. Boston 1958, p. 77
2.2.14 John, R. et al.: *Voprosi raketnoy tekhniki* **8** (1960) p. 19
2.2.15 Peters, Th.: *Naturwiss.* Bd. **41** (1954) pp. 571—572
2.2.16 Jurczyk, K.: German patent specification No. 954,816
2.2.17 Rother, W.: *Schweisstechnik* **2** (1961) pp. 77—78
2.2.18 Osborn, A. B.: *J. of. Sci. Instr.* **7** (1959) pp. 317—319
2.2.19 Morton, H. S.: U.S. patent specification No. 2,906,858 (1959)
2.2.20 Briggs, R. L.: U.S. patent specification No. 2,769,079 (1956)
2.2.21 Briggs, R. L.: U.S. patent specification No. 2,770,708 (1936)
2.2.22 Gage, R. M.: U.S. patent specification No. 2,806,124 (1957)
2.2.23 Gage, R. M.: U.S. patent specification No. 2,858,411 (1958)
2.2.24 Kane I. S.—Hill, C. W.: U.S. patent specification No. 2,858,411 (1958)
2.2.25 U.S. patent specification No. 2,062,099 (1958)
2.2.26 Gage, R. M.: U.S. patent specification No. 2,868,950 (1959)
2.2.27 Reed, T.: U.S. patent specification No. 2,874,265 (1959)
2.2.28 Oyler, G.: U.S. patent specification No. 2,890,322 (1959)
2.2.29 Anderson, J. E.—Reed, T.: U.S. patent specification No. 2,951,143 (1960)
2.2.30 Giannini, G.—Ducati, A. C.: U.S. patent specification No. 2,941,063 (1960)
2.2.31 Ducati, A.—Blackmann, V. H.: U.S. patent specification No. 2,941,063 (1960)
2.2.32 Giannini, G. M.—Ducati, A. C.: U.S. patent specification No. 2,929,952 (1960)
2.2.33 Giannini, G.—Blackmann, V. H.: U.S. patent specification No. 2,944,140 (1960)
2.2.34 Blackmann, V. H.: U.S. patent specification
2.2.35 Kugler, T.—Miklossy, K.: Czechoslovak patent specification No. 3,097.292 (1963)
2.2.36 Miklossy, K.: ČSSR patent application No. 4627-61
2.2.37 Miklossy, K.: ČSSR patent application No. 7084-63
2.2.38 Miklossy, K.: ČSSR patent application No. 7083-63
2.2.39 Ayrton: The Electric Arc. London, The Electrician, 1902
2.2.40 Kaptsov, N. A.: Elektricheskie yavlenia v gazokh i vakuume (Electric Phenomena in Gases and in a Vacuum) Moscow. 1950
2.2.41 Nottingham, W. B.: *Trans. Am. Inst. El. Engrs.* **42** (1923) p. 302
2.2.42 Nottingham, W. B.: *Phys. Review* **28** (1926), p. 764
2.2.43 Simon, Th.: *Phys. Zeitschr.* **6** (1905) p. 297
2.2.44 Browne, E. T.: *J. of Electrochem. Soc.* **102**, No. 1, p. 27
2.2.45 Finkelnburg, W.—Maecker, H.: Elektrische Bögen und thermisches Plasma (Electric Arcs and Thermal Plasma) *Handbuch d. Physik*, Bd. XXII, Berlin, Springer Verlag, 1956

2.2.46 WEIZEL, W.—ROMPE, R.—SCHÖN, M.: *Zeitschr. f. Phys.* **119** (1942), p. 366
2.2.47 BAUER, A.: *Zeitschr. f. Phys.* **138** (1954), p. 35
2.2.48 ERA Report GXT/165
2.2.49 KLAUSS, E.: Die Probleme des elektrischen Lichtbogen- und Widerstandsofens (Problems of the Electric-Arc and Resistance Furnaces) Berlin, Springer Verlag, 1951
2.2.50 FOLWER, R. H.—NORDHEIM, L.: *Proc. Roy. Soc.* A **119** (1928), p. 173
2.2.51 SNUCKMAN, A. N.: *Zhurnal tekh. fiz. AN SSSR* **28** (1958) No. 8
2.2.52 MAXFIELD, F. A.—BENEDICT, R. R.: Theory of Gas Conduction and Electronics, New York, McGraw Hill, 1941
2.2.53 COBINE, J. D.—GALLAGHER, C. J.: *AIEE Transac.* **70** (1951), p. 804
2.2.54 GOLDMAN, K.: *Brit. Weldg. J.* (1963), No. 10, p. 516
2.2.55 BAUER, A.—SCHULZ, P.: *Zeit. f. Phys.* **139** (1954), pp. 197—211
2.2.56 MÜLLER, G.—FINKELNBURG, W.: *Zeit. f. Phys.* **144** (1956), p. 244
2.2.57 BUSZ-PEUKERT, G.—FINKELNBURG, W.: *Naturwissenschaft* **10** (1955), p. 294
2.2.58 COBINE, J. D.: Gaseous Conductors New York, Mc. Graw Hill, 1941
2.2.59 HÖCKER, K. H.—BEZ, W.: *Zeit. f. Naturforschg.* **9a** (1954) pp. 72—81, *Zeit. f. Naturforschg.* **10a** (1955) pp. 706—714
2.2.60 MÜLLER—FINKELNBURG, *W. Naturwiss.* **10** (1955), p. 294
2.2.61 BUSZ-PEUKERT, G.—FINKELNBURG, W.: *Zeit. f. Phys.* **140** (1955), p. 540
2.2.62 BUSZ, G.—FINKELNBURG, W.: *Zeit. f. Phys.* **139** (1954), p. 212
2.2.63 ENGEL, A.—STEENBECK, M.: Elektrische Gasentladungen, ihre Physik und Technik (Electric Discharges in Gas, their Physics and Technology) Berlin, 1932
2.2.64 STEENBECK, J. M.: *Phys.* **7, 33** (1932), p. 809
2.2.65 KIRCHSTEIN, B.—KOPPELMANN, F.: Wiss. Veröff. a. d. Siemens Konzern 16/III (1937), p. 56
2.2.66 STEENBECK, M.: Wiss. Veröff. a. d. Siemens Konzern 19 (1940), p. 59
2.2.67 PETERS, T.: *Zeit. f. Phys.* **144** (1956), p. 586
2.2.68 KING, L. A.: *Nature* **174** (1954), p. 4439
2.2.69 SCHIRMER, H.: *Zeit. f. Phys.* **142** (1955), No. 1, p. 116
2.2.70 IRVING, R. R.: *The Iron Age*, (1963), January 10th
2.2.71 REED, T. B.: *Journ. Appl. Phys.* **32** (1961) p. 821
2.2.72 BROWN, G. H.—BOYLER, C. N.—BIERWIRTH, R. A.: Theory and Application of Radio-Frequency Heating. New York, van Nostrand Inc., 1947
2.2.73 *Revue generale d'électrique* **17**, No. 96, pp. 36—39
2.2.74 REED, T. B.: *Journ. Appl. Phys.* **32** (1961), pp. 2534—2535
2.2.75 KING, A.: Colloquium Spectroscopicum Internationale VI. Amsterdam. London, Pergamon Press, 1956
2.2.76 MAECKER, H.: *Zeit. f. Phys.* **157** (1959), p. 1
2.2.77 MIKLÓSSY, K.: *RIEE Report* No. Z-883
2.2.78 MAECKER, H.: *Zeit. f. Phys.* **129** (1951), p. 360
2.2.79 VARGA, T.: *Techn. Rundschau* **55** (1963), No. 10

2.2.80 SERGEYEV, V. L. et al.: *Inzh. Fiz. Zhurnal* VI. (1963), No. 1

2.2.81 BROWNING, J. A.—KLASSEN, G. A.: *Mech. Engng.*, August 1961, p. 43

2.3.1 KULHÁNEK, L.: *Research Report* No. 60—135, Welding Research Institute, Bratislava

2.3.2 MIKLÓSSY, K.: *RIEE Report* No. Z-891

2.3.3 MIKLÓSSY, K.: *RIEE Report* No. Z-981

2.3.4 MAČÁT, J.: *RIEE Report* No. Z-958

2.3.5 MATEJEC, K.: Candidate's Thesis, Bratislava, Technical University, 1962

2.3.6 MIKLÓSSY, K.: *RIEE Report* No. Z-893

2.3.7 MIKLÓSSY, K.: *Svarochnoye proizvodstvo* (1962), No. 11, p. 23

2.4.1 LINDE: *Welding Journal* **34** (1955), pp. 1097—1098

2.4.2 BROWNING, J. A.: *Welding Journal* (1959), No. 9

2.4.3 Linde prospects No. 51—405; 51—402; 51—406

2.4.4 BROWNING, J. A.: Private communication

2.4.5 Plasmadyne Corp. Prospects S-series plasma systems

2.4.6 Plasmadyne Corp. Prospects P-series plasma systems

2.4.7 Metal-spraying by plasma gun. *Chemical Processing*, June 1962

2.4.8 KULAGIN, I. D.—NIKOLAYEV, A. V.: Obrabotka materialov dugovoy plazmenoy struyey (Machining Materials by an Arc Plasma Jet). Moscow, Imetizdat 1960

2.4.9 Ustanovka UDR 58 dla gazodugovoy rezki metallov (UDR 58 apparatus for Metal Cutting with a Gas-Stabilized Arc) *Technical leaflet* No. 151 *VNIIAVTOGEN* (Soviet National Research Institute for Electric Welding Apparatus)

2.4.10 Gazorezateľnaya mashina SGU-1-58 (SGU-1-58 Gas-Stabilized Cutting Machine) Sovnarkhoz Moskva *Information Leaflet* No. 172]

2.4.11 VASSILIEV, K. V.: Gazoelektricheskaya rezka metallov (Metal Cutting by Gas-Stabilized Electric Arc). Moscow, Mashgiz, 1963

2.4.12 VON ARDENNE, M.: *Die Technik* **18** (1963), pp. 263—274

2.4.13 VON ARDENNE, M. et al.: *Schweisstechnik* **14** (1964), No. 10

2.4.14 HANTZSCHE, H.: *Schweisstechnik* **13** (1963) No. 9, pp. 388—392

Part 3

Technical Application of Plasma

3.1 Metal Cutting with Plasma Torches

3.1.1 Introduction

The idea of using the electric arc to cut metal is relatively old. It was the high cost of electric current which stood in the way of its realization on a large scale until, more recently, electric power became cheap enough to make the arc an attractive metal cutting tool.

Actual arc cutting of metals was preceded by development work aimed at substituting the electric arc for oxy-acetylene in preheating before oxygen cutting. An arc burning between a carbon electrode and the workpiece surface – the two including an angle of 45° – was used to heat the place of the cut. The oxygen jet was directed exactly into the anode spot of the arc. With the carbon kept at least 4 mm off the workpiece, the material was not carburized. Very good cuts were obtained under favourable conditions, but they were not fully reproduceable since the arc was very sensitive to contamination by the material ("stuck" anode spot).

In the arc-atomic method of preheating before oxygen cutting, the material to be cut was heated indirectly. The electric arc – independent of the material to be cut – burned between two tungsten electrodes in a hydrogen atmosphere, and the hydrogen dissociated in the arc rapidly transferred its dissociation heat on contact with the material. Alternating current was supplied by a special transformer (35 to 50 A, 50 to 60 V) with 2·2 to 2·5 kW output. The hydrogen consumption was 1,700 to 2,600 $dm^3hr.^{-1}$. Although the resultant cuts were good, this method did not abolish the dependence on gas supplies and the inconvenience associated with the handling of gas cylinders. Moreover, the operating cost was substantially increased by the high rate at which the tungsten electrode was consumed because of its contamination with iron oxides.

The methods described above made too high demands on the operator to become popular.

An emergency method sometimes used for cutting copper and high-alloy steels was an electric arc burning between a carbon electrode—the cathode—and the material acting as anode. The heat supplied to the material by the anode drop was actually the only energy utilized for the purpose, while the heat liberated on the cathode and in the arc column was lost. The appearance of the fused kerf was poor and the cost for supplementary processing high, particularly in the case of high-alloy steel, where the surface layers were carburized and therefore became very hard.

The appearance of the kerf was improved and the rate of advance in fusion cutting enhanced by axial air blowing. A copper-coated carbon electrode was fixed in the centre line of a tube through which compressed air was supplied at a pressure of 5 to 6 atm. As soon as the arc was ignited, a current relay actuated the solenoid valve which opened the air inlet, and the air stream blasted the molten metal from the kerf. However, no improvement was achieved in the appearance of the kerf, and productivity remained low. Moreover, the carburization of the steel made further operations on the edges of the kerf rather difficult. Nevertheless, this method also spread to grooving jobs, not only on materials which do not lend themselves to oxy-acetylene cutting, but on structural steels too.

Temporarily, oxy-acetylene cutting with pulverized steel or silicon supplied into the cut superseded the electric arc in the cutting of high-alloy steels. Both the quality of the cut and the productivity of labour were greatly improved by this change-over. Still, even then the edges had to be finished mechanically before welding, because iron oxides contaminated the cut surfaces.

The later half of the fifties saw the construction of the first plasma torches with a stabilized electric arc. These torches were suitable both for cutting and for other applications in which a freely burning arc did not prove satisfactory.

The plasma of a stabilized arc issues from the plasma torch

at a high temperature and constricted to a small cross-section. The torch offers the following advantages:

- A stable plasma jet. The stabilizing gas, which blows the arc in the constricted space of the nozzle, ensures a stable direction of the plasma jet and also limits the undesirable travel of the anode spot on the surface of the material. This stabilizing effect, produced even by a relatively small stream of gas, is of prime importance for the quality of the cut surface.
- High concentration of energy in the plasma jet. This characteristic permits a narrow kerf to be produced at a high cutting speed, and consequently the heat-affected zone can be kept exceptionally thin.
- The kinetic energy concentrated in the plasma jet causes the molten metal to be promptly blown out from the kerf; hence the cut is clean and free of dross.

3.1.2 Metal Cutting with Stabilized-Arc Plasma

The basic principle of metal cutting with a stabilized arc is the rapid fusion of the material, down to a depth that equals the thickness of the metal, under the action of a plasma jet directed at the place to be cut. The separating gap is melted by the continuous movement of the torch above the material in the direction of the cut. The quality of the cut is affected by the heat-flow distribution in the kerf; uniform heat supply throughout the thickness of the material produces a cut of excellent quality. The optimum cutting speed (for the given torch characteristics) is dependent upon the uniform distribution of the heat flow at the plasma-to-material interface, which—in turn—is found by trial and error.

According to the way in which the heat energy is conveyed to the workpiece, we distinguish plasma cutting with a transferred and with a non-transferred arc.

In cutting with a non-transferred arc, the process is analogous to gas cutting. The plasma obtains all its heat from the zone between the inner electrode and the nozzle which acts as the anode. The plasma jet flowing at a high speed from the nozzle differs from the combustion products of the gas torch by an axial temperature several times higher—and therefore a higher energy content—largely due to ionization heat (and where polyatomic gas is used, also to dissociation heat). Ionization and dissociation heat is transferred to the material as the particles recombine on its surface. The intensive transfer of ionization and dis-

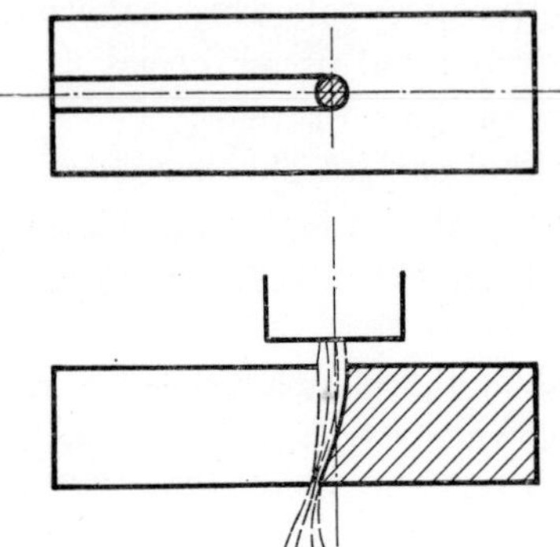

Fig. 101 Contact between plasma and workpiece

sociation energy is accompanied by rapid cooling of the plasma and consequently the transition of the plasma phase into a gaseous phase in the bottom part of the cut. The heat transfer from the gas to the surface of the material is slower. The differences in the heat supply, caused by the differing heat-transfer mechanisms in the top and bottom parts of the cut, are balanced by the bevelling of the contact surfaces towards the centre line of the jet.

The contact between the plasma and the workpiece is schematically represented in Fig. 101. The front half of the jet is cooled by the material, and owing to the high flow velocity of the plasma, the temperatures and heat contents of the front and rear parts of the jet are not immediately equalized. Therefore with increasing depth of penetration of the plasma, the contact surface (between the plasma and the material) is shifted into deeper plasma zones. The resultant slope of the contact face improves the heat transfer from the rear parts of the jet which contain enough energy to melt the metal.

By the cutting speed we control the amount of thermal energy transferred in the top part of the cut, and hence also the energy transferred deeper down in the cut. If the speed is too high, then the upper edge of the contact face reaches too far into the plasma jet and receives an excessive amount of heat. The kerf is thus very wide on top and narrows downward, forming a V-shaped cross-section. Too low cutting speeds also result in an excessive kerf width at the top; however, owing to the excess energy remaining in the plasma, the kerf moreover widens in the direction towards the bottom edge.

The optimum cutting speed is achieved by advancing the torch head at a rate at which the distribution of the heat flow from the plasma into the material is almost uniform throughout the thickness of the material. Under such conditions we obtain perpendicular edges, i.e. the kerf is approximately equally wide at the top and the bottom. Plasma torches with non-transferred arc are usually employed for cutting electrically non-conductive materials, and only exceptionally for metal cutting. Conductive materials are more conveniently cut with a transferred arc, which has a better total efficiency and permits the handling of thicker materials.

The transferred arc burns between the internal electrode (cathode) and the material to be cut (anode). With the arc and the cathode region placed in the cutting gap, the heat conditions greatly differ from those prevailing in cutting a non-transferred arc. The factors affecting the quality of the cut include—in addition to the cutting speed—the characteristic of the current source and the intensity at which the arc is blown by the stabilizing medium.

The kerf in which the arc burns may be considered an extension of the stabilizing nozzle: the arc continues to be stabilized by the rapidly flowing relatively cold gases of the boundary zones. There is no well-defined anode spot, the contact between the arc and the material is diffuse, and there is only a certain active anode region in which most of the current passes into the anode. The location of this region depends upon the cutting speed, the intensity at which the arc is blown, and the static characteristic of the current source. If the source has a soft characteristic i.e. a sufficient voltage reserve, then the active anode region is easily blown into the lower third of the kerf. Joule's heat generated by the current flowing through the arc column compensates for all the losses due to the cooling of the arc by the material, while the heat supplied by the anode drop compensates then for the lower amount of heat energy supplied into the bottom third of the kerf.

Fig. 102 shows the cutting process on thicker materials as described in this paragraph.

The cutting mechanism and the transition of the current into the anode material were investigated by experiment; oscillographic

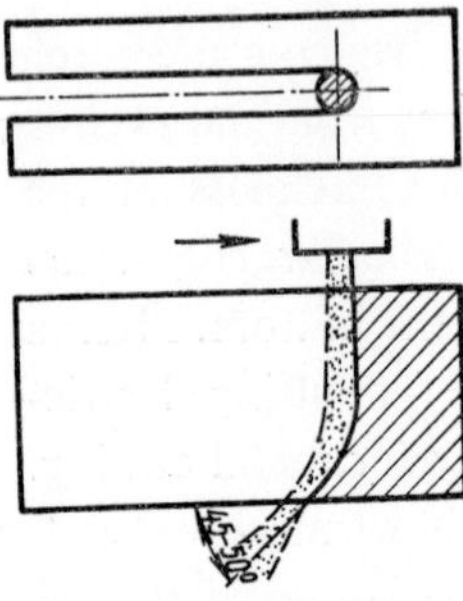

Fig. 102 Cutting process on thicker materials

records were made of the current flowing through a set of three brass plates placed in parallel one upon another. The plates were each 3 mm thick and interlaid with 10 mm thick insulating straps at their edges.

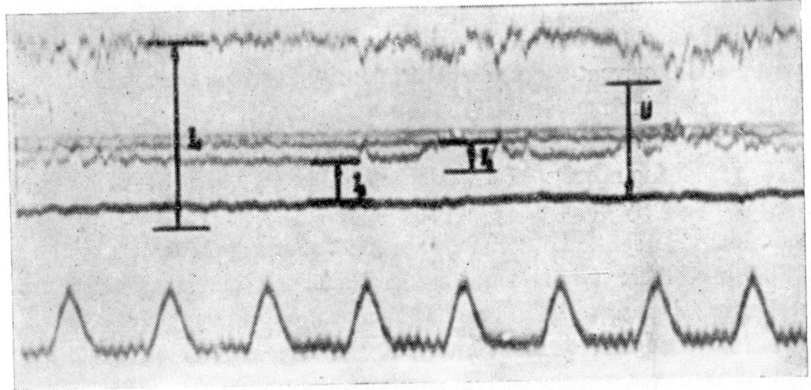

Fig. 103 Oscillogram of current and voltage

One of those oscillograms, recorded at a total current of 310 A and 200 V, is reproduced in Fig. 103. It shows that the cutting current flows simultaneously through all plates. In the set-up described, the current was far higher on the first plate I_1 than on I_2 and I_3; this, however, is understandable because no continuous cutting kerf, coherent throughout the depth of the cut, could be formed. The distance between the plates could not be reduced, since at smaller intervals metal droplets flowing from the cut formed conductive bridges and short-circuited the plates.

When thin materials (about 20 to 25 mm thick) are cut, the high cutting speed shortens the section where the contact surface is perpendicular or slightly inclined (cf. the upper part of the cut shown in Fig. 101), and its slope sets in as early as a few millimetres below the upper edge. Provided optimum cutting speeds are employed, the slope of the contact surface is smaller than on thick metal (Fig. 102) and depends on the kind of material to be cut. On stainless steel the slope angle is 15–20°, on aluminium 25–30°.

Fig. 104 schematically shows the appearance—viewed perpendicularly to the cutting plane—of the edges produced by cutting with

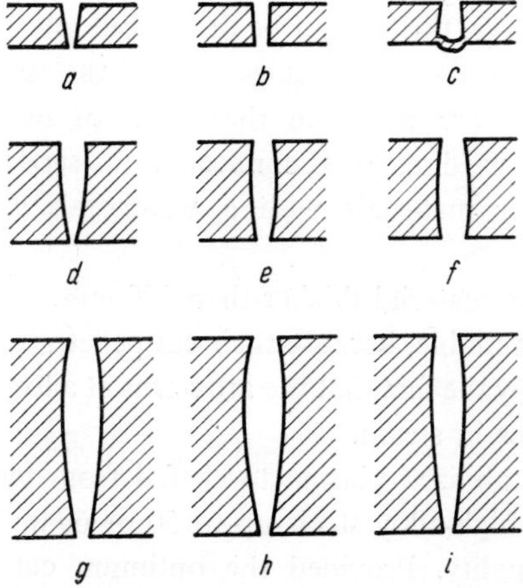

Fig. 104 Appearance of cut edges at various speeds and thicknesses

a water-stabilized plasma torch at various thicknesses and speeds [Ref. 3.1.1].

Excessive speeds produce kerfs with marked bevels (diagram a) because not even in thinner materials can the arc be drawn out to the lower edge of the cut. The same cross-section is produced by a current source with a voltage insufficient to lengthen the arc to the requisite value (a torch operated with a single welding set). If the cutting speed is correct and the current source has a sufficient voltage, kerf surfaces nearly perpendicular and parallel (diagram b) can be obtained on thin materials. A fused lower edge proves that the active anode region has been kept stable in the lowest position, i.e. that the current-carrying plasma is distributed throughout the depth of the cut and thus maintains a uniform heat supply to the material.

Insufficient cutting speed causes excessive heat supply to the region of the lower edge; the material is fused over a greater width than that of the directed plasma jet, and therefore it is not completely blown out from the cut. As the torch passes on, the molten metal joins again and forms bridges (diagram c).

If two welding sets connected in series are used to cut thicker materials, a moderate widening of the kerf in the middle of the cut face cannot be avoided. However, fed from a soft source with a sufficient voltage reserve, the arc can be drawn out into the neighbourhood of the bottom edge. Therefore, unless one of the special 400-V sources is available which are made in the USA for feeding metal-cutting apparatus, three welding sets connected in series should be used for cutting thicker materials with a water-stabilized VS-45 plasma torch.

For cutting material thicker than 100 mm, gas-stabilized plasma torches are more suitable, because their cut penetrates to a geater depth. This is explained by the fact that the arc burns at a lower voltage gradient in gas than it does in steam.

A water-stabilized plasma torch fed from two series-connected welding sets cuts high-alloy steels up to 50 or 60 mm thick with a kerf of satisfactory quality. Provided the optimum cutting speed for the given thickness is employed, the kerf will have the shape indicated in diagrams (a) and (i). On aluminium, the shape is better than on steel. Although in gas-stabilized torches the composition of the gas mixture and its throughput can be varied over a comparatively wide range, the widening of the kerf under the top edge and its constriction at the bottom cannot be avoided; in other words, not even gas-stabilized torches produce a fully uniform heat-flow distribution throughout the face of contact, on thicker materials.

According to literary sources and prospectuses [Refs. 3.1.2, 2.2.88], American PT-2 and F-80 gas-stabilized plasma torches, with inputs as high as 100 – 140 kW, produce relatively perpendicular cuts with slight deviations on materials up to 50 mm thick; on aluminium, uniform edges are also obtained on thicker material.

The nozzle diameter governs both the width of the kerf and the intensity at which the arc is blown, that is the electric-field gradient in the plasma of the nozzle region. If thin sheets are cut, the total length of the arc is small, hence the arc voltage is low. A higher electric output could only be obtained by excessively increasing the current; this, however, would result in an undesirably widened kerf. A smaller nozzle diameter and consequently higher flow rates of gas are therefore used on thin sheet. This raises the electric gradient in the arc, i.e. the total arc voltage. A higher electric output is thus attained without excessively raised current. For thicker materials we increase the nozzle diameter. The outlet velocity of the gas, referred to the nozzle employed (no arc burning), is reduced. The lower potential gradient thus permits the arc in the kerf to be lengthened into the region of the bottom edge without any change in the characteristic of the source.

In the PT-2 and F-80 torches, the gas throughput is far higher than the 25 to 80 m/s which the publications discussing the first types of plasma cutting torches recommended as gas velocity in the nozzle (no arc burning). With the initial small gas throughput, the quality of the cut was unsatisfactory, especially in thinner alloy steels, where the cost for mechanical deburring of the bottom edge had to be added to the cutting costs. The electric output and hence the cutting speed were low. A higher gas throughput was required to raise it. Moreover, the nozzle diameter was also gradually reduced, particularly for cutting thinner materials.

Linde's high-output torches type PT-2 are for 30 to 50 kW output fitted with a nozzle of 1/8 in. (3·18 mm) diameter in its narrowest place. Calculated from the throughput quantity of the argon-hydrogen mixture recommended for cutting alloy steel, the outlet velocity in the nozzle neck amounts to 198 m/s (no arc burning). For cutting aluminium, the calculated outlet velocity amounts to 120 m/s, since the recommended gas throughput is lower. With a 1/4-in. (6·35 mm) nozzle used to cut stock of either of both materials, about 100 mm thick, the calculated outlet velocity drops to 49·5 m/s.

In the F-80 plasma torch of Thermal Dynamics, using a mixture of nitrogen and hydrogen for stabilizing the arc, the nozzle diameter for low output is 3/32 in. (2·38 mm). Both the output and the nozzle diameter are increased for cutting thicker materials; an output of about 200 kW and a nozzle of 3/16 in. (4·76 mm) is required for cutting stock about 100 mm thick. The corresponding gas velocities in the nozzle range between 190 and 103 m/s.

The width of the cut is affected by the cutting speed, the distance ("stand-off") of the torch from the workpiece, and the nozzle diameter. With the optimum set-up, the cut is 1·5 to 2 nozzle diameters wide, and the tolerance equals the nozzle radius.

With the arc column in the nozzle severely constricted and strongly blown with gas, the total arc voltage increases to such a value that plasma torches with small-diameter nozzles require special current sources with high no-load voltages. Linde's PHC-250 source has a no-load voltage of 400 V; for reasons of safety, the use of such sets must be limited to remote-controlled machine cutting. They produce high-quality cuts, that is kerfs with almost perfectly parallel surfaces. On materials of lesser thickness the maximum included kerf angle is 2°.

In water-stabilized plasma torches, the throughput of the stabilizing medium cannot be controlled except by indirect means—by the selected diameter of the water vortex, the length of the vortex channel, and the current flowing through the arc, which affects the heat supplied to the water surface. A longer channel, however, reduces the heating effect produced on the material, as proved by calorimetric measurements [Ref. 3.1.21]. The longer channel requires a higher arc voltage and therefore calls for special current sources. A further disadvantage of the long channel in Maecker's vortex-chamber design (cf. Fig. 45) is a comparatively great mass of water that has to be kept rotating by the tangential introduction of a narrow water jet. Consequently the rotation is decelerated, hence the surface of the water vortex becomes defective, and water sputters from the torch. This causes excessive cooling of the cutting plasma and spots of incomplete cutting.

The ceramic cylindrical nozzle according to Fig. 82d permitted the length of the water chamber to be reduced to 3 mm, while the channel length remained 9 mm, i.e. the surface evaporating water was not altered. This eliminates the danger of an excessive water layer in the mouth of the nozzle and occasional droplets spattered into the plasma. The heat of the arc evaporates almost all the water which penetrates into the channel because of the slower rotation on the surface of the nozzle. Moreover, the difference between the diameters of the front and rear nozzles has been increased to 2 mm, consequently the water from the chamber runs off into the water-collecting chamber (cf. Fig. 85), and the amount of water penetrating into the channel is thus very small. Measurements of the steam quantities generated [Ref. 3.1.21] bear out that the production of steam in a cylindrical nozzle with 9 mm total channel length is fully sufficient (about 3 g as against 1·2 to 1·6 g per second obtained from a channel of equal length in a Maecker-type nozzle). This design enhances reliability in operation and improves the quality of the cut. The optimum quality is obtained with a current input in the range of 380 to 420 A.

3.1.3 Heat Balance of Metal Cutting with Plasma

3.3.1 Energy balance

The total energy balance is based on the input at the terminal of the plasma torch, without considering the efficiency of the power source. Welding generators have a relatively low efficiency – 60 to 68% – whereas silicon rectifiers reach 85 to 88%.

Of the input P [W] supplied to the torch terminals, part of the heat – Q_{z1} [W] – is lost in the conversion of electrical into heat energy. The gas-and-plasma beam issuing from the nozzle of the torch entrains (per unit time) the energy

$$Q_P = P \cdot \eta_z \qquad [W] \tag{3.1.1}$$

where η_z are the measured efficiencies listed in Chap. 2.2.11. We neglect the small losses due to radiation and convection into the surrounding medium before the beam strikes the material.

In addition to the useful heat Q_u required for melting the material in the kerf, a certain amount of heat Q_m must be supplied to cover the losses due to heat conduction by the material, which are caused by the thermal gradient in the unmelted part of the workpiece. The total heat supplied $Q_c = Q_u + Q_m$ has to be computed. The useful heat per unit mass and unit time Q_u can be calculated from the cutting rate v, the width of the kerf d, the thickness of the material cut δ, the specific gravity of the metal γ, the specific heat c, the specific latent heat of liquefaction c' and the temperature difference ΔT between the initial temperature and the melting temperature.

$$Q_u = vd\gamma\delta(\Delta T \,.\, c + c') \qquad [\text{cal} \,.\, \text{g} \,.\, \text{s}^{-1}] \tag{3.1.2}$$

The amount of heat removed by the material – Q_m – is found by the subtraction of the useful heat Q_u from the total heat input Q_c, disregarding the heat radiated from the surface of the material into the surrounding medium.

The heated gases escaping from the kerf carry off energy amounting to

$$Q_{z2} = Q_p - Q_c. \tag{3.1.3}$$

Having ascertained the total amount of heat Q_c required for melting the kerf, we can draw up the energy balance.

3.1.3.2 Calculation of the total heat Q_c required for the fusion cutting of a plate

The plasma jet in the cutting kerf of a plate of thickness δ can be considered a linear source of heat with a constant heat output Q_c uniformly distributed over the length δ along the axis OZ (Fig. 105). This source moves at a constant velocity v in the plane X_0Y_0. For the calculation we assume the dimensions of the cut plate to be infinite in the plane X_0Y_0. The boundary surfaces $z_0 = z = 0$ and $z_0 = z = \delta$ transfer heat to the surrounding medium. Rosenthal [Ref. 3.1.3] and Rykalin

[Ref. 3.1.4] have computed the distribution of the temperature field which is formed when a linear heat source moving at uniform velocity heats a plate. In this calculation the field of the linear source is considered a superposition of the fields of point sources into which the linear source can be resolved.

The principle of superposition can only be used for problems soluble by linear differential equations. In computing the heat conduction,

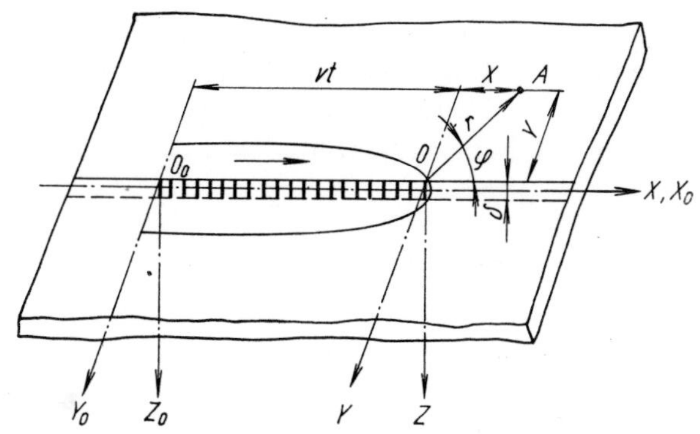

Fig. 105 Linear heat source moving at constant velocity above workpiece

we may use this principle provided the coefficients $\varkappa$, c and γ are independent of temperature. At high temperatures, in the region of the melting point, this condition is not satisfied. We by-pass it by mean values for the entire temperature interval under consideration substituted in the first approximation for the coefficient of thermal conductivity $\varkappa$, the specific heat c and the specific gravity γ of the material. The coefficient of thermal diffusivity $a = \dfrac{\varkappa}{c \cdot \gamma}$ is then likewise independent of temperature. In order to use the principle of superposition, we must, however disregard the latent heat of changes in the structure and aggregation of the material.

If a very intensive source moves at a constant velocity, the temperature field in the plate can in a steady state be considered approximately equal to the field produced by an infinitely long-lasting action of the source ($t \rightarrow \infty$).

Under these assumptions, the temperature field of the linear source in the plate (in a coordinate system moving along with this source) is described by the relation

$$T_{(r,x)} = \frac{Q_c}{2\pi\varkappa\delta} \exp\left(-\frac{vx}{2a}\right) K_0\left(r\sqrt{\frac{v^2}{4a^2} + \frac{b}{a}}\right); \qquad (3.1.4)$$

x is here the distance – projected onto the axis of the motion – between an arbitrary point A in the temperature field and the source; r is the distance of point A from the momentary position of the source O; K_0 is Bessel's function of zero order, second kind.

The quantity b in the argument of Bessel's function K_0 is related to the heat transfer coefficient:

$$b = \frac{2\alpha}{t\gamma\delta}. \qquad (3.1.5)$$

In view of the insignificant value of α we may disregard losses by heat transfer to the medium, and the argument (u) of function K_0 is then

$$u = \frac{vr}{2a}. \qquad (3.1.6)$$

If we disregard the losses to the surrounding medium and express the temperature distribution for unit thickness of the plate ($\delta = 1$), Eq. (3.1.4) assumes the form

$$T_{(r,x)} = \frac{Q_c}{2\pi\varkappa} \exp\left(-\frac{vx}{2a}\right) K_0\left(\frac{vr}{2a}\right). \qquad (3.1.7)$$

For the isotherm T_t of the melting point of the material, Wells [Refs. 3.1.5, 3.1.6, 3.1.7] has solved Eq. (3.1.7) for the unknown Q_c. That isotherm indicates the width of the kerf d in the case of butt-welding and steel cutting by an oxy-acetylene flame.

In solving Eq. (3.1.7), we introduce instead of the variable Q_c the quantity $M = \frac{Q_c}{Q_u}$, i.e. the ratio between the amounts of heat supplied to the material and useful heat consumed in melting the metal in a kerf of width *d*. This relation is the ideal efficiency of the cutting process, disregarding all losses.

$$M = \frac{Q_c}{vd\varrho cT}. \qquad (3.1.8)$$

By expressing Q_c for the isotherm T_t in terms of Eq. (3.1.7) and substituting it into Eq. (3.1.8), we take into account the original thermal distribution in the material outside the kerf. The magnitude of M as a function of $\frac{vd}{4a}$ for $0{\cdot}1 < \frac{vd}{4a} < \infty$ is given by the approximate formula

$$M = 2\left[\frac{1}{5\left(\frac{vd}{4a}\right)}\right] + 1 \qquad (3.1.9)$$

By substituting this value into Eq. (3.1.8) we obtain

$$Q_c = 8\varkappa T\left(\frac{1}{5} + \frac{vd}{4a}\right). \qquad (3.1.10)$$

From Eq. (3.1.9) we can easily derive the efficiency of the kerf-melting process as a function of the fraction $\frac{vd}{va}$. This efficiency—i.e. the ratio $\frac{Q_e}{Q_u}$, total heat input over useful heat—cannot exceed 50%. The value of η_t is given by

$$\eta_t = \frac{5\frac{vd}{4a}}{2\left(1 + 5\frac{vd}{4a}\right)} \qquad (3.1.11)$$

3.1.4 Technology of Metal Cutting by means of Plasma Torches

Plasma torches are mostly designed for machine cutting. Only small types of torches are used for cutting by hand, and this technique is limited to emergencies (repairs, erections) or cutting of risers, where the quality of the cut is of no consequence. For with its head guided by hand, the torch will never travel over the material sufficiently uniformly for a high-quality cut.

3.1.4.1 Mechanical metal-cutting sets

Mechanical cutting sets consist of the following constituent parts:

a) **The Plasma Torch** (for description cf. Divs. 2.3, 2.4).

b) **The Traversing Gear** which propels the torch over the material to be cut at a controlled velocity.

Carriages such as are manufactured for oxy-acetylene cutting or automatic welding are employed for straight cuts. The range of controllable velocities is adjusted to suit the cutting rates of the particular plasma torch. – For the plasma torches manufactured in Czechoslovakia, the best choice is the tractor of the automatic welding set SUM-1000; its speed range covers – without any special adjustments – the full interval of the cutting rates attainable with these torches.

For shape cutting, the plasma torch is attached to an ordinary flame-cutting machine whose traversing velocity has to be increased if the torch has a rated output of more than 30 kW. The sloping contact surface formed in plasma cutting causes trouble when the direction is abruptly changed, i.e. when a shape with a small radius or a sharp corner is cut. Therefore programme-controlled machines have been designed for the purpose, in the USA; before reaching the point of the sudden change in direction, they slow down the motion of the torch and simultaneously reduce the electric input so that the quality of the cut should not suffer [Ref. 3.1.10]. In tube cutting, the workpiece held in the jaws of the cutting machine traverses, and the torch is fixed in a stationary holder. For cutting tube shapes, the holder with the torch also moves; its motion in relation to the tube is axial and coordinated with the rotary motion of the latter. Fig. 106 shows shape-cutting on a tube with a ZIS-Leuna 247 plasma torch.

c) **The Mechanism for Vertical and Horizontal Adjustment** of the torch is fixed to the carriage. It consists of two racks placed at right angles, similarly to those used in oxy-acetylene sets. The plasma torch is fixed to the vertical rack as shown in Fig. 89. This design permits the torch head to be quickly adjusted.

Since the quality of the cut is largely dependent on the interval between the head and the material, the operator must adjust this distance by hand if cutting material with an uneven surface. In order to avoid this necessity, the Vs-45 set is fitted with a tracer mechanism.

Fig. 107 shows the device for vertical and horizontal adjustment together with the tracing mechanism for uneven surfaces and a VS-45 plasma torch fitted to the SUM-1000 carriage.

d) **The Cutting Bench with the Carriage Runway** does not differ form the current designs used for flame cutting. However, an exhaustor has to be installed for permanent operation. Exhaustion from below

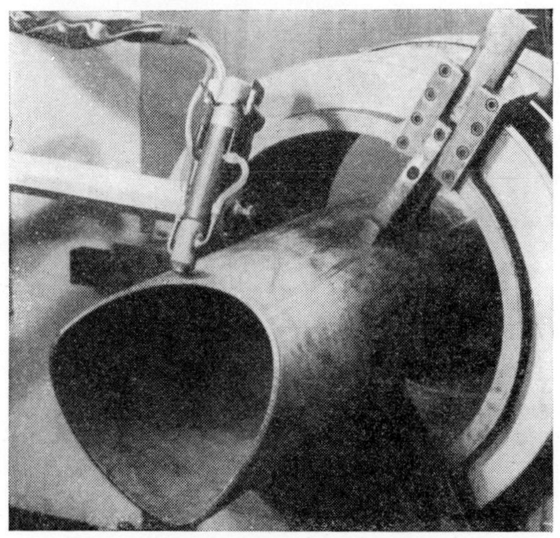

Fig. 106 Shape-cutting on tube with ZIS LEUNA 247 torch (By courtesy of ZIS, Halle)

of metal vapours and dust produced in cutting is the most advantageous solution; as a rule, it takes the form of a sheet-metal trough installed below the cutting bench and connected with a suction fan. Such a cutting bench with exhaustor together with a Thermal-Dynamics set for circular cuts is shown in Fig. 108.

e) **The Water Circulation System.** Gas-stabilized plasma torches with high outputs call for copious cooling with water. As a rule, direct supplies from the tap cannot ensure the requisite gauge pressure (4 to 7 atm). Some types of gas-stabilized plasma torches even require distilled water cooled in a heat exchanger. Water-stabilized torches need a constant water pressure of approximately 3 atm to ensure the requisite flow velocity in the various chambers. A centrifugal pump with a suction strainer is employed to circulate the water from a tank and produce the necessary pressure.

The drain hose of a water-stabilized plasma torch must not be too long (~ 1·3 m), yet the head between the torch and the water tank

Fig. 107 VS-45 plasma torch mounted on SUM-1000 carriage

must be at least 0·5 m. For the satisfactory drainage of water from the torch, the tank should be installed immediately behind the carriage roadway.

f) **The Remote Control Desk.** The process control and check on the parameters of the cutting process is concentrated at the control desk which also contains elements for partly automated operation.

The working cycle for machine cutting with a gas-stabilized plasma torch consists of the following operations:

1) Opening the solenoid valve for argon supply to the torch;

2) switching-in the ionizer which initiates the pilot arc;

3) switching-on the main current circuit, and simultaneously switching-off the ionizer, opening the hydrogen supply and switching-on the carriage motor;

4) if the operation is discontinued and the main current circuit

Fig. 108 Cutting bench with Thermal-Dynamics set (By courtesy of Thermal Dynamics Corp.)

switched off, the previous procedure is reversed: the ionizer initiating the pilot arc is switched on, and the hydrogen supply and carriage traverse are stopped;

5) on discontinuing operation, the solenoid valve controlling the argon supply acts with an automatic delay of a preset length, usually half a minute at most.

In operation 3 and 4, simultaneous switching is ensured by a current relay which acts as soon as the current in the main circuit exceeds a value of 70 to 80 A.

The control panel of a water-stabilized plasma torch operates on the same principle, but instead of the controls for gas valves and

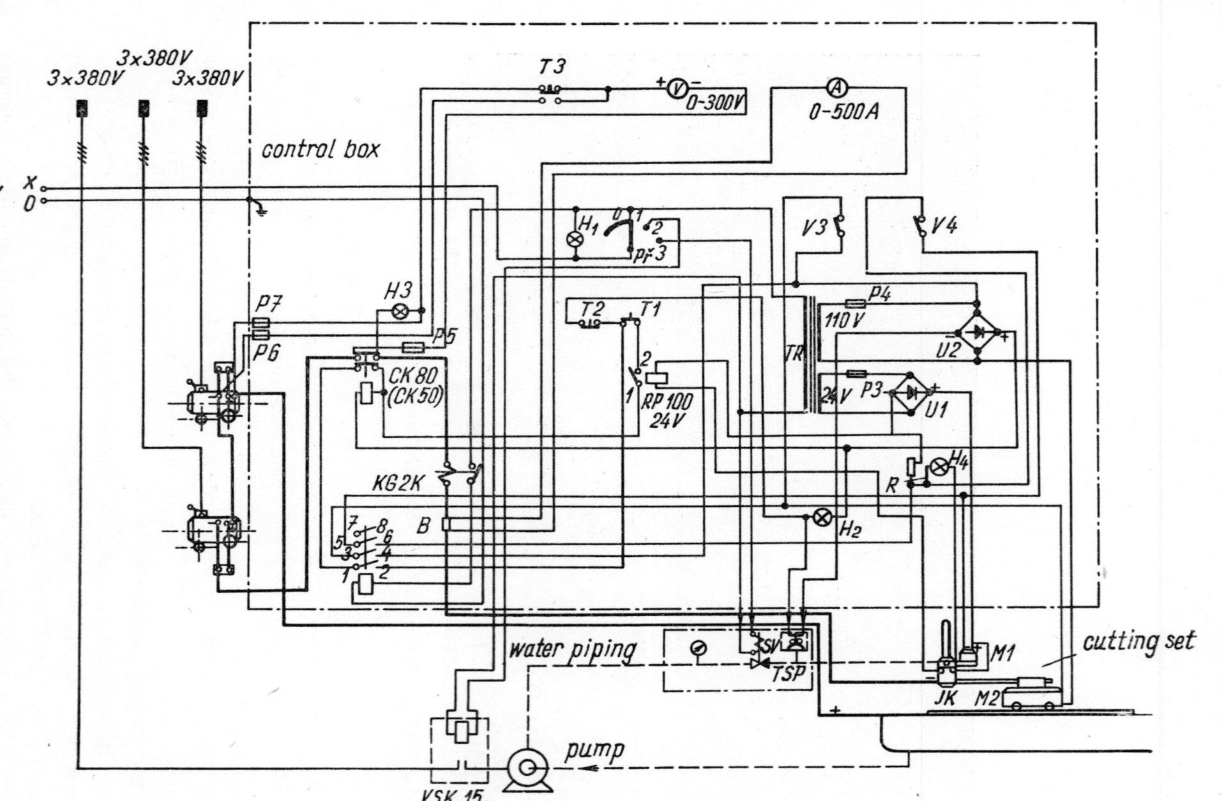

Fig. 109 Wiring diagram of control panel for water-stabilized VS-45 plasma torch

the ionizer there are those for the water valve, the water pump and the electrode feed.

To permit the tractor (and in the water-stabilized torch also the electrode feed) to be started without an arc burning, the control panel is fitted with by-pass switches for the current relay. Fig. 109 shows the wiring diagram for the control panel of a water-stabilized plasma torch.

g) **Accessories.** Apart from water-and gas-supply hoses and connecting cables, the equipment for gas-stabilized plasma cutting includes the appropriate number of gas cylinders with pressure regulators, and flow meters for setting the correct amounts of gas. Sets of spare nozzles for replacement and sometimes special nozzles for various gas mixtures are supplied with the torch.

The accessories of a water-stabilized plasma torch include a device for grinding electrode tips and drilling a hole for the ignition wire.

Table 3.1.1 VUS-ARG-ORS TYPE SET WITH 3 MM NOZZLE DIAMETER FOR PLATES OF ALL THICKNESSES [REF. 3.1.11]

Plate thickness [mm]	Current [A]	Voltage [V]	Flow rate [litres per min.]		Cutting Rate [cm . min^{-1}]
			argon	hydrogen	
Aluminium and its alloys					
10	300	65	13	9	280
25	400	72	17	11	125
30	380	74	18	12	85
40	430	80	24	16	48
50+	450	90	29	19	42·5
65+	480	102	38	25	33·7
105+	415	130	54	36	8·6
Chrome-nickel steel					
10	400	50	17	11	56
20	420	60	20·5	13·5	53
30	440	80	24	16	45
50	520	87	28	18	34
60	540	90	30	20	24

+ Two welding sets

3.1.4.2 Parameters for cutting practice

Tables 3.1.1 to 3.1.4 contain recommendations for nozzle diameters, electric power inputs and rates of gas flow for various types of plasma torches, and also optimum rates of cutting as found by experiment.

Table 3.1.2 VS-45 SET WITH 3.5 MM NOZZLE DIAMETER FOR PLATES OF ALL THICKNESSES [REFS. 3.1.11 AND 3.1.12]

Plate thickness [mm]	Current [A]	Voltage [V]	Cutting rate [cm . min^{-1}]
Aluminium and its alloys			
12	400	120	320
15	400	130	210
20	380	135	166
30	380	145	100
40	360	150	75
50	350	150	45
65*	400	212	51
105*	340	240	16·2
Chrome-nickel steel			
8	400	120	320
12·5	400	120	150
20	400	135	87·5
35	400	140—150	48
60	360	160	20—22
60*	400—420	190—210	30—31
70*	400	210—220	22—23

* Three series-connected welding sets

The parameters of Soviet-made plasma torches and of the ZIS-Leuna 247 are similar to those quoted for the VUS-arg-ORS in Table 3.1.1. The inputs are identical, but the gas flow rate is slightly higher. The values quoted from Ref. [3.1.11] are the lowest flow rates for the kinds and thicknesses of material indicated. However, exaggerated economy does not pay, since it narrows down the range of cutting rates at which cuts free of dross can be obtained. In workshop practice

Table 3.1.3 Type PT-2 Linde "Plasmarc" Torch [Ref. 3.1.2]

Plate thickness [mm]	Nozzle dia. [mm]	Electrode dia. [mm]	Input [kW]	Gas flow rate [litres . min^{-1}]			Cutting rate [cm . m^{-1}]
				35% H_2 65% Ar	N_2	H_2	
Aluminium and its alloys							
25·4	3·175	7·937	55	47·2	—	—	51
5·1	3·968	7·937	65	47·2	—	—	51
76	4·762	7·937	75	94·4	—	—	38·1
102	4·762	7·937	80	94·4	—	—	30·3
127	4·762	7·937	85	94·4	—	—	25·4
51	3·968	7·937	70	—	66·1	28·3	51
76	6·35	12·7	80	—	66·1	28·3	38·1
102	6·35	12·7	80	—	66·1	28·3	30·3
127	6·35	12·7	80	—	66·1	28·3	20·3
Chrome-nickel steel							
12·7	3·175	7·937	50	94·4	—	—	127
1	3·175	7·937	50	94·4	—	—	63·5
2	3.968	7·937	85	94·4	—	—	45·7
3	4·762	7·037	90	94·4	—	—	40·6
4	4·762	7·932	90	94·4	—	—	20·3
5	6·35	7·937	100	106·5	—	—	12·7
12·7	3·175	7·937	50	—	61·3	—	178
25.4	3·968	7·937	55	—	61·3	4·7	63·5
25.4	4·762	7·937	100	—	82·5	7	254
51	6·35	12·7	140	—	82·5	7	61
76	6·35	12·7	150	—	82·5	7	40·6
102	6·35	12·7	150	—	82·5	7	20·3

it is rather difficult to keep accurately to the indicated parameters, and it is advisable to double the amounts of gas at least [Ref. 3.1.18].

The 100-kW plasma torch developed at the M. v. Ardenne Institute in Dresden has virtually the same parameters as quoted in Table 3.1.4. The nozzle diameter is 4 mm.

3.1.4.3 Cutting structural steels

There is no economic advantage in cutting structural steels with plasma torches. The stabilizing gases cost more than oxygen and fuel gas.

Table 3.1.4 TYPE F-80 THERMAL DYNAMICS PLASMA TORCH [REF. 2.2.88]

Plate thickness [mm]	Nozzle dia. [mm]	Gas flow rate [litres . min^{-1}]			Input [kW]	Cutting rate [cm . min^{-1}]
		N_2	H_2	Ar		
Aluminium and its alloys						
12·7	2·38	21·15	21·15	—	30	76·2
25·4	2·77	47·2	21·15	—	50	76·2
51	3·57	56·6	28·3	—	100	89
Chrome-nickel steel						
12·7	2·38	42·4	2·36	—	30	63·5
25·4	3·57	71	9·4	—	100	139·5
51	3·57	71	9·4	—	100	38·1
102	4·76	94·4	9·4	—	200	15·2
25·4	3·175	—	9·4	68·4	30	63·5
76	5·55	—	28·3	51·7	100	63·5
127	5·55	—	33·1	61·3	150	25·4
200·3	6·746	—	33·1	61·3	150	10·2
Structural steel						
12·7	2·38	47·2	—	—	30	88·9
25·4	2·77	52	—	—	50	76·2
51	3·57	85	4·7	—	100	38·1
76	4·76	94·4	9·4	—	150	25·4
102	4·76	94·4	9·4	—	150	17·7
127	4·76	94·4	9·4	—	200	12·7

Not even the higher cutting rates attained on thinner materials are sufficient to cover the higher cost. Table 3.1.5 compares the cutting rates achieved by means of a 50-kW plasma torch (with its arc stabilized by an argon-hydrogen-mixture [Ref. 3.1.14]) and those of oxy-acetylene cutting.

If the input of the plasma torch is increased to 150 – 200 kW for cutting thicker steels (about 100 mm), then the cutting rates are approximately the same as with oxygen torches (Table 3.1.4); the capital cost, however, and consequently the cost per linear metre is

Table 3.1.5

Plate thickness [mm]	Cutting rates [cm . min^{-1}]	
	Plasma torch Zis Leuna 247	Oxy-acetylene flame cutting
10	80	45
20	45	36
40	30	26
60	17	22
80	15	19
100	8	17·5

higher. Plasma torches with air-stabilized arc have therefore been developed to reduce the cutting costs on structural steels. The exothermic

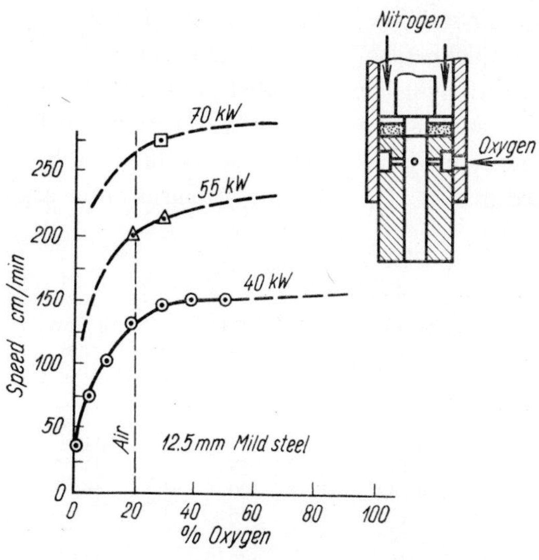

Fig. 110 Cutting speeds with mixture of nitrogen and oxygen (mild steel)

reaction between atmospheric oxygen and iron increases the cutting rate, and the demand on the output of the source is therefore lower.

The effect of pure oxygen added into the steam of nitrogen in the stabilizing nozzle is shown in Fig. 110, in which maximum cutting speeds for various torch inputs are plotted against the percentage of oxygen in the plasma [Ref. 3.1.17]. The diagram proves that up to an oxygen content of 20% the oxidation of the iron greatly accelerates the cutting rate. Employing air is therefore a sound proposition, although

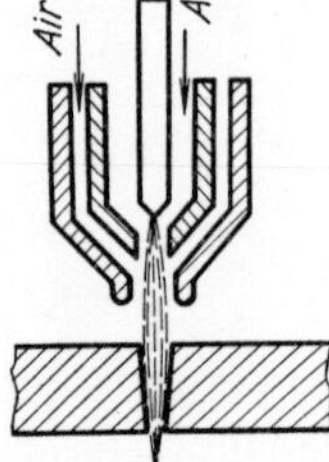

Fig. 111 Plasma torch with combined argon- and air-stabilization of arc

a comparison between the costs of cutting with an oxy-acetylene flame and with an air-stabilized plasma torch [Ref. 3.1.16] indicates that only up to a thickness of 35 or 38 mm the latter is advantageous.

In a lightweight hand-operated plasma torch used in the USSR [Ref. 3.1.17], a small amount of argon is supplied to the region of the tungsten cathode, and in addition the arc is stabilized by air as schematically indicated in Fig. 111. The pilot arc burns in the argon atmosphere

Table 3.1.6

Plate thickness [mm]	Nozzle dia. [mm]	Input [kW]	Air flow rate [l . min^{-1}]	Cutting speed [cm . min^{-1}]
6·35	3·175	55	118	508
12·7	3·175	55	118	254
12·7	4·762	100	165	330
25·4	3·175	55	118	127
25·4	4.762	100	165	228
51	4·762	100	165	76·2

at a current of 30 to 50 A, and the current flowing through the torch during cutting is about 300 A.

Cutting speeds for the Linde PT-2 type of air-stabilized plasma torch are tabulated below.

Water-stabilized plasma torches cut structural steels at slightly lower cost per linear metre than do oxy-acetylene flames. However, the high hydrogen content in the plasma results in a somewhat poorer quality of the cut (the edges are never entirely perpendicular and the kerf is wider), and the tendency towards formation of burrs (dross) is very pronounced.

3.1.4.4 **Quality of cuts**

Gas- and water-stabilized torches yield cuts of virtually the same appearance.

Fairly smooth and dross-free cuts may be obtained on austenitic steels and aluminium. Fig. 112 illustrates surfaces of cuts performed on austenitic steel 12·5, 20 and 58 mm thick with a water-stabilized torch. Carbon and chromium steels are more inclined to form dross than are chrome-nickel steels.

Aluminium up to 100 or 110 mm may be cut with high-quality edges, whereas copper is less easily cut with water-stabilized torches: material 25 to 30 mm thick forms a drossy burr at the lower edge of the cut. However, if copper is cut on an iron backing (a sheet of 2 – 3 mm), a surface of good quality is obtainable up to a thickness of 40 mm: all the material is lapped onto the iron backing, and when the latter is removed, the copper plate has a clean edge.

Sheets can readily be cut in a packet, provided they are perfectly flat or else thin enough to be drawn together in a vice to the complete exclusion of air gaps. Thus clamped, the sheets are not welded together and the cut is good. If, however, an air gap is left between two sheets, molten metal flows into it and forms a weld. Fig. 113 reproduces the appearance of a cut through a packet of 10 1-mm sheets of 18-8 stainless steel [Ref. 3.1.18].

No difficulties are encountered in cutting bevels. With water-stabilized plasma torches, cuts can be bevelled up to about 38°, hence by slanting the head, edges can be prepared for groove welds of 60 or 75°. The slant has to be about 8° more than the intended bevel, i.e. about 38° for a 30° bevel, or 45° for a 37·5° bevel. A slightly higher speed is employed than corresponds to the thickness, i.e. to the length of the path which the plasma takes in the material. It is advisable to increase the speed only after starting the cut, when the beginning of the kerf has already been formed.

The depth of the heat-affected zone of the cut has been studied on ground sections, both macroscopically and microscopically. On the

Fig. 112 Surfaces of cuts performed with a water-stabilized arc on austenitic steel

whole, the effects are slight and the depth is inversely proportional to the cutting speed. In austenitic steels the fused zone has a dendritic structure and its depth ranges from 0·15 mm (in plates of 15 to 20 mm) to 0·25 mm (for 60 mm plate thickness). The fused layer is followed by

Fig. 113 Cut through a packet of sheets (By courtesy of M. v. Ardenne Forschungsinstitut, Dresden)

the transitional zone with slightly enlarged austenite grain. On 60-mm steel, this transitional layer is 2 to 2·2 mm thick, at the most. In 65 mm duralumin, the melted zone is 0·2 mm deep, and the heat-affected zone with enlarged grain up to 2·5 mm. The structure consists of an α solid solution, and eutectic separations of intermetallic phases occur on the grain boundaries which are far coarser in the unaffected zone.

Copper exhibits only an inconsiderable coarsening of grain to a depth of 0·3 or 0·5 mm.

In high-carbon steels, the surface layer of the cut is hardened. In a ball-bearing steel cut for experimental purposes, the parent metal consists of globular pearlite with a gradual transition into the surface-hardened zone which exhibits a fine martensite structure; the depth of the hardened zone is 0·6 mm at the most. In high-speed steels, the coarse purely martensitic structure of the surface-hardened zone passes into finer martensite with residual austenite, and through transitional structures into globular pearlite. The pure-martensite layer is 0·1 mm thick, and the depth of the entire heat-affected zone is 1·4 to 1·5 mm.

The typical structural appearance under the surface of the cut through an austenitic steel is reproduced in Fig. 114. No microscopic cracks were found in any of the steels examined.

Hardness was studied in the macroscopic section of an air-hardening chromium steel. A thin hardened zone formed under the

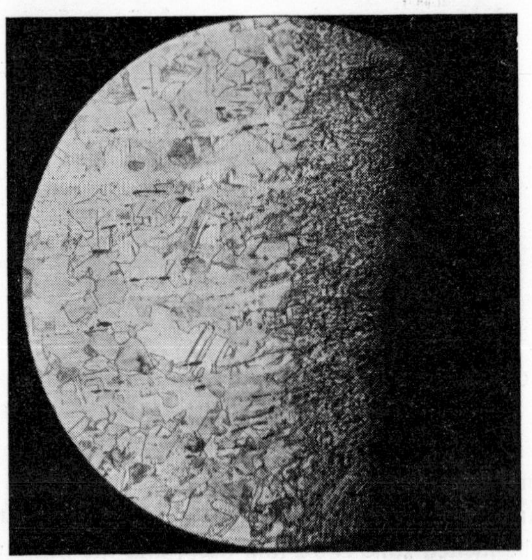

Fig. 114 Ground section through austenitic steel (magnified 100×.)

surface exhibits a Vickers hardness of 525 to 530, the VHN of the parent metal being 245.

In the case of water-stabilized plasma torches, the oxide layer formed on the cut faces of alloy steels is of a negligible depth. The effect of this layer on the quality of welds was studied in a series of experiments performed on a 15-mm plate of an 18/8 chrome-nickel steel and on 18-mm austenitic chromium-manganese steel (18Cr9Mn–4NiO . $2N_2$). The test welds were performed on unmachined cuts, with check tests made on machined edges. The chromium-manganese steel was included in the tests because the burned-out manganese makes the brown oxide layer on the cut edges more easily distinguishable than it is in chromium-nickel steels.

In both materials the cuts were bevelled, and on the check test pieces the bevel was subsequently machined. The welds were in all cases preformed by electric-arc welding with a stainless austenitic electrode and without preheating, by the procedure usually employed on austenitic stainless steel. Ground sections from all welds were thoroughly investigated by both macroscopic and microscopic methods.

Particularly in the transitional zone between deposited and parent metal, the metallographic examination revealed virtually no difference in the appearance of machined and unmachined bevel faces. In both cases the transition between added and parent metal was gradual, without any considerable coarsening of the steel grain. The grain size in the parent metal and the transitional zone was equal, classified as 5 to 6 according to ASTM. The amount of oxide inclusions was identical in both cases and conformed to the usual purity standard for austenitic stainless steels.

3.1.5 **Economic Comparison of Various Cutting Methods**

The costs per linear metre of oxy-acetylene, oxy-propane-butane and steel-powder cutting are quoted from Ref. [3.1.20]; the gas consumption and cutting speeds of the VUS-arg-ORS and VS-45 torches are those quoted in Tables 3.1.1 and 3.1.3, respectively [Ref. 3.1.11]. Calculations were based on the following prices:

argon	Kčs 42·– per m^3
hydrogen	Kčs 4·– per m^3
electric power	Kčs 0·25 per kWh
500-mm carbon electrodes 13 mm in diameter	Kčs 2·20 each
steel powder	Kčs 4·50 per kg
oxygen (in cylinders)	Kčs 5·85 per m^3
acetylene (in cylinders)	Kčs 15·40 per m^3
propane-butane (in cylinders)	Kčs 4·34 per m^3

Consumption of the tungsten electrode and water is disregarded.

The results of the calculations are confronted in Tables 3.1.7 and 3.1.8.

Table 3.1.7 ALUMINIUM AND ITS ALLOYS

Material thickness [mm]	28	40	52	65	65*	105*
ORS-arg set [Kčs per lin. metre]	1·127	2·544	3·455	5·661		30·6
VS-45 set [Kčs per lin. metre]	0·233	0·390	0·621	1·580	1·117	2·35

* VS-45 operating with three welding sets in series.

Table 3.1.8 CHROME-NICKEL AND STRUCTURAL STEELS

Material thickness [mm]		12·5	20	35	60	60*	70*
VS-45 [Kčs/lin. m]		0·161	0·296	0·386	1·386	1·161	1·770
Material thickness [mm]		10	20	30	40	50	60
ORS arg	[Kčs per lin. m]	1·50	2·059	2·70	3·26	4·22	6·41
oxygen + steel powder	[Kčs per lin. m]	3·00	3·62	4·75	5·65	6·05	
oxy-acetylene	[Kčs per lin. m]	0·52	1·04	1·87	2·64	3·52	
oxy-propane-butane	[Kčs per lin. m]	0·50	0·81	1·18	1·58	2·17	

* VS-45 operating with three welding sets in series.

From the economic analysis summarized in Tables 3.1.7 and 3.1.8 it is evident that the plasma torch with water-stabilized arc is far cheaper in operation, particularly as the price of gases does not include the costs for intra-factory transport, exchange of empties, and losses due to leakage. The difference is even more pronounced if higher inputs

(3 sets in series) are employed. The drawback of the water-stabilized torch is in the less convenient ignition and the slightly higher consumption of electric power per linear metre of cut.

3.1.6 Safety Precautions

Apart from the usual safety precautions in welding shops – in Czechoslovakia imposed by standard ČSN 05,0630 – plasma cutting calls for special care. The no-load voltage of sources required for cutting thick materials is above the limit of 100 V d.c., and remote control is therefore obligatory. The controls (hand-wheel for vertical adjustment of the torch head) have to be at least 0·2 m off the head. Moreover, the supply cables must be more thoroughly insulated than in current welding practice. The sandard quoted above also obliges the makers of special welding gear with sources above the limit voltage to set out full instructions for operators.

Apart from the hazards involved in electric currents, the operators of metal-cutting plasma torches are exposed to intensive ultraviolet radiation, noise and metal vapours stemming from the cut.

The intensity of ultra-violet radiation from the free part of the arc (between the head and the workpiece) is many times higher than in welding, and even in short-term operation the operator's eye-sight and skin must be appropriately protected.

The noise caused by the plasma flowing from the torch nozzle must not be underestimated. The intensity of the noise caused by gas-stabilized plasma torches is found to be 96 to 125 dB, the maximum being reached at a frequency of 16 kc.p.s. Water-stabilized torches reach the highest noise level (102 dB) at 8 kc.p.s., and at higher frequencies the value drops to 94 dB [Ref. 3.1.22]. These measurements are in good agreement with data presented in Ref. [3.1.23] for a gas-stabilized torch of U.S. make (78 to 110 dB with the maximum between 4·8 and 10 kc.p.s.). The frequency involving the maximum noise level decreases as the thickness of the cut material increases. For permanent operation the operator must have his ears protected by sound-damping devices.

Metal is vaporized in the cutting process, and the vapours –

oxidized in the atmosphere – form microscopic dust particles. The standard quoted earlier in this Chapter imposes the obligation of adequately ventilating the workshop and installing an exhaustor under the grating of the cutting bench, to draw off the metal particles. So far no Czechoslovak standard limits the permissible metal-vapour and oxide concentration. The internal directives of the Medical Health Service permit the maximum concentrations tabulated below:

Table 3.1.9

	average concentration during shift [mg per m^3]
chromium and chromium compounds	0·05
aluminium oxide	2
iron oxides	10
manganese and manganese compounds	2
nickel	0·5
silica	2
copper	10

No measurable amount of nitrous gases were found; repeated measurements in the media of both types of plasma torches did not yield more than trace concentrations.

3.2 Plasma Torches Used for Surfacing

3.2.1 Introduction

Surfacing by means of a plasma torch is carried out by either spray coating or fusion surfacing. In either case, a layer of different material and with different properties forms on top of the base materials. The coating material and the method of its deposition is selected to suit the surface properties required.

Recent designs of plasma torches intended for surfacing permit the average temperature of the plasma jet to be controlled in a very wide range (cf. Chap. 2.4.1). Moreover, the gas mixture can be selected to provide the most appropriate medium for the surfacing material. These advantages permit any desired material to be applied to a surface, metal carbides as well as high-melting metals (e.g. tungsten, tantalum, molybdenum), oxidic materials, and even plastics.

Surface layers deposited by spray coating do not form any metallurgical link, not even if a metallic coating material is used: the cohesion is due to mechanical forces only.

Fusion surfacing (weld surfacing), though technically more difficult, yields surface layers metallurgically bonded with the base metal; the coat is more compact and therefore resistant to abrasive wear.

Neither of these surfacing methods is based on a new principle; yet giving better results at lower costs than the techniques used so far, they deserve to be extensively employed.

3.2.2 Applying Protective Coatings by Plasma Torches

Spray coating is a process in which a surface coat of arbitrary thickness is obtained by spraying the previously prepared surface of the base material with droplets of a molten material. This method, originally used for metallizing only, was elaborated by the Swiss expert M.U. Schoop in 1910 (hence schoop-plating or schooping).

No metallurgical bond is formed between the sprayed coating and the base material. The perfect cohesion of the surface coating with the roughened surface of the base material depends on mechanical forces produced by shrinkage stresses; these result from the deformation and rapid cooling of the sprayed particles, wedged into the uneven surface of the base material. The spraying material can enter into the process in wire or powder form.

In the first industrial spray guns, the so-called metallizers, gas was burned in a stream of oxygen, and the material to be sprayed was supplied by a wire, axially fed into the middle of the flame at a controllable rate. The droplets of molten metal were accelerated by a stream of compressed air which flowed through channels concentrically surrounding the supply channels for the gas-and-oxygen mixture.

In addition to these gas-heated metallizers, electric-arc spray guns with substantially higher spraying outputs were later developed. The arc burns between two metallizing wires which are introduced in parallel into the spray gun. In the melting part of the gun, a guiding tube bends them so that their points—the electrode tips—include an angle of 30 to 35°. A tube supplying compressed air runs axially between the metallizing wires. The air, flowing at a high velocity, atomizes the molten metal and flings the droplets onto the surface to be metallized.

While metal spraying is not a recent development, it is only a few years since spray coating with ceramic materials and high-melting metals has been practised. The impulse for developing heat-resisting coatings was given by the requirements of the aircraft industry during the Second World War. Anticorrosive silicate-base layers were developed as high-temperature protective coatings (high-temperature enamels). They were in particular used to protect various parts made of expensive alloys and exposed to high temperatures in operation.

The successes achieved were an incentive to intensify the research of ceramic coatings after the War. The application of oxide coatings and cermets called for the development of suitable spray guns. Since electric-arc appliances were limited to conductive materials, oxy-acetylen guns seemed the obvious choice. The principle of the "ceramizer", the spray gun in which a ceramic rod supplies the coating material, is the same as used in the metallizer: an oxy-acetylene flame fuses the rod and a stream of compressed air accelerates the molten droplets.

Modern powder-type ceramizers (Metco, Lurgi) use part of the oxygen to convey the powder into the centre of the flame and accelerate the particles by the combustion products moving at a high velocity. They dispense thus with compressed air, obtain a better heat transfer from the combustion gases to the powder, and prevent the temperature of the jet from dropping.

The ceramic rods are compacted and sintered. Rather high demands are made on their accuracy: they must be straight and of equal diameter throughout their length which is at least 350 mm. These stringent demands result in a high percentage of waste and accordingly high prices. On the other hand, rod-fed spray guns produce better coatings than powder-fed ceramizers. They perfectly melt the coating material, whereas coatings made from ceramic powders contain a fair amount of insufficiently fused and stuck-up particles and are therefore less resistant to abrasive wear.

Ceramic materials in rods cost up to a hundred times the price of powders; moreover their high brittleness and the frequently interrupted spraying (because of the short lengths) are drawbacks in operation. Attempts were therefore made to press ceramic "wires" from powders of 10 to 400 μm grain size with a plastic binder. The perfect combustion of this latter in the spraying process is essential lest the coating be contaminated. A mixture of polyethylene and polyisobutylene was for instance suggested, the binder content of the "wire" being less than 30%. However, literature mentions no practical applications of this method although a US patent application was published as long ago as 1949.

The first paper giving a full account of ceramic coatings made from rod materials was published by Ault [Ref. 3.2.3]. It describes the properties of sprayed coatings of aluminium oxide, zirconium oxide and zirconium silicate. Commercially available rod materials include Rokide Z (98% zirconium dioxide) and Rokide ZS (65% zirconium dioxide and 34% silica). Mayer [Ref. 3.2.4] describes the spraying of powder-base aluminium oxide coatings and their properties. Levy

[Ref. 3.2.5] has investigated the thermal-insulation properties of ceramic coatings; he reports that such coatings exhibit a high thermal reflectance, in addition to their low thermal conductivity. As to corrosion protection, no satisfactory results have been reported so far. Corrosion is only inhibited and not prevented, because the coatings are fairly porous. Good results have been obtained with ceramic coatings on parts exposed to abrasive wear. However, the temperature of the oxyacetylene flame is insufficient for melting some materials which are of particular interest because of their high abrasive resistance; such are various metal carbides, borides and nitrides as well as oxides of high-melting metals.

The utilization of plasma torches for spraying protective layers has brought about a rapid progress in coating technique. Higher operating temperatures and the possible use of inert atmospheres are the main advantages derived from this innovation.

Commercially available plasma spray guns are designed for arc-stabilization by an argon-hydrogen or nitrogen-hydrogen mixture (with a maximum of 10% hydrogen) and coating materials supplied in powder form. The gas mixture used to stabilize the arc simultaneously transfers the requisite heat to the particles of the spraying material and accelerates them to the appropriate velocity.

Plasma sprayers were originally expected to bring the solution to all open questions of coating technique, the most important problems being

- coating with all technically important materials, including materials with the highest melting points, provided they melt without decomposition;
- protection of droplets from oxidation,
- compact coatings firmly connected to the base material;
- relatively low temperatures of the base material during coating.

However, systematic and thorough studies showed that the initial optimism was unfounded and coating practice involves more complex problems than expected. Nevertheless, plasma torches have undeniable advantages over oxy-acetylene spray guns. The choice of coating materials is far wider and the coats are less porous. The full elimination of porosity has not proved feasible, so far, although the particles sprayed with plasma are perfectly melted. As the coating is cooled, it contracts and the strains involved result in fine cracks, usually located at the grain boundaries. Since some of the pores are open, the coating is not gasproof and does not completely prevent corrosion. In some materials the composition of the coating is not identical with that of the initial compound. Inert gases used to stabilize the arc are no sufficient protection against the oxidation of the coating material. However, this drawback can be overcome and pure metallic coatings even of tungsten, molybdenum or tantalum can be obtained by spraying in a protective atmosphere.

3.2.3 Spray-Coating with Plasma — Technique and Equipment

The technique is similar to that of coating with oxy-acetylene or electric-arc metallizers. The process consists of two basic operations – the preparation of the surface to be coated and spraying proper.

The surface preparation is of paramount importance for the perfect adhesion of the coating to the parent material. The surface of the parent metal is cleaned and appropriately roughened by degreasing and steel-grift blasting. – Thorough degreasing in a hot alkaline bath or trichlorethylene is indispensable if the part to be coated has already been used and is contaminated with oil.

In current metallizing practice, the surface of parts requiring high reliability or thicker coatings is often roughened by threading or dovetailing. In spraying with plasma guns grit-blasting will mostly be sufficient, because plasma produces coatings with very good adhesion, and coats of special metals, alloys or oxides are thin and their adhesion is superior to that of thicker coats. – As a rule, blasting is done with sharp steel grit of No. 16 to 24 grain, preferably in a blasting booth to keep down the loss of grit. Good-quality steel grit can be used 60 to 90 times.

The spray-coating technique using plasma torches is practically the same as with oxygas guns. The lightweight design of surfacing plasma torches facilitates their handling. The differences between the two processes can be summarized in the following points:

- The beam temperature differs. The mean temperature of the plasma beam is about four times that of the combustion gases from an oxy-acetylene gun. Consequently high-melting materials can also be spread and – generally speaking – even on a less roughened surface of the base material, the coating adheres better than it ever does with oxy-acetylene spraying.

- The discharge rate of the plasma is higher than the flow velocity of the combustion gases. This and the higher gas temperature result in a far more rapid heat transfer to the particles. As the latter impinge on the base material, their deformation is more thorough because they are more completely molten and have received a higher acceleration. Consequently the coating is finer and more compact. The higher temperature and velocity of the plasma result in a more efficient utilization of its enthalpy as compared with that of the combustion gases; hence the plasma torch has a higher spraying output per unit time than an oxy-acetylene gun with the same heat output.
- Selecting the proper atmosphere for the spraying process, we can successfully apply materials which are very easily oxidized, decarbonized or react with the combustion gases of the oxy-acetylene flame.
- Precision feeding of powdered materials or wires is more difficult in the plasma torch than in the oxy-acetylene gun. Owing to the higher prime cost for plasma equipment, its total cost per unit output is slightly higher, although the direct operating costs of a plasma torch with nitrogen-stabilized arc are lower than those of an oxy-acetylene gun with the same thermal output.

The plasma spraying equipment consists of the torch; the cooling circuit with a circulating pump; the gas cylinders with the cylinder-pressure regulators and gas supply tubing; the storage bin with a device for metering the powder and continuously supplying it into the plasma; the d.c. supply source; and the box containing the electric control circuits and the gas flow meters. The most sensitive part of the spraying plasma torch is the nozzle – also acting as the anode – with its front part adapted to mix the powder into the plasma. The aperture for the powder supply must be in a place through which no current flows.

The length of the arc depends on the stabilizing gas, the gas mass flow and the nozzle diameter. In universal torches (for various kinds of stabilizing gases) the cathode is adjustable; thus the optimum length of the discharge path for various gases or gas mixtures can be set without exchanging the nozzle. Fig. 115 indicates the transition of the current

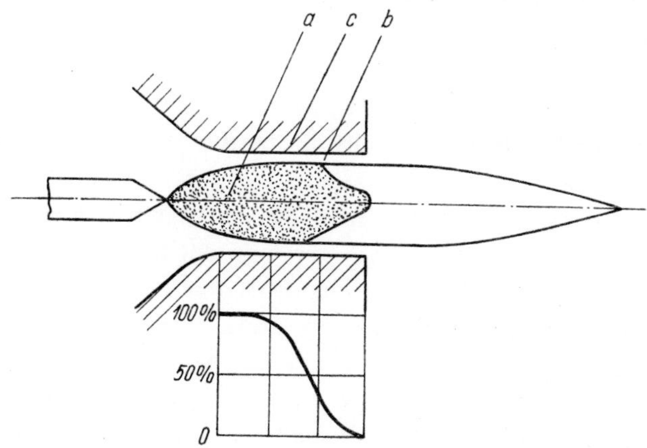

Fig. 115 Transition of current from plasma into anode nozzle

from the plasma into the individual sec..ons of the nozzle-anode. From the centre of the plasma (*a*) the current can flow into the nozzle wall (*c*) only through the relatively cold boundary layer (*b*). The flow of the current can be mediated by rapidly transferred small secondary arcs or diffusion of electrons across the boundary layer.

An important issue is the service life of the nozzles which depends on the uniform stressing of their active zone and the observance of the permissible stress limit in all of their sections. Since hydrogen – because of its good thermal conductivity – increases the temperature of the anode, it must not be added in a proportion exceeding 10%. In water-stabilized plasma torches, where the plasma contains 66% of hydrogen, the nozzle cannot be used as anode. A water-cooled copper disk is employed instead; by its rapid rotation it enlarges the active surface of the anode (the circumference of the disk) and the anode spot of the arc is mechanically transferred; this ensures the short stay of the anode spot in any single place.

These anode disks – 100 mm in diameter, revolving at 3000 r.p.m., and cooled by water streaming at the rate of 40 cm^3 per sec – have a relatively long service life. Aluminium disks were used for spraying aluminium oxide, lest copper oxides contaminate the coating. However, the maximum current load permissible on aluminium disks of the size mentioned is 330 to 350 A; the edge of the disk is rapidly worn if this limit is exceeded.

3.2.4 Factors Affecting the Quality of Spraying

With a plasma torch, spraying is more difficult than with a conventional gun: the greater intricacy of the spraying device involves a greater number of breakdowns, and the quality of the coating is affected by a greater number of variables. The operating charcteristic of the plasma torch is influenced by the input, the kind and flow rate of the stabilizing gas, and also the geometry of the plasma jet. Therefore most papers describe the results obtained with only a single type of torch, which reduces the variables to the input used and the gas feed per unit time.

The temperature in the centre line of the plasma jet depends, apart from the input, on the thermal properties of the stabilizing gas, i.e. mainly on its enthalpy. This problem is disccussed in greater detail in Chap. 2.2.9, where the differences in the radial temperature distribution of different gases were also emphasized. More important than the mean temperature of a plasma is the enthalpy or heat content of the gas. It is this quantity which determines the rate of heat transfer onto the individual particles of the powdered material as this is carried in the plasma. The minimum axial temperature of the plasma has to be far higher than the melting point of most coating materials. Given the flow volume and the efficiency at which electrical energy is transformed into heat energy in the given apparatus, the gas enthalpy can be calculated with a fair degree of accuracy.

However, at equal input and equal flow volume of the gas, hence equal enthalpy of the plasma, its discharge velocity (outlet velocity) will depend on the nozzle diameter. The difficulties involved become quite obvious if we take into consideration that the optimum parameters of the coating process are also substantially affected by other design features—the diameter and location of the aperture for feeding the powder into the plasma, the type of powder feeder—and by material properties such as the quality of the powder and the kind of the base material. Therefore experience derived from the operation of one type of torch calls for a certain amount of reserve before application to a different type of sprayer.

The powder feed is of great importance for a high-quality coating. Neither pressure injectors nor vibration feeders are capable of feeding the powder into the plasma at the absolutely uniform rate required for a uniform coating. The best results have been obtained with mechanical designs such as rotary-vane or worm-type feeders. A due proportion must be maintained between the amounts of the stabilizing gas and of the carrier gas for the powder. The quantity of the carrier gas should be as low as possible lest the plasma is excessively cooled. Some feeders work to full satisfaction with a large amount of carrier gas, yet they are absolutely unreliable at lower flow rates.

It is a matter of common knowledge that the quality of the coating is materially affected by the powder quality; nevertheless, this problem has not been fully studied so far. Most authors use commercially available materials and only observe grain sizes and their distribution in unit volume. The optimum grain size for a given spraying distance has been studied relatively thoroughly. As the heat output of the plasma gun increases, the spraying distance has to be lengthened to prevent the base material from being overheated. Small droplets solidify very rapidly once they have left the hot zone of the plasma; their flight path must therefore be shorter than that of larger drops. The permissible maximum grain size depends upon the kind of the material. The individual particles must be supplied with the heat for bringing their temperature up to the melting point plus the latent heat of liquefaction. Moreover, their stay in the plasma must be lengthened by the time required for the heat to penetrate from the surface of the particle to its core. It follows that the termal conductivity of the material greatly affects the permissible maximum grain size. Ceramic materials have a low thermal conductivity. The unsatisfactory results experienced in coating with magnesium and silicon oxides are due to the narrow temperature range between the melting and boiling points of these materials, together with their low thermal conductivity. The surface of the particles attains the boiling temperature and much heat is lost by its vaporization. Yet the delay of the particles in the plasma is insufficient to heat and melt their core.

Besides the size of the particle it is its shape that affects its transport characteristics. Fine-grained powders tend to agglomerate and clog up the supply hose. Sharp-edged powders cause trouble in pneumatic transport through small-diameter hoses: they form bridges of mutually wedged grains and the hose is frequently clogged. An approximately spherical grain size is favourable; such powders are conveniently conveyed and charged, their feed is easily controlled, and the quality of the coating layer is reproducible.

Since the drops move at a very high velocity before impinging on the base material, the conversion of their kinetic energy into thermal energy retards their cooling and helps to weld them up. The complete fusion of the individual drops requires that every drop be hit by the following one while it is still in a plastic state.

All this shows that the fusion of particles in the plasma depends on many factors, some of which are usually unknown or at least very inaccurately known. For instance, far from accurate data are the result if we use known torch parameters to compute the maximum diameter at which particles are still melted throughout their cross section. Experimental data seem more useful; they are obtained by spraying the material onto glass plates quickly drawn through the plasma at a standard distance from the nozzle [Refs. 3.2.7, 3.2.8]. The results are evaluated by microscopical viewing of the shape which the drops have assumed after striking the glass.

The spraying technique is another factor influencing the quality of the coating. Most important are the distance between the base material and the nozzle, and the angle at which the particles strike the surface. Their effect is fairly well known, and all published reports discuss it. The effect of the velocity at which the workpiece travels upon the quality of the coating seems not to be very important, except in extreme cases. The effect of the atmosphere in which the coating is applied has been discussed earlier in this Chapter.

The base material affects the quality of the coating by its

composition, the mode of its surface preparation, the degree to which the surface has been roughened, and the temperature it attains during the spraying procss.

Research into spray coating by plasma torches is by no means completed. The precise analysis of the effects wrought by all quantities and factors involved still calls for a great amount of experimental work.

Mash et al. [Ref. 3.2.9] have found that these quantities affect most of all the spraying efficiency, whereas the density of the coat varies no more than slightly. In Fig. 116, the curves of spray-coating efficiency are plotted against the throughput of stabilizing gas, and in Fig. 117 against the enthalpy of the plasma which varies with the amount of gas supplied (argon in this case). It appears from these curves that the optimum for carbides lies at lower flow rates of the stabilizing gas than does the optimum for oxides, i.e. they require a higher enthalpy and lower gas flow velocity. Obviously, maximum efficiency calls for a certain optimum combination of plasma enthalpy, dwelling time of particles in the plasma and particle velocity. Further results reported

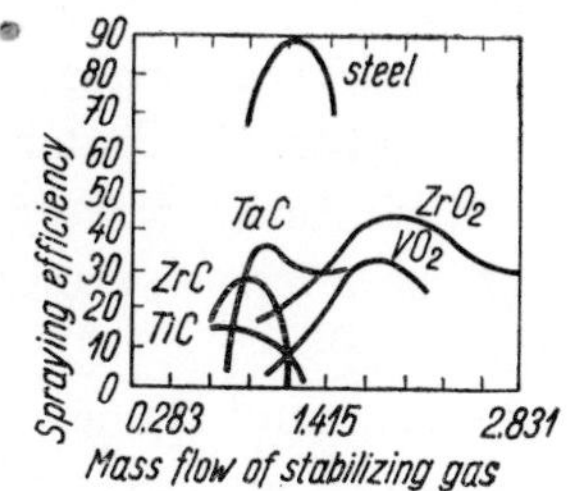

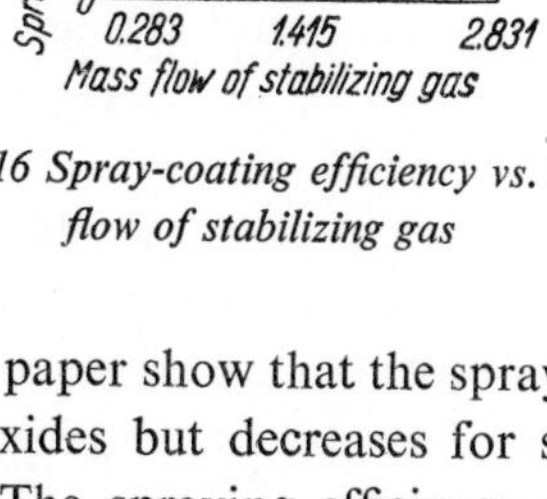

Fig. 116 Spray-coating efficiency vs. mass flow of stabilizing gas

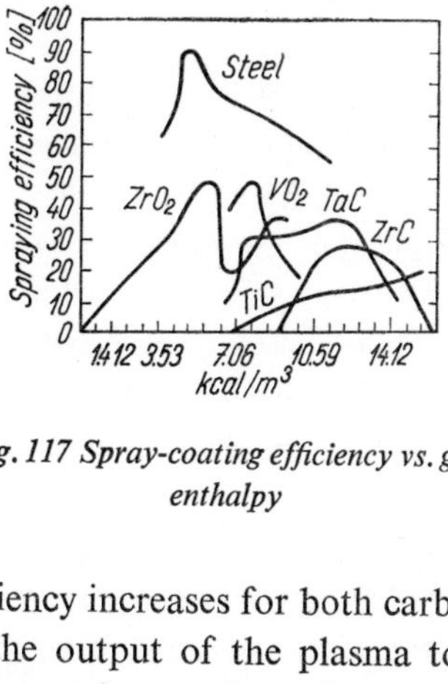

Fig. 117 Spray-coating efficiency vs. gas enthalpy

in the paper show that the spraying efficiency increases for both carbides and oxides but decreases for steel as the output of the plasma torch rises. The spraying efficiency rises because the number of welded-on particles increases. Once the optimum enthalpy for a given kind of powder material has been attained, any further rise in torch output has a negative effect: As soon as the complete fusion of all particles is achieved, any further supply of energy causes overheating and losses by evaporation (particularly clearly perceptible in the steel curves).

The optimum grain size in spray-coating with medium output gas-stabilized plasma torches is 45 to 47 μm for oxidic materials [Refs. 3.2.9, 3.2.11, 3.2.12], while smaller grain sizes – 12 to 43 μm – give good results with carbides [Refs. 3.2.9, 3.2.14]. The spraying distance recommended is 12 to 15 cm.

At the present time plasma torches with a water-stabilized arc are being tried out for spraying oxidic materials. They may also be used for metallizing, provided no high demands are made on the purity of the coat. If metals are sprayed with water-stabilized plasma torches, oxides are formed on the grain boundaries in approximately the same amount as experienced in electric-arc metallization. However, the quantity of oxides can be considerably diminished if a liquid producing a neutral or reductive atmosphere is used for stabilizing the arc. Trials with a mixture of denatured alcohol and water (50 : 50) gave good results, also when carbides or tungsten were sprayed. An oil emulsion can be employed, too.

The greater sturdiness of water-stabilized torches and the need to provide them with an electrode-feeding device have already been mentioned. The weight of the torch is also increased by the rotary electrode employed in securing a non-transferred arc. Water-stabilized plasma torches are therefore less appropriate for manual spraying than gas-stabilized units. Their spraying output, however, is higher (cf. Table 3.2.1).

The distance of the rotary anode (i.e. the edge of the disk) from the centre line of the plasma (cf. Fig. 79) has to be set very carefully. The current passes from the hot parts of the plasma onto the rotary anode by a small secondary arc. If the disk is too near, the plasma jet declines from the centre line of the torch, as the streaming medium is reflected from the obstacle that obstructs its path. If the distance between the edge of the anode disk and the plasma is excessive, the anode is not sufficiently heated. The anode spot sticks to the cold surface, and for some time it moves together with the edge of the disk. The column of the secondary arc is thus lengthened until it reaches a critical value and shortens by a jump. If the changes in the secondary-arc length are periodically repeated, the main jet of the plasma begins to oscillate.

Table 3.2.1

Type of torch	Stabilizing medium	Torch input [kW]	Spraying output [kg/h]	Material sprayed
Thermal Dynamics	N_2 + max. 10% H_2	40	3·0	tungsten
Avco	N_2 + 5—10% H_2	50	4·5 1·3	tungsten alumin. oxide
Metco	N_2 + max. 10% H_2	23 50	1·5 2·2	alumin. oxide
Plasmadyne	argon	40	3·2	tungsten
R.I.E.E.	water	53	3·8	alumin. oxide

In both alternative cases of incorrectly set anode distances the spraying efficiency drops because the powder is not conveyed in the right direction.

Unlike in plasma torches with gas-stabilized arc, no ancillary nozzle can be used to introduce the powder into the plasma, because this device would be too short-lived. However, powder feed through a copper tube placed vertically above the electrically neutral plasma jet (ahead of the anode disk) gives good results if it is carefully directed.

Oxidic coatings applied with water-stabilized torches have a somewhat better quality than is obtained with gas-stabilized torches. This is due to slightly better mutual fusion of the particles, which in appearance shows as glossier surface lustre. Aluminium oxide was used as material in experimental studies of this subject. A rough comparison of porosities was carried out by measuring the dielectric strength of the coats applied to test pieces, since this quantity is a function of the number and size of pores. Aluminium-oxide coatings were applied to an iron tube 80 cm long, and this test piece was coated with a 20-cm silver stripe in its middle. The conductive film thus produced was one electrode for measuring the dielectric strength, and the base metal (i.e. the iron tube) was the other. The breakdown voltage was measured with alternating current of 50 c.p.s. frequency. The gas-stabilized torch was operated on nitrogen with 10% hydrogen added.

The breakdown voltage, measured on coats of 0·15 to 0·25 mm, was 6,000 to 8,000 V on the layers deposited by a water-stabilized torch, and 4,500 to 6,000 V on layers sprayed with gas stabilization. The application of two layers totalling more than 0·3 mm in thickness reduced the dielectric strength in both instances. In this case, the cracks formed owing to cooling stresses are probably deeper and result in the formation of more numerous intercommunicating pores.

The operation of the water-stabilized torch is simpler, because there is no setting of the optimum gas flow volume. The amount and issuing velocity of the plasma are a function of the current flowing through the arc, and with the current intensity constant, they can only be altered by changing the geometry of the water channel. In practice, the reproducibility of the coating quality is ensured by the stability of

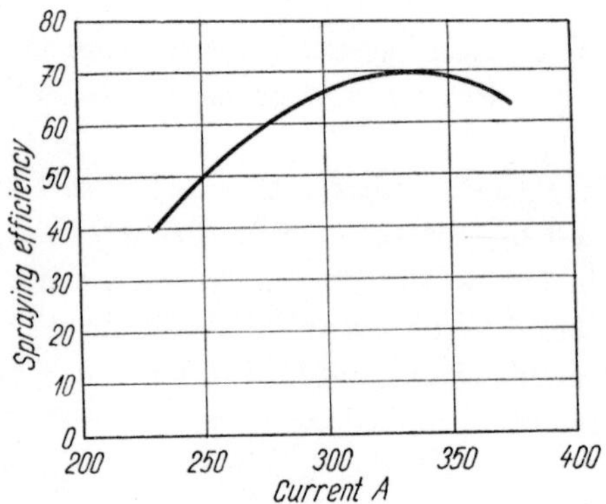

Fig. 118 Efficiency of water-stabilized torch in spraying aluminium oxide vs. current intensity

the correct current intensity (ascertained by trial and error) and of the amount of powder fed into the plasma.

The spraying efficiency depends upon the current. This efficiency of a torch with a stabilizing channel of 4 mm, spraying aluminium oxide of 60 to 75 μm grain size, is plotted against current intensity in Fig. 118. The maximum lies at 320 A, and above this value the efficiency slowly drops. If we compare this result with the static characteristic of the plasma torch (Fig. 70), we find that the highest efficiency is attained if the stabilizing channel is entirely filled by the arc column. A larger core of the plasma results in the good fusion of most particles. The optimum

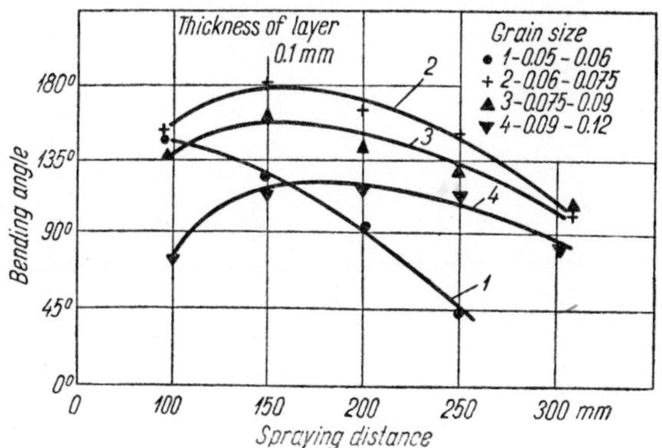

Fig. 119 Maximum permissible bending angles for various grain sizes—thickness of layer 0.1 mm

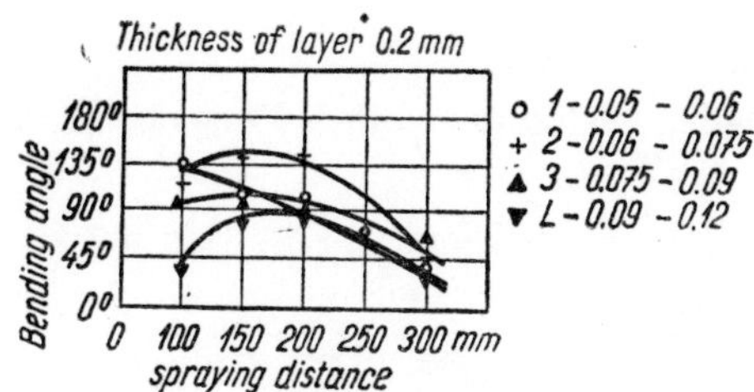

Fig. 120 Maximum permissible bending angles for various grain sizes—thickness of layer 0.2 mm

grain size of aluminium oxide was found to range between 60 and 75 μm [Ref. 3.2.13].

The adherence of the coating was tested by bending over a mandrel of 12 mm diameter. The bending angles at which the coats became defective are plotted—for various grain sizes—as a function of the spraying distance in Figs. 119 and 120 (for 0·1- and 0·2-mm layers, respectively).

The following optimum conditions for spraying aluminium oxide by means of a water-stabilized plasma torch have been derived from experimental results:

length of stabilization channel	10 mm
diameter of water vortex in stabil. channel	4 mm
d.c. input	51 – 55 kW
current	320 – 340 A
voltage	160 V
water pressure	3 atm gp
water flow through torch	8 l/m
diameter of rotary anode	100 mm
speed of rotary anode	3000 r.m.p
spraying output	3·8 to 4 kg/h
grain size of aluminium-oxide powder	60 – 75 μm
optimum spraying distance	150 mm

The advantage of spray-coating by means of water-stabilized plasma torches lies in the low operating costs.

3.2.5 Spray-Coating Materials and their Properties

Surfaces may be spray-coated for the following purposes:

- protection against heat;
- enhanced surface protection against abrasive wear;
- inhibition of surface corrosion;
- decoration.

The first two of these purposes are the most important ones. Because of their porosity, sprayed coats do not, as mentioned before, afford a perfect protection against corrosion hazards at elevated temperatures. Sometimes, however, corrosion on a surface protected by a sprayed coating is so substantially inhibited that the cost of the treatment seems fully justified. In the lower-temperature bracket (up to 300°C) plastics are used to seal the pores in the coatings, which is then a good corrosion protection. Hot-spraying of plastics by means of plasma torches permits various base materials to be quickly and cheaply surfaced with a coating resistant to corrosion (at lower temperature) or decorative in appearance.

Commercially available materials for deposition by means of plasma torches are supplied in powder form. The kinds most frequently used are discussed in the following paragraphs.

Metals

Tungsten. Although having the highest melting point of all the metals (3,380°C) tungsten is comparatively easy to deposit. It yields highly adhesive coatings of superior quality. The optimum grain size of the powder ranges between 10 and 40 μm. A lamellar structure of the coat, consisting of deformed oblong particles, indicates satisfactory fusion. If the coating is deposited in the presence of air, it contains a fairly large amount of oxygen; together with reaction products of the plasma, this forms shells deposited between the lamellae of the coating. Tungsten is for instance sprayed onto the nozzle walls of rocket engines. Various tungsten parts are produced by depositing a coat onto a metal or graphite pattern; the content of impurities in the deposit is reduced by spraying in a chamber filled with inert gas. The pattern is then removed and the workpiece is sintered and machined. Annealing at sintering temperature removes up to 90% of the gaseous contaminants contained in the workpiece [Ref. 3.2.7]. Tungsten parts produced by spraying are readily forgeable after annealing; their structure, purity, density and mechanical properties are virtually the same as those of tungsten products manufactured by the traditional method—cold pressing of powders and sintering of the compacts [Refs. 3.2.8, 3.2.17]. Spraying onto patterns, however, facilitates the production of more intricate shapes, and the products are cheaper. A method of spraying onto a removable metallic mandrel has been developed for workshop practice [Ref. 3.2.17].

Molybdenum. This metal, melting at 2,630°C, is intensively oxidized at temperatures above 700°C. Spraying by means of plasma torches yields high-quality coatings of excellent adhesion to steel surfaces; these need not be roughened, though thorough cleaning is essential. Photomicrographs of the transitional zone show a very thin layer consisting of an iron-molybdenum alloy or of oxides of both metals. The coating has good mechanical properties, is hard, and resistant to abrasion. Molybdenum is not employed by itself, but it is a very valuable binder metal for high-quality ceramic coatings such as are used in the lower temperature range.

Chromium. Chromium coatings with appropriately sealed pores are corrosion resistant and are also employed for surfacing bearings operating at elevated temperatures. The melting point is 1,895°C.

Nickel and Cobalt. In spray-coating techniques, both these metals are employed as binders mixed into ceramic powders. Nickel melts as 1,454°C and cobalt at 1,491°C.

Stainless Steel. Since austenitic stainless steels are mostly available in wire form, electric-arc metallizers are more convenient than plasma torches for spraying them. The melting point is 1,400°C. An intermediate metallized layer of stainless steel is often employed to improve the adherence of ceramic coats to steel bases. It reduces the wide difference existing between the coefficients of thermal expansion of the base metal and the ceramic surface coating. Moreover, being comparatively rough, the metallized interlayer favours the reliable adherence of the subsequent ceramic layer.

Oxides

Oxide coatings (mostly aluminium oxide) applied onto a metallized interlayer of stainless steel greatly improve the high-temperature corrosion resistance of the base metal.

Aluminium oxide melts at 2,050°C. Coatings produced by spraying aluminium-oxide powder are highly resistant to abrasive wear, constant in both oxidizing and reducing atmospheres up to 1,500°C, and exposed to baths of molten metal they show but little reactivity. Both thermal-insulating properties and dielectric strength are rather satisfactory (the coefficient of thermal conductivity measured on aluminium-oxide coatings is $0{\cdot}1$ kcal . m^{-1} . h^{-1} . deg^{-1} as compared with 11 kcal . . m^{-1} . h^{-1} . deg^{-1} for sintered corundum). The low price of powdered aluminium oxide invites its application on a large scale. Water-stabilized plasma torches can be used to advantage for spraying it. Pure aluminium oxide is liable to change in structure and specific gravity when hot-sprayed [Refs. 3.2.4, 3.2.8]. The high-temperature modification – αAl_2O_3

(corundum)—with a specific gravity of 4 kg/dm^3 changes into γAl_2O_3 of only 3·6 which is stable up to 1,000°C. This phenomenon is explained by the rapid cooling of the particles impinging on the base material. If the coating is heated above 1,000°C, the gamma oxide changes again into the alpha modification with the higher specific gravity; consequently the deposit contracts, cracks and increases in porosity. The addition of stabilizers (titanium dioxide or calcium oxide) inhibits the formation of the gamma oxide.

Titanium oxide melts at 2,125°C. Titanium dioxide coatings are hard and abrasion-resistant; they have the least porosity of all spray-coating materials and an excellent adhesion to the base material. Titanium oxide is added to other ceramic powders, especially to aluminium oxide, in order to improve the physical properties of the deposit. It is also a constituent part of various metal and ceramic powder mixtures.

Zirconium dioxide. The melting point is 2,710°C. The coatings are applied for similar purposes as aluminium oxide. Up to 2,250°C they are stable, hence the frequent application as heat-resistant protection, mainly in high-temperature nuclear reactors. For these purposes, however, the material has to be purified of comtaminants, particularly hafnium. Pure zirconium oxide stabilized with calcium oxide has a small effective cross-section for thermal neutrons.

Ceric oxide melts at 2,600°C. As a heat-resistant coating it is attributed catalytic properties in the combustion process, in addition to good thermal properties. It is sprayed onto the inner surfaces of combustion chambers.

Calcium metazirconate. It yields dense, hard, abrasion-resistant coats of excellent adhesion to the base material and remains stable in oxidizing atmospheres up to 1,300°C. The melting point is 2,333°C. Metal baths do not wet these coats, hence they are employed to great advantage on the inside of melting pots.

Zirconium silicate coatings have properties similar to those of calcium metazirconate and are employed for the same purposes. The melting point is 2,550°C.

Aluminium oxide mixed with titanium oxide. An addition of 3% of titanium oxide improves the structural properties of aluminium oxide; the melting point of this mixture is 1,995°C. A mixture in the proportion 50 : 50 is also used; it improves the adhesion of the coating to the base material and the resistance to rapid temperature changes. The stability in oxidizing atmospheres, however, is reduced to 1,110°C. The melting point is 1,921°C.

Mixtures of oxides with metals. Such mixtures are used for coatings exposed to rapid temperature changes. The most frequent compositions are aluminium oxide with nickel (60 + 40%), aluminium oxide with chromium (70 + 30%) and zirconium dioxide with cobalt (60 + 40%). The metallic components of these mixtures diffuse into the base material, which improves the adhesion of the coat at higher temperatures. Moreover, the admixture reduces the difference in the thermal expansion of the base metal and coating. Such deposits can stand very considerable thermal shocks. The best resistance is exhibited by coatings of zirconium dioxide with 40% cobalt. The hardness and abrasive resistance of the oxide remains unaffected, and the corrosion-inhibiting effect is enhanced.

Carbides

Tungsten carbide melts at 2,870°C. The high temperature of the plasma permits pure tungsten carbide to be deposited without any binder. The coatings are dense, homogeneous, and well fused. For their excellent resistance to abrasive wear, tungsten carbide coats are deposited on components exposed to the most stringent conditions. The nozzles of rocket engines, for instance, are thus protected against premature wear by rapidly streaming hot combustion products.

Tungsten carbide with 12% cobalt powder added yields coatings with improved adhesion and resistance to thermal shocks, in addition to good abrasive resistance. Other metal carbides are rarely used by themselves for spray coating.

Zirconium carbide (melting point: 3,540°C) and **boron carbide** (2,450°C) are mainly employed as constituents in cermets. Their application in the pure state does not meet with any major difficulties, although a certain amount of oxidation by atmospheric oxygen calls for the meticulous observance of optimum spraying conditions. Trials made with carbides of hafnium, titanium and tantalum bear out that these materials cannot be sprayed in the presence of air: oxidation causes the loss of half the carbon from hafnium carbide, 30% from titanium carbide, and 80% from tantalum carbide. Stetson and Hauck ascertained carbon losses from carbides sprayed in an argon atmosphere as a function of the oxygen content in the argon [Ref. 3.2.14]. An oxygen content of 1% or 0·1% in the shielding gas reduced the carbon content from the original value of 20·7 to 17·8 or 20·5% respectively. Correspondingly, the oxygen content in the coating rose from 0·25% in the original powder to 2·5% in the first, and 0·27% in the second case. Tantalum carbide proved even more sensitive: an oxygen content of 0·1% in the argon caused the carbon content in the coating to drop from the original value of 6·4 to 5·6% with the oxygen content rising from 0·07 to 0·24%. The carbide coats deposited under these conditions are of satisfactory quality, with very good adhesion, 90% minimum density, and resistant to impacts. They find application in rocketry.

Plastics

The spraying of penton-, nylon- and teflon-type plastics as well as of most epoxy resins became feasible by new anode designs for plasma torches, in which a water-cooled powder-supply tube prevents the plastic powder from being prematurely fused in the tube orifice. Plasmadyne were the first to market a plasma torch for spraying plastics. Such plasma-deposited coatings are employed as corrosion protection at low temperatures, thermal and electrical insulation, as well as for decorative purposes. The dielectric properties of plastics are not appreciably affected by hot spraying. The breakdown voltage of a 0·25 mm coating is 20 kV [Ref. 3.2.21]. Plastics can be sprayed onto steel, aluminium, glass, ceramics or other plastics, and the minimum thickness of the coating is 0·05 mm. Coats up to 6 mm thick are currently produced, and even thicker layers can be obtained.

The maximum grain of the powder is 150 μm. A tolerance of 0·025 mm is acceptable for the thickness of the overlay on horizontal or cylindrical surfaces, and the corresponding value for the inner surface of tubes is 0·075 mm. Fig. 121 shows an adapted Plasmadyne torch, coating the outer surface of a cylinder with a plastic.

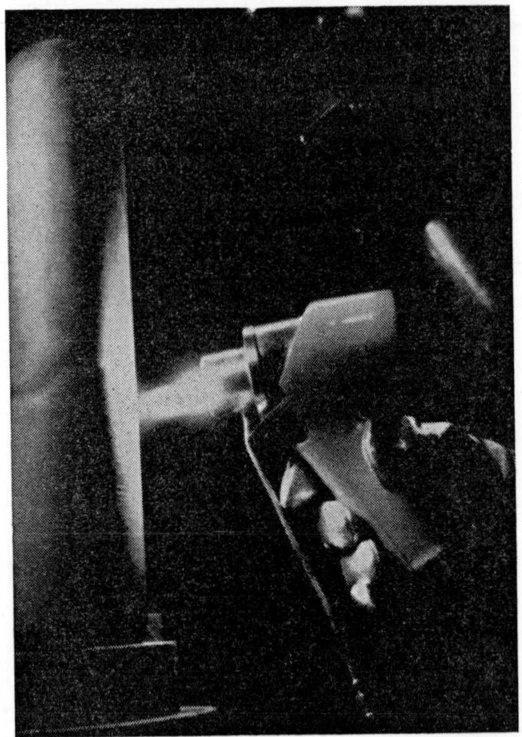

Fig. 121 Plasmadyne torch adapted for spraying with plastics (By courtesy of Plasmadyne Corp.)

Hot spraying of plastics has the advantage of obviating the need of curing in an oven.

3.2.6 Hard Facing with Metals by Plasma Torches

Hard facing with metals by means of an oxy-acetylene flame or an electric arc has now been practised for many years, particularly in mechanical engineering. This method consists in welding ("building up") a deposit of high-quality material onto a machine part made of cheap carbon or low-alloy steel, and thus protecting its surface against intensive wear, heating or chemical influences to which it is exposed in operation.

Hard facing with an oxy-acetylene flame is slower than with an electric arc, but it yields a smooth built-up surface with no cracks or pits and with a minimum of dilution by the parent metal (1%).

Building-up with an electric arc is quicker, but the concentrated heat supply increases the tendency to crack, particularly in parts of intricate shape. Moreover, hard facing has to be done in several successive layers, since the dilution by parent metal is up to 30% in the first layer. The surface (welding beads) is more uneven than in gas-welded build-up, hence finish machining is more expensive. The deposited metal has to solidify successively from the base material toward the surface; otherwise shrinkage cavities and gas holes are formed in the deposit. In flame welding, the heat supply to the built-up surface is more easily controlled then in arc-welding.

In order to accelerate hard facing by flame while preserving the advantages of this method, the "thermospray" procedure divides the process into two operations. An oxy-acetylene spray gun developed for coating with ceramic materials (cf. Chap. 3.2.2) is fed with a hard-alloy powder and applies a coat to the previously blasted surface of the base metal. The powder feed is then stopped, and the entire workpiece is heated to the melting temperature of the coating. This can be done by the spray gun or by inductive heating, and a protective atmosphere may be used in this operation. The remelting of the coat eliminates the pores, enhances the compactness of the built-up metal, and secures the diffuse connection between base and weld metal. It is a special advantage of this method that the dilution* is very low, hence a comparatively thin layer, well connected by diffusion, secures the excellent qualities of the hard facing alloy. The main drawback of the "thermospray" method consists in the heating of the entire workpiece to a high temperature, which in some cases is inadmissible.

The development of electric-arc surfacing was aimed at reducing the dilution and mechanizing the surfacing process. New methods were evolved: electric-arc surfacing with a filler rod; atomic-hydrogen (arc-

* In this connection, the term dilution is used to express the contamination of the weld metal by the parent metal; it is the ratio [per cent] of melted base metal to the total weld deposit.

atom) surfacing in which the filler rod is melted by the recombination heat of hydrogen; surfacing with filler material in a stream of argon; the Sigma process in which the arc burns between two rod-type electrodes in a stream of argon. Surfacing by submerged arc welding—with a single arc or three arcs—was introduced for the automatic deposition of materials available in wire form. Apart from increasing productivity, the utilization of three arcs has the advantage that only the first arc heats the base metal, and the following two remelt the filler material already deposited. An automatic submerged-arc surfacing welder developed in the USSR uses a strip electrode—1 mm thick and 100 mm wide—to deposit stainless chrome-nickel steels on large surfaces. Mention should also be made of vibratory surfacing which permits the deposition of low- and medium-alloy steels in thin layers.

This development has now been extended by a surfacing method which uses specially adapted plasma torches and combines some of the

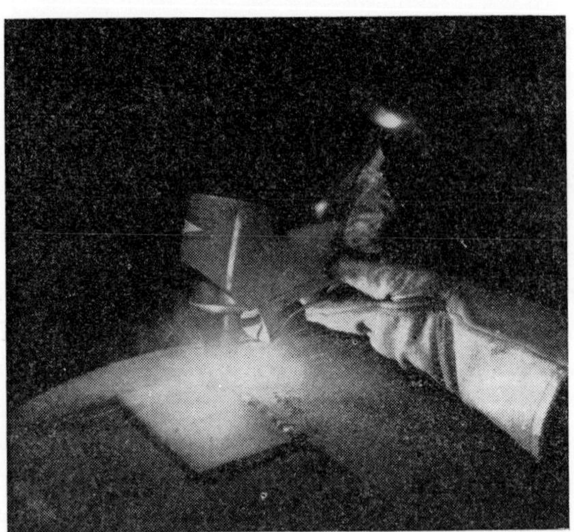

Fig. 122 Hand-operated Plasmadyne torch adapted for weld-surfacing, and appearance of deposit (By courtesy of Plasmadyne Corp.)

advantages of electric-arc and flame surfacing. The principle has been described in Chap. 2.2.7, and devices for manual (Plasmadyne) and automatic surfacing (Linde) in Chap. 2.4.1. A wide and smooth surface layer 0·25 to 4·5 mm thick is the main advantage of surfacing by plasma welding; the productivity of hand-operated torches is higher than obtained by any other manual method. Fig. 122 shows a hand-operated Plasmadyne torch and also the appearance of the deposit. The wiring of the current sources (cf. Figs. 63 and 64) facilitates the control of the transferred part of the arc and hence of the heat supply to the surface of the

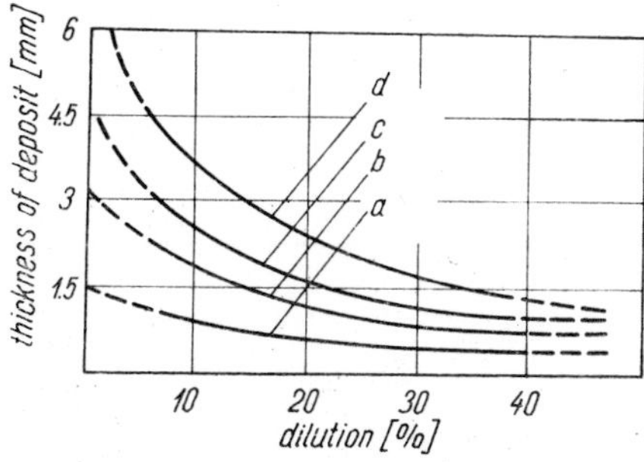

Fig. 123 Interdependence of output per hour, thickness of deposit and percentage of dilution in the deposition of cobalt-tungsten-chromium-boron hard-facing alloy by means of a Linde plasma torch

base metal. The output in mechanical surfacing by means of the Linde torch ranges between 0·45 and 5·5 kg per hour. The dilution of the deposit depends on the melting point of the hard facing alloy and on the process characteristics. An alloy of cobalt, chromium and tungsten is, for instance, more contaminated than the same alloy with an addition of boron which lowers the melting point of the hard facing metal. The percentage of dilution increases with the surfacing output per hour and with the decreasing thickness of the deposited layer.

Fig. 123 illustrates this relationship for a cobalt-chromium-tungsten-boron hard-facing alloy deposited with a Linde torch. The curves are for various outputs: (a) – 0·45, (b) – 0·90, (c) – 2·72 and (d) – 5·44 kg/h. These graphs show that with outputs such as are usual in operation, dilution can be limited to about 5% if the thickness is chosen accordingly. Fig. 124 presents ground sections through a deposited layer of a cobalt alloy and shows how easily the process is controlled and adapted to various purposes.

The width of the deposit is also controllable in a relatively wide range, from a few millimetres up to 1 3/4 in. (45 mm). An automatic device, controlled by a current relay in the circuit of the transferred arc, oscillates the torch head at right angles to the direction in which the bead is being deposited. The amplitude and frequency of the oscillation

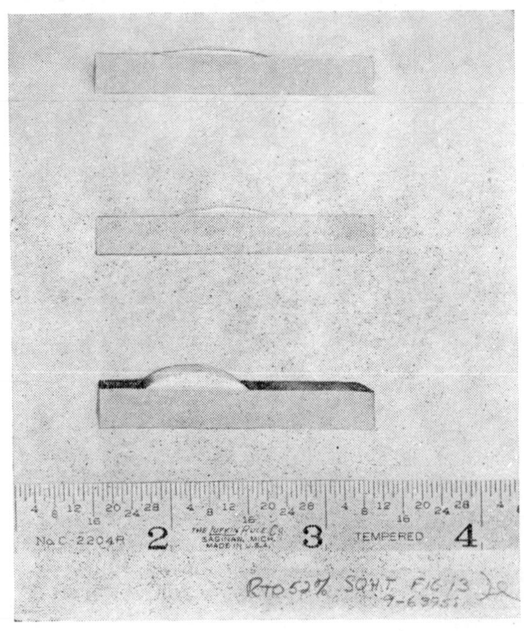

Fig. 124 Ground sections of cobalt-base deposit (By courtesy of Union Carbide International Co., N.Y.)

are preset to suit the selected width (cf. the description of the device in Chap. 2.2.4).

For the time being, it is not an economic proposition to use a plasma torch for weld-surfacing with low- or medium-alloyed or chrome-nickel steels, such as are available in wire form. The most appropriate application of plasma torches is hard facing with cobalt-base alloys containing more than 25% chromium, 4 to 18% tungsten, and other elements like molybdenum, niobium, vanadium, nickel and boron [Ref. 3.2.23]. In addition to cobalt-based alloys, there are powdered nickel- and iron-based alloys suitable for application by means of plasma torches, as are heterogeneous built-up coatings containing tungsten carbide with a nickel-, cobalt,- or iron-based binder alloy. The carbide and binder-metal powders are mechanically mixed, and the proportion of the tungsten carbide can be up to 85% by weight.

The dendritic microstructure of coatings built up by the plasma technique is finer then obtained in oxy-acetylene-welded layers, although the quality of the coat is very good in either case. The coarser structure of the oxy-acetylene-deposited layer is due to the slower cooling, since the whole workpiece has previously been heated.

Plasma torches are also used to advantage where the quantity production of thin built-up layers is required. The range of application is the same as with the methods hitherto used [Ref. 3.2.24]. The industrial application of most building-up methods using plasma torches is favoured by their lower costs, particularly where technologists make full use of some special opportunities for further improvements in quality and economies in expenses which this technique offers [Ref. 3.2.23].

3.3 The Plasmodynamic Conversion of Heat into Electrical Energy

3.3.1 Introduction

By plasmodynamic conversion, electric power can be obtained from the energy of an electrically conductive fluid flowing through an electric or magnetic field. Most frequently this fluid is an ionized gas, although in principle conductive vapours of liquids can be used to this end.

Among the new principles of energy conversion (Fig. 125), the plasmodynamic method is most important, because it seems best

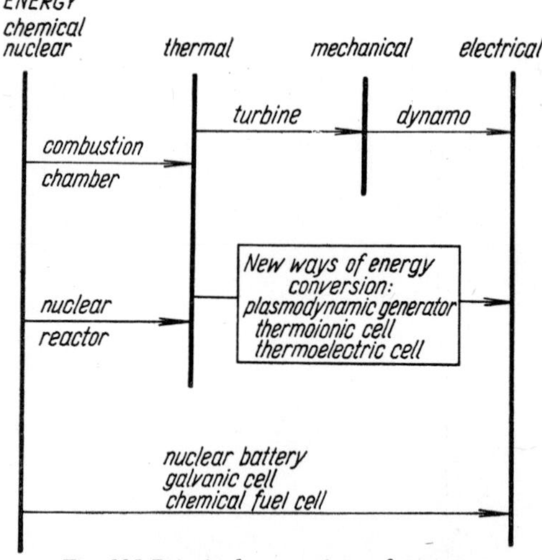

Fig. 125 Principal conversions of energy

qualified for power generation in large plants. Since the initial form of the energy is heat, the efficiency of the plasmodynamic conversion is limited by the Second Law of Thermodynamics. The plasmodynamic conversion of energy is sometimes called "direct", because thermal energy is converted into electric power without the need of any intermediate transformation into mechanical energy.

The principle of plasmodynamic conversion has been known as early as the thirties of last century, when Arago and Faraday performed their basic experiments with electricity. In 1831, Faraday studied mercury flowing through a glass tube between the poles of a magnet and suggested that the motion of electrically conductive sea water in the magnetic field of the Earth should be used to generate electricity. However, his experiments with the tides at the mouth of the river Thames failed: the values of the conductivity of sea water, terrestrial magnetism and the flow rate of water in nature are too small and their interaction does not produce the requisite conditions for the exploitation of Faraday's idea in this form.

Although the principle of the plasmodynamic generator is identical with that of the usual rotary electric generators, solid electric conductors have for a full century held a position of exclusivity in the generation of electric power. Not until the middle of our century did rapidly extending knowledge in the field of plasma physics and astrophysical studies in a new field—magnetohydrodynamics—bring about a radical change.

Plasmodynamic generators can be classified in two main groups: the conductive fluid flows through an external electric field in one of them, and through a magnetic field in the other. The difference between these two types clearly stands out in the equation for the current density:

$$\mathbf{j} = \sigma(\mathbf{E} + \mathbf{v} \times \mathbf{B}) + \varrho_e \cdot \mathbf{v}, \tag{3.3.1}$$

where σ is the electric conductivity of the fluid, $\mathbf{E}$ electric field intensity, $\mathbf{v}$ the flow rate of the fluid in the laboratory system in which $\mathbf{E}$ is measured, and ϱ_e the charge density. The first member on the right-hand side, expressing the conduction current, prevails in magneto-plasmodynamic generators, while the second member, which stands for convection current, is decisive in electro-plasmodynamic or convection-type generators.

Magnetoplasmodynamic or as they are often called magnetohydrodynamic generators are of two different kinds: induction- and conduction-type generators. In conduction generators, the plasma flows between electrodes which have a similar purpose as the brushes in rotating electrical machinery. Induction generators have no electrodes, and the principle of their operation is similar to that of normal rotating induction generators. In both cases the plasma flows through an external magnetic field. If the magnetic field is intensified, the electric conductivity of the plasma becomes anisotropic. Where this phenomenon exerts a dominant influence on the properties of the generator, the latter is called a Hall-type magnetoplasmodynamic generator.

It is a distinguishing feature of electroplasmodynamic generators that the flowing gas transfers the electric charge against the direction of the electric field, in the same way as it moves in electrostatic generators with a moving insulation strip. Large-size electroplasmodynamic generators were constructed already before 1940; lately, however, they have been eclipsed by magnetoplasmodynamic generators. As our attention must be concentrated on the latter, we limit our discussion of electroplasmodynamic generators to a brief description.

3.3.2 Electroplasmodynamic Generators

Electroplasmodynamic generators are characterized by their ability to produce electric power of high potential and low current. Fig. 126 schematically represents a 1·2 MV generator by Pauthenier [Ref. 3.3.1] who studied the subject in considerable detail. Blower B circulates the working gas in a closed circuit consisting of tubes T_1 and T_2. In ionizer I the gas acquires an electric charge which it surrenders to collector C. The collector is part of a metal sphere which is charged with a high positive or negative potential, depending on the voltage which is led into the ionizer. The maximum potential to which the sphere can be charged solely depends on the equilibrium between the charge fed to it and the leakage current.

The ionizer employed by Pauthenier was a system of metal fibres, about 0·1 mm in diameter, stretched along the centre line of thin tubes; the fibre was loaded with a voltage of about 10 kV.

The current generated was in the range of 10 μA, and the output a few watts. Both the current and the output rose a thousand times if instead of the ionized gas solid particles of 10^{-3} mm diameter were used as charge carriers.

In an apparatus elaborated by AEG designers [Ref. 3.3.2], steam was ionized in ionizer *I* (cf. Fig. 127) and flowed through an

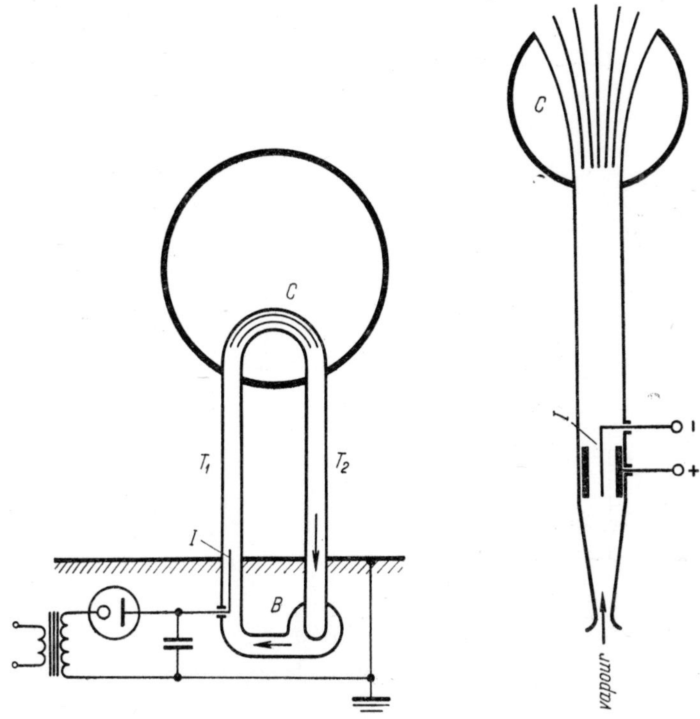

Fig. 126 Pauthenier's electroplasmodynamic generator

Fig. 127 AEG electroplasmodynamic generator

insulating tube into collector *C* which was located in a metal sphere and shaped as a diffuser.

Several authors suggested generators with steam or vapour flowing in two different streams of opposite polarity. Braun's design (Fig. 128) dates from the year 1917. Steam is generated in boiler Z and led into a pair of nozzles which make it stream round the tips H_1 and

H_2 in an electric field. Positive or negative ions are thus formed which surrender their charges on grid-type collectors. A design proposed by Semenov in 1937 (Fig. 129) is based on the same principle.

Generators operating on mercury vapours or gases of high dielectric strength gave the highest current intensities. Not even

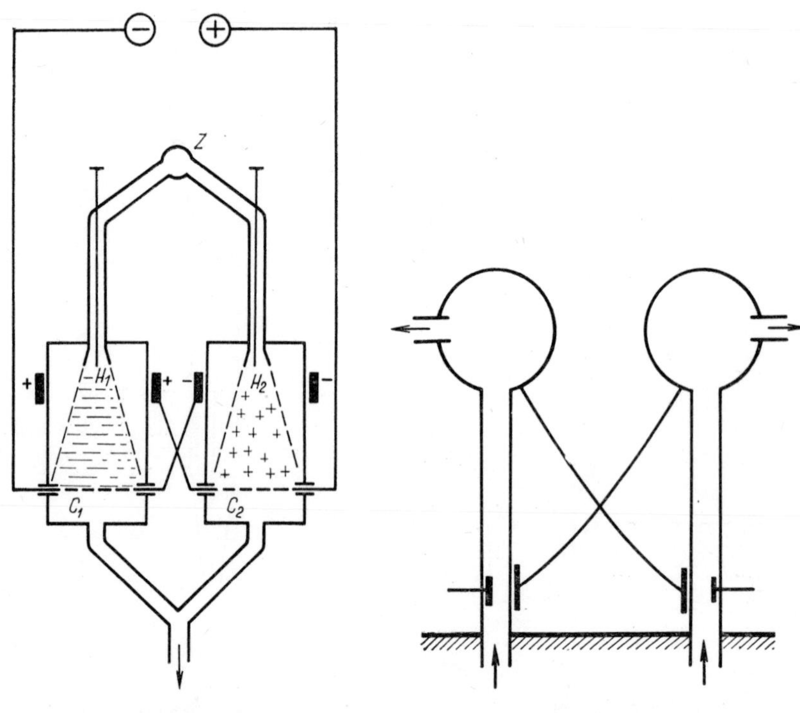

Fig. 128 Braun's electroplasmodynamic generator

Fig. 129 Semenov's electroplasmodynamic generator

in these cases, however, did the current density exceed a few tens of A m^{-2}, and the specific output was also low. Thus electroplasmodynamic generators can be used to obtain very high voltages, but they are unsuitable for generating purposes in electric power stations.

3.3.3 Simplified Description of the Magnetohydrodynamic Conversion of Energy

Electromotive force $\mathbf{v} \times \mathbf{B}$, induced in a plasma that flows perpendicularly to the direction of a magnetic field, causes an electric current to flow through an external closed circuit. Fig. 130 indicates the elementary

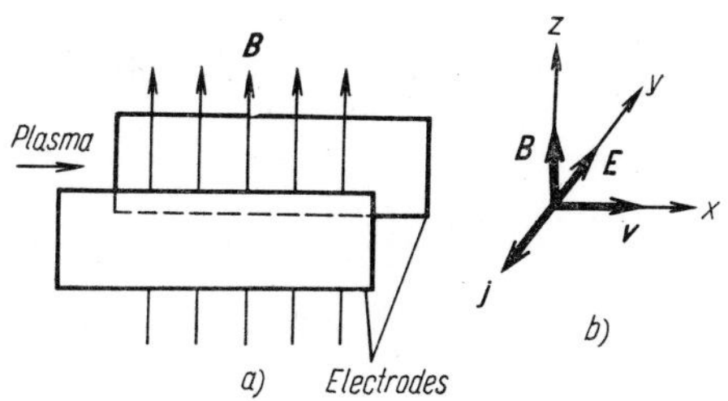

Fig. 130 Magnetohydrodynamic conduction generator—Diagram and direction vectors of principal quantities

configuration of a conduction generator and the direction vectors of the principal quantities. For the sake of simplicity we shall assume in this Chapter that the magnetic field strength $\mathbf{B}$, the velocity $\mathbf{v}$ of the plasma and the electric field $\mathbf{E}$ in the space between the electrodes lie along the axes of an orthogonal coordinate system; at the same time we consider σ, the electric conductivity of the plasma, a scalar quantity. In discussing the plasmodynamic conversion we shall assume the permittivity and permeability of the plasma to be constant and their numerical values to agree with those for a vacuum.

The electric power supplied by the generator is drawn from the energy of the plasma – either its kinetic energy (enthalpy i = const, $v \neq$ const) or potential energy ($i \neq$ const, v = const). In the first alternative, the gas is decelerated as in an impulse turbine, while in the second case the gas temperature is reduced as it is in a reaction turbine. In practice both instances are combined. The mechanism of energy transfer from the flowing plasma into the external load is the result of a complex interaction between charged particles and the external electromagnetic field. Neutral particles transfer energy to charged gas particles by collisions.

As a current of density $\mathbf{j}$ flows through the gas between the electrodes, the Lorentz force $\mathbf{j}\times\mathbf{B}$ acts upon a unit element of the gas; it retards the gas and gives rise to a pressure gradient in it

$$\nabla p = \mathbf{j}\times\mathbf{B}. \tag{3.3.2}$$

The action of the Lorentz force in the unit volume of gas generates a power

$$\mathbf{v}\,.\,\nabla p = \mathbf{v}\,.\,(\mathbf{j}\times\mathbf{B}) = -\mathbf{j}(\mathbf{v}\times\mathbf{B}). \tag{3.3.3}$$

As in every conductor through which flows an electric current, Joulean energy dissipation also arises in the working medium of the magnetoplasmodynamic generator; it amounts to j^2/σ per unit volume of the gas. Therefore, the generator output power per unit volume of plasma is

$$P = -\mathbf{j}\,.\,\mathbf{E} = \mathbf{j}\,.\,(\mathbf{v}\times\mathbf{B}) - \frac{j^2}{\sigma}, \tag{3.3.4}$$

and the current density

$$\mathbf{j} = \sigma(\mathbf{E} + \mathbf{v}\times\mathbf{B}) = \sigma\,.\,\mathbf{E}^*, \tag{3.3.5}$$

where $\mathbf{E}^*$ refers to the coordinate system which moves together with the gas. $\mathbf{E}$ is the electric field measured in a stationary laboratory system at a general external load. Eq. (3.3.5) is the general form of Ohm's law for a conductor moving in a magnetic field.

Joulean heat does not involve energy losses in the same sense as in classical electrical machinery, because it remains in the gas and can be used for conversion into electric energy just as every other kind of heat. However, Joulean heating increases the entropy of the entire thermodynamic system as do friction losses.

In electrical machines of classical types, $|\mathbf{E}| \to |\mathbf{v}\times\mathbf{B}|$; nevertheless, thanks to the high conductivity of copper they are able to yield high currents. The conductivity of the plasma in magnetoplasmodynamic generators is lower by a factor of about a million; hence the value $|\mathbf{v}\times\mathbf{B}| - |\mathbf{E}|$ must be fairly high. The relationship between the two quantities is expressed by what is called the load factor K, defined as

$$K = \frac{\mathbf{E}}{\mathbf{E}_0}, \tag{3.3.6}$$

where $\mathbf{E}_0$ refers to the open-circuit electric field. It follows from Eq. (3.3.5) that

$$\mathbf{E}_0 = -\mathbf{v} \times \mathbf{B}. \tag{3.3.7}$$

Using Eq. (3.3.4), we also have

$$K = \frac{\mathbf{E}\,.\,\mathbf{j}}{\mathbf{E}_0\,.\,\mathbf{j}} = \frac{P}{P + \dfrac{j^2}{\sigma}} = \frac{R_{\text{ext}}}{R_{\text{ext}} + R_{\text{int}}}, \tag{3.3.8}$$

and K thus indicates the ratio of the external loading resistance R_{ext} to the total resistance $R_{\text{ext}} + R_{\text{int}}$, R_{int} being the internal resistance of the generator.

Under the assumption that the vectors $\mathbf{v}$, $\mathbf{B}$ and $\mathbf{E}$ are mutually perpendicular, the specific output power of the generator is

$$P = -Ej = -E\sigma(vB + E) = \sigma v^2 B^2 (1 - K)\,K. \tag{3.3.9}$$

Since the expression $(1 - K)\,K$ attains its maximum value at $K = \dfrac{1}{2}$, the highest specific output power obtainable from the generator is

$$P_{\text{max}} = \frac{1}{4}\,\sigma v^2 B^2, \tag{3.3.10}$$

and the maximum output power of the entire generator is

$$P_{\text{total}} = \frac{1}{4}\,\sigma v^2 B^2 A l; \tag{3.3.11}$$

A is the cross-sectional area of the generator chamber in a plane which is perpendicular to the gas velocity, and l the length of the chamber. We see that the output power rises in proportion to the volume of the generator ($P_{\text{total}} \sim L^3$); the thermal losses, however, depend on the area of the chamber walls ($Z_t \sim L^2$), hence

$$\frac{Z_t}{P_{\text{total}}} = \mathrm{f}(P_{\text{total}}^{-1/3}). \tag{3.3.12}$$

While the thermal losses relatively decrease with increasing output (i.e. with the size of the generator), they are so high a proportion of the output of small generators that it is virtually impossible to construct a small-size magnetoplasmodynamic generator of satisfactory efficiency.

Let us now consider the thermal and pressure gradients between the inlet and outlet sections of the generator. We designate the quantities at the inlet and outlet sections by the indices 1 and 2, respectively. By integrating Eq. (3.3.2), we obtain for the one-dimensional case $\left(\nabla = \frac{\partial}{\partial x}\right)$

$$p_1 - p_2 = jBl. \tag{3.3.13}$$

The length and output power of the entire generator depend on the pressure gradient $\Delta p = p_1 - p_2$ by the following simple relations:

$$l = \frac{\Delta p}{\sigma v B^2(1 - K)} \tag{3.3.14}$$

and

$$P_{\text{total}} = vAK\,\Delta p. \tag{3.3.15}$$

In Fig. 131, curves plotted according to the equations derived show the effects of the various quantities on the principal parameters of the generator.

Let us further assume that the cross-section of the chamber and the velocity of the gas flowing through it do not greatly vary, and that A and v are the mean values of these quantities. Moreover, let γ_m be the specific gravity of the gas at the mean operating pressure p_m and the mean temperature T_m—these quantities in terms of the thermal and pressure gradients being

$$\left.\begin{aligned} p_m &= \frac{1}{2}(p_1 + p_2) = \frac{1}{2}(\Delta p + 2p_2) \\ T_m &= \frac{1}{2}(T_1 + T_2) = \frac{1}{2}(\Delta T + 2T_2) \end{aligned}\right\}; \tag{3.3.16}$$

from the equation of state for the perfect gas:

$$\gamma_m = \frac{1}{R}\cdot\frac{\Delta p + 2p_2}{\Delta T + 2T_2}. \tag{3.3.17}$$

The amount of gas flowing through the chamber in unit time is $Av\gamma_m$. If all the heat removed from the gas as it passes through the chamber is transformed into electric energy, then the electric output power is

$$P_{total} = Av\gamma_m c_p \Delta T = Av \frac{\Delta p + 2p_2}{\Delta T + 2T_2} \cdot \frac{c_p}{R} \Delta T =$$

$$= Av \frac{\Delta p + 2p_2}{\Delta T + 2T_2} \cdot \frac{\varkappa}{\varkappa - 1} \cdot \Delta T. \tag{3.3.18}$$

Here, c_p is the specific heat of the gas at constant pressure, and $\varkappa$ is Poisson's constant. We equate Eq. (3.3.18) to Eq. (3.3.15) and compute the thermal gradient across the generator:

$$\Delta T = \frac{2T_2}{\frac{\varkappa}{\varkappa - 1} \frac{\Delta p + 2p_2}{K \Delta p} - 1}. \tag{3.3.19}$$

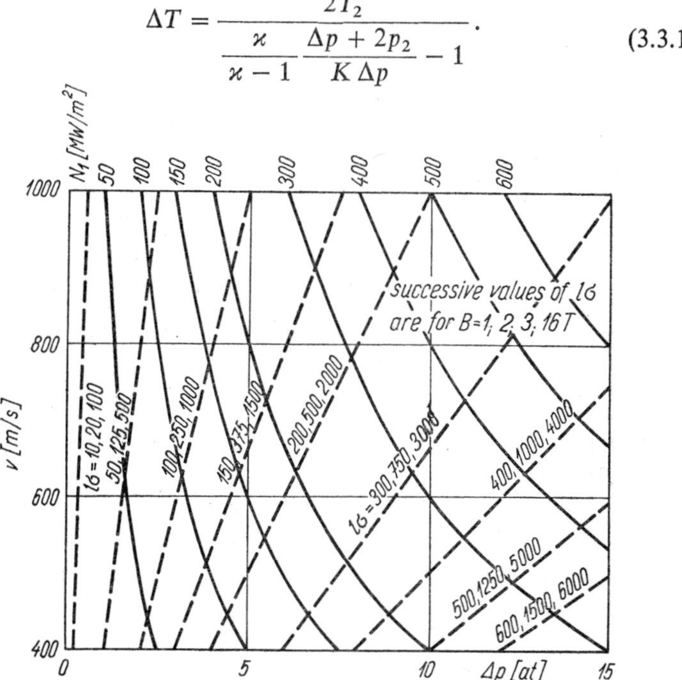

Fig. 131 Principal parameters of magnetoplasmodynamic generator and their dependence on gas velocity and pressure

v — gas velocity, p — pressure gradient of generator, l — length of chamber, σ — specific conductivity of plasma, B — magnetic flux density in channel, N_1 — specific generator output per square metre of channel section

The maximum possible efficiency yields the ideal Carnot cycle with

$$\eta_C = \frac{T_1 - T_2}{T_1} = \frac{\Delta T}{\Delta T + T_2} = \frac{2}{1 + \dfrac{\varkappa}{\varkappa - 1} \cdot \dfrac{\Delta p + 2p_2}{K\,\Delta p}}. \quad (3.3.20)$$

The results obtained by solving this equation are graphically represented in Fig. 132. The calculation has been greatly simplified

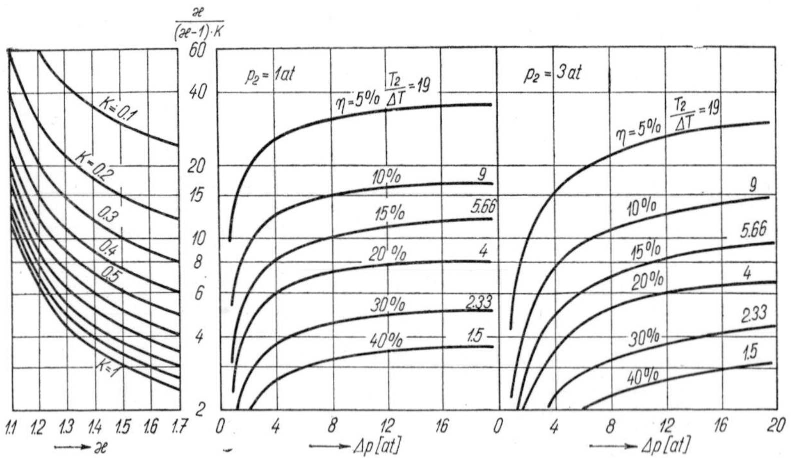

Fig. 132 Generator efficiency vs. pressure gradient

and the results have a qualitative rather than a quantitative significance. Nevertheless, important conclusions may be drawn from it.

The substance with the highest possible Poisson constant proves to be the most advantageous working gas. In this case, the generator has a high efficiency and a loading factor approaching the optimum value of $K = \frac{1}{2}$; its specific output power is high, its dimensions for a given total output are small, hence the prime cost is low. Actually, the value of $\varkappa$ cannot be affected by the choice of the working gas, except if the generator is part of a closed thermodynamic cycle. Then it is advantageous to use argon or helium, whose Poisson constants have very high values; moreover, these gases have thermal and nuclear properties which prove valuable when the gas is heated in a nuclear reactor. In an open cycle the working gas consists of combustion products. Their composition can hardly be influenced, and in this case the optimalization of the magnetoplasmodynamic conversion by Poisson's constant is therefore of no practical significance; in the given range of operating temperatures $\varkappa$ will assume a value of less than 1.4.

With rising exit pressure, the efficiency of the magnetoplasmodynamic conversion rapidly decreases, particularly at low pressure gradients in the generator. The lowest possible pressure in the generator chamber is also desirable from the designer's point of view.

3.3.4 The System of Magnetohydrodynamic Equations

Computations of magnetoplasmodynamic generators are generally based on the assumption that the working medium can be considered continous. A plasma satisfies this condition, if the mean free path of its particles is so small that the characteristic quantities of the plasma and the field do not undergo any considerable changes over this length. The flow of the conductive medium in the magnetic field is then described by three kinds of equations, related to the electromagnetic field, to hydrodynamics and to thermodynamics, respectively. Together they constitute what is called the system of magnetohydrodynamic equations; unlike their ordinary forms, they contain terms which take into account the connections between the various groups of equations. As in every study of phenomena occurring in the magnetic field, data on permittivity, permeability and electric conductivity of the medium must also be added.

In recent years attempts are being made to derive the magnetohydrodynamic equations directly from Boltzmann's equation, that is from the notion of the particle character of plasma. Using the first of these methods, we shall discuss the basic problems involved in the adjustment and utilization of magnetohydrodynamic equations for calculating the plasmodynamic conversion of energy.

Let as assume that the plasma has a scalar electric conductivity σ, and that the magnitudes of its permittivity ε and permeability μ are the same as in a vacuum. We shall demonstrate in Chap. 3.3.6 that the first of these assumptions has a limited validity; the second is always satisfied under the conditions studied.

The relation between the quantities of the electromagnetic field is then in a general way expressed by Maxwell's equations

$$\operatorname{curl} \mathbf{H} = \mathbf{j} + \frac{\partial \mathbf{D}}{\partial t} \tag{3.3.21}$$

$$\operatorname{curl} \mathbf{E} = -\frac{\partial \mathbf{B}}{\partial t} \tag{3.3.22}$$

$$\operatorname{div} \mathbf{B} = 0 \tag{3.3.23}$$

$$\operatorname{div} \mathbf{D} = \varrho_e \tag{3.3.24}$$

and

$$\mathbf{D} = \varepsilon \mathbf{E}; \qquad \mathbf{B} = \mu \mathbf{H} \tag{3.3.25}$$

$$\mathbf{j} - \varrho_e \mathbf{v} = \sigma \,.\, \mathbf{E}^* = \sigma(\mathbf{E} + \mathbf{v} \times \mathbf{B}) \tag{3.3.26}$$

Let us first discuss the magnitude of the displacement current $\frac{\partial \mathbf{D}}{\partial t}$ and the convection current $\varrho_e \mathbf{v}$. We introduce the characteristic length L over which the magnitude of the variable quantities undergoes a significant change. In our considerations, the value of $\mathbf{E}$ is in the range of $vB = \mu v H$. Then $\frac{\partial \mathbf{D}}{\partial t}$ ranges about $\mu \varepsilon v^2 \frac{H}{L}$, and curl $\mathbf{H}$ about $\frac{H}{L}$; because of $\varepsilon\mu = \frac{1}{c^2}$ where c is the velocity of light, the value of curl $\mathbf{H}$ is by a factor of c^2/v^2 greater than the displacement current. Similarly, Eq. (3.3.24) shows that the value of ϱ_e ranges about $\varepsilon \mu v \frac{H}{L}$, and the convection current is thus smaller than curl $\mathbf{H}$ by a factor of v^2/c^2. In all our cases, the medium moves at a velocity which is far smaller than that of light; consequently, we can neglect the displacement and convection currents, and the Eqs. (3.3.21) and (3.3.26) are thus simplified to

$$\operatorname{curl} \mathbf{H} = \mathbf{j} \tag{3.3.27}$$

and

$$\mathbf{j} = \sigma \mathbf{E}^* = \sigma(\mathbf{E} + \mathbf{v} \times \mathbf{B}). \tag{3.3.28}$$

If we connect Eq. (3.3.24) with the law of conservation of the electric charge $\operatorname{div} \boldsymbol{j} = -\frac{\partial \varrho_e}{\partial t}$ then

$$\operatorname{div} \boldsymbol{j} + \frac{\partial}{\partial t} \operatorname{div} \mathbf{D} = 0,$$

and after permutating the sequence of differentiation in the second member of the equation and neglecting the displacement current, we have instead of Eq. (3.3.24)

$$\operatorname{div} \boldsymbol{j} = 0. \tag{3.3.29}$$

In the theory of classical electrical machinery, Maxwell's equations are connected with those of mechanics (e.g. in studying the rotor motion in an electric motor). The difficulty in solving magnetohydrodynamic problems lies in the fact that the "mechanical" part of the equations consists of hydrodynamic equations which in many cases have not yet been solved mathematically.

Such is in the first place the equation of motion that expresses the dynamics of fluids. It is based on Navier-Stokes' equation for a viscous compressible fluid, which for a constant dynamic viscosity has the form

$$\varrho \frac{d\mathbf{v}}{dt} = -\operatorname{grad} p + \frac{1}{3} \eta \operatorname{grad} \operatorname{div} \mathbf{v} + \eta \nabla^2 \mathbf{v} + \mathbf{F}. \tag{3.3.30}$$

The left-hand side of the equation expresses the force of inertia acting upon an elementary volume of a fluid having the density ϱ. Since the fluid velocity is a function of both time and coordinates, its total derivative with respect to time is expressed by the vector equation

$$\frac{d\mathbf{v}}{dt} = \frac{\partial \mathbf{v}}{\partial t} + (\mathbf{v} \,.\, \nabla) \,.\, \mathbf{v}, \tag{3.3.31}$$

where the first and second members on the right-hand side stand for local and convective acceleration, respectively.

The meaning of the individual terms on the right-hand side of Eq. (3.3.30) is as follows: The first is the force generated in the unit volume by a variable pressure, and the second the force due to supplementary friction as the fluid is compressed. The third member takes into account the friction between the layers of the flowing fluid, and the fourth summarizes the action of all volume forces due to other than hydrodynamic causes. Generally speaking, the forces $\mathbf{F}$ are caused by the action of external fields upon the unit volume of fluid. The principal forces $\mathbf{F}$ are: the gravitational force $\varrho \cdot \mathbf{g}$ ($\mathbf{g}$ being the acceleration of terrestrial gravity); the electrostatic force $\varrho_e \cdot \mathbf{E}$; and the electromagnetic force $\mathbf{j} \times \mathbf{B}$, generated by the interaction of the electric currents flowing through the conductive fluid and the magnetic field. The gravitational and electrostatic force may be neglected, as a rule; the electromagnetic force, however, is an integral constituent of the magnetohydrodynamic equation of motion.

Therefore the equation of motion of a viscous compressible fluid has the following form:

$$\varrho\left[\frac{\partial \mathbf{v}}{\partial t} + (\mathbf{v} \cdot \nabla) \cdot \mathbf{v}\right] = -\operatorname{grad} p + \frac{1}{3}\eta \operatorname{grad} \operatorname{div} \mathbf{v} + \eta \nabla^2 \mathbf{v} + \mathbf{j} \times \mathbf{B}. \tag{3.3.32}$$

For an incompressible fluid, div $\mathbf{v} = 0$, and the second member on the right-hand side of the equation vanishes. If viscosity may be neglected, the second and third members equal zero. The equation of motion assumes the simplest form if it stands for the one-dimensional steady flow of a non-viscous incompressible fluid, and the electric current flows perpendicularly to the direction of the magnetic field

$$\varrho v \frac{\mathrm{d}v}{\mathrm{d}x} = -\frac{\mathrm{d}p}{\mathrm{d}x} + jB. \tag{3.3.33}$$

Using component symbols we can write the equation of motion (3.3.32) in a very clear form

$$\varrho \frac{\mathrm{d}v_i}{\mathrm{d}t} = \frac{\partial \tau_{ij}}{\partial x_j} + F_i = \frac{\partial \tau_{ij}}{\partial x_j} + (\mathbf{j} \times \mathbf{B})_i; \tag{3.3.34}$$

a few basic concepts (discussed in greater detail in Chap. 1.3.2) are sufficient to explain this expression. To start with, these are rectangular Carthesian coordinates, the components of the quantities in the direction of the individual axes being designated by the indices 1,2 and 3 ($i, j = 1, 2, 3$). If the same index occurs twice in the same member, this means the summation over all prescribed values of this index according to Einstein's rule of summation, e.g.

$$a_i b_i = \sum_{i=1}^{3} a_i b_i = a_1 b_1 + a_2 b_2 + a_3 b_3 .$$

The quantity τ_{ij} is the ij^{th} component of the stress tensor described by

$$\tau_{ij} = \tau'_{ij} - p\delta_{ij} . \tag{3.3.35}$$

In this equation τ'_{ij} stands for the ij^{th} component of the viscous stress, p for the total pressure and δ_{ij} is Kronecker's symbol which equals unity at $i = j$, and zero at $i \neq j$. (Cf. 92). Since the tensor of the viscous stress can be expressed by the equation

$$\tau'_{ij} = \eta \left(\frac{\partial v_i}{\partial x_j} + \frac{\partial v_j}{\partial x_i} \right) - \delta_{ij} \frac{2}{3} \eta \frac{\partial v_k}{\partial x_k} , \tag{3.3.36}$$

we can write Eq. (3.3.34).

$$\varrho \frac{\mathrm{d} v_i}{\mathrm{d} t} = - \frac{\partial p}{\partial x_j} \delta_{ij} + \eta \frac{\partial}{\partial x_j} \left(\frac{\partial v_i}{\partial x_j} + \frac{\partial v_j}{\partial x_i} \right) - - \delta_{ij} \frac{2}{3} \eta \frac{\partial}{\partial x_j} \frac{\partial v_k}{\partial x_k} + F_i . \tag{3.3.37}$$

Rewritten in its vector form, this equation is identical with Eq. (3.3.32).

Other important relations are the equation of state for the ideal gas

$$p = \varrho R T , \tag{3.3.38}$$

where R is the gas constant, and the equation of conservation of the flowing fluid mass

$$\frac{\mathrm{d}}{\mathrm{d} t} (\varrho \, \mathrm{d} V) = 0 , \tag{3.3.39}$$

where $\mathrm{d}V$ stands for an elementary volume. This equation is equivalent to the equation of continuity valid for both an ideal and a viscous fluid, which is used more frequently:

$$\frac{\partial \varrho}{\partial t} + \operatorname{div}(\varrho \mathbf{v}) = 0. \qquad (3.3.40)$$

If the fluid is incompressible, its density ϱ is constant and the equation of continuity assumes the form

$$\operatorname{div} \mathbf{v} = 0, \qquad (3.3.41)$$

which expresses the condition of incompressibility of the fluid already used in deriving Eq. (3.3.33).

The last equation in the system of fundamental magnetohydrodynamic relations is the equation of energy, also called the equation for heat flow. It introduces the law of conservation of energy into the system of the magnetohydrodynamic equations. The derivation of the energy equation (cf. e.g. [Ref. 3.3.32]) is based on the consideration that the time change of the internal and kinetic energies of the fluid must equal the sum of conducted heat gained in unit time by the fluid volume under consideration plus the output of external and pressure forces. Radiated heat can be neglected, because the temperature of the working fluid is relatively low. The form in which the energy equation usually appears in hydrodynamics in only supplemented by a member accounting for Joule's heat. It is then written

$$\varrho \frac{\mathrm{d}}{\mathrm{d}t}\left(c_v T + \frac{v_i v_i}{2}\right) =$$
$$= \frac{\partial}{\partial x_j}\left(\lambda \frac{\partial T}{\partial x_j}\right) + \frac{\partial}{\partial x_j}(v_i \tau_{ij}) + v_i F_i + \frac{j^2}{\sigma}. \qquad (3.3.42)$$

Symbol λ is the coefficient of thermal conductivity and F_i the volume force component referred to the unit volume. Since we consider the electromagnetic force only, $F_i = (\mathbf{j} \times \mathbf{B})_i$. The symbol c_v stands for specific heat at constant volume, and according to Mayer's relation it is related to c_p by the equation

$$c_v = c_p - R. \qquad (3.3.43)$$

Let us denote the power of pressure and viscosity forces by N_t and re-write Eq. (3.3.42) in vector form. We have then

$$\varrho \frac{\mathrm{d}}{\mathrm{d}t}\left(c_v T + \frac{v^2}{2}\right) = \operatorname{div}(\lambda \operatorname{grad} T) + \mathbf{v}(\mathbf{j} \times \mathbf{B}) + \frac{j^2}{\sigma} + N_t. \quad (3.3.44)$$

The energy equation (3.3.42) can also be given various other forms of which we only quote those used most frequently. By using Eqs. (3.3.38), (3.3.40) and (3.3.43), we give Eq. (3.3.42) the form

$$\varrho \frac{\mathrm{d}}{\mathrm{d}t}(c_p T) - p \operatorname{div} \mathbf{v} = \frac{\mathrm{d}p}{\mathrm{d}t} + \frac{\partial}{\partial x_j}\left(\lambda \frac{\partial T}{\partial x_j}\right) -$$

$$- \varrho v_i \frac{\mathrm{d}v_i}{\mathrm{d}t} + \frac{\partial}{\partial x_j}(v_i \tau_{ij}) + v_i F_i + \frac{j^2}{\sigma}. \quad (3.3.45)$$

By scalar multiplication of the equation of motion (3.3.34) with velocity v_i, substitution for $\varrho v_i \frac{\mathrm{d}v_i}{\mathrm{d}t}$ in Eq. (3.3.45) and adjustment, we obtain the energy equation in the form

$$\varrho \frac{\mathrm{d}}{\mathrm{d}t}(c_p T) = \frac{\mathrm{d}p}{\mathrm{d}t} + \frac{\partial}{\partial x_j}\left(\lambda \frac{\partial T}{\partial x_j}\right) + \tau_{ij} \frac{\partial v_i}{\partial x_j} + \frac{j^2}{\sigma}, \quad (3.3.46)$$

or

$$\varrho \frac{\mathrm{d}}{\mathrm{d}}(c_p T) = \frac{\mathrm{d}p}{\mathrm{d}t} + \operatorname{div}(\lambda \operatorname{grad} T) + \frac{j^2}{\sigma} + N_v, \quad (3.3.47)$$

and in case of $\frac{\partial p}{\partial t} = 0$, we have

$$\varrho \frac{\mathrm{d}}{\mathrm{d}t}\left(c_p T + \frac{v^2}{2}\right) = \varrho \frac{\mathrm{d}h}{\mathrm{d}t} = \mathbf{j} \cdot \mathbf{E} + Q, \quad (3.3.48)$$

where h is the stagnation enthalpy and Q expresses the energy dissipation jointly due to heat conduction and pressure force. The electric output $\mathbf{j} \cdot \mathbf{E}$ has been defined by Eq. (3.3.4). The energy equation has the simplest form in the case of the one-dimensional flow of a conductive fluid in the direction of the x axis:

$$\varrho v \frac{\mathrm{d}h}{\mathrm{d}x} = jE + Q. \quad (3.3.49)$$

The entire system of magnetohydrodynamic equations includes then nine variables—$\mathbf{v}$, p, T, ϱ, $\mathbf{j}$, $\mathbf{H}$, $\mathbf{B}$, $\mathbf{E}$, and σ—if c_p, λ and η are considered constants. The system consists of nine equations: the equations of motion, energy, state, continuity and of the electromagnetic field (3.3.22), the right-hand equation (3.3.25), and Eqs. (3.3.27), (3.3.28) and (3.3.29). Most frequently, the solution of this system is sought for the case of one-dimensional flow. By simplifying assumptions depending on the purpose of the calculation and by the successive elimination of unknown quantities, the number of equations can be reduced to two. Cases of two-dimensional flow are mathematically so difficult as to be virtually unsolvable except by electronic computers.

One-dimensional approximation offers a wide choice of starting premises. Among fundamental studies of unidimensional flow in magnetoplasmodynamic generators, mention should first of all be made of Neuring's work published in 1960. His subject is the steady one-directional flow of an incompressible and non-viscous fluid in a homogeneous magnetic field within a channel of slightly varying cross-section, and of a compressible non-viscous fluid in a channel of constant cross-section. Rosa (1961) studied flow at constant gas velocity, Sarychev (1962) a case of isothermal flow, Nowacki (1961) also flow at constant pressure. Sutton (1961) and Dahlberg (1963) investigated the effect of a strong magnetic field upon the properties of a magnetoplasmodynamic generator, and the flow of a compressible inviscous plasma was the subject of Crown's studies (1961). Swift-Hook and Wright (1963) concerned themselves with flow at constant Mach number, and Ralph (1961), on the contrary, was out for its optimum value. Numerous further calculations have been made, and comparisons of theoretical results with measured values are more and more frequently published.

So far, we have assumed that the electric conductivity of the plasma is a constant quantity. In more accurate computations, however, the initial system of equations has to be supplemented by an equation that takes into account the dependence of electric conductivity upon the composition of the working fluid and on the quantities of state or, moreover, on the current density, the external magnetic field, etc. These problems so greatly affect the function of the magnetoplasmodynamic generator that we have to discuss them in greater detail, in the following Chapter.

3.3.5 Electric Conductivity of Partly Ionized Plasma

Two basic conditions to be satisfied by the working fluid of a magnetohydrodynamic generator are electric conductivity the magnitude of which has to surpass several S/m, and temperature not exceeding 3,000°K. Consequently substances with a low ionization potential, so-called seed materials, have to be introduced into the working gas. The vapours of alkali metals, mainly cesium and potassium, are found very useful to this end.

For greater simplicity let us assume, at first, that the working fluid is a mixture of two gases – an electron gas and a carrier gas (ions and neutral atoms) – which move independently of each other. The gas as a whole has the velocity $\mathbf{v}$. Since the mass of electrons is by several orders of magnitude smaller than that of ions and atoms, we shall further assume that $\mathbf{v}$ is also the velocity of the carrier gas. The relative velocity of the electron gas with respect to the carrier gas is $\mathbf{v}_d$.

Outwardly, the gas is electrically neutral and we equate the electron density to the density of positive ions ($n_e = n_i$). If e is the charge of an electron, the density of the electric current is

$$\mathbf{j} = -n_e e \mathbf{v}_d . \tag{3.3.50}$$

On the average, the electron is in ν_e collisions per second with particles of the carrier gas, hence the time between two collisions of the electron is $\tau_e = 1/\nu_e$. If after each collision the electron velocity decreases to the velocity of the carrier gas, then a unit volume of electron gas suffers in one second a loss of momentum amounting to $n_e m_e \mathbf{v}_d \nu_e$, where m_e is the mass of the electron. The effect is the same as if a retarding force of $n_e m_e \mathbf{v}_d \nu_e$ were acting upon the unit volume of electron gas. If we neglect the effect of the gravity force and the pressure gradient of the electron gas, the forces affecting the electron gas are in equilibrium when

$$n_e e(\mathbf{E}^* + \mathbf{v}_d \times \mathbf{B}) = -n_e m_e \mathbf{v}_d \nu_e . \tag{3.3.51}$$

The electric force acting upon the electrons is on the left-hand side. $\mathbf{E}^*$ is the intensity of the electric field in a system moving with the gas at a velocity of $\mathbf{v}$ ($\mathbf{E}^* = \mathbf{E} + \mathbf{v} \times \mathbf{B}$). By using Eq. (3.3.50), we can write Eq. (3.3.51) in the form of

$$\mathbf{E}^* - \frac{1}{n_e e} \mathbf{j} \times \mathbf{B} = \frac{m_e \nu_e}{n_e e^2} \mathbf{j}. \quad (3.3.52)$$

If there is no magnetic field in the plasma, this equation expresses Ohm's law $\mathbf{j} = \sigma_0 \mathbf{E}^*$, and the specific electric conductivity of the gas is then

$$\sigma_0 = \frac{n_e e^2}{m_e \nu_e} = \frac{n_e e^2}{m_e} . \tau_e . \quad (3.3.53)$$

The mean electron velocity according to Maxwell's distribution is

$$\bar{v}_e = \sqrt{\frac{8kT}{\pi m_e}}$$

and by it we express the collision frequency ν_e, provided the electron moves in a gas that consists of a mixture of k species with the densities $n_1, n_2, \ldots, n_k$ and the effective cross-sections for collision with an electron $Q_1, Q_2, \ldots, Q_k$

$$\nu_e = \bar{v}_e \sum_k n_k Q_k .$$

Since we have assumed thermodynamic equilibrium, the temperatures of the electrons and of the gas are equal ($T_e = T$). The conductivity is therefore

$$\sigma_0 = \sqrt{\frac{\pi}{8}} \cdot \frac{n_e e^2}{\sqrt{m_e kT}} \cdot \frac{1}{\sum_k n_k Q_k} = 0.628 \frac{n_e e^2}{\sqrt{m_e kT}} \cdot \frac{1}{\sum_k n_k Q_k} . \quad (3.3.54)$$

Chapman and Cowling [Ref. 3.3.13] have accurately computed the electric conductivity of a partially ionized gas; their resultant equation is

$$\sigma_0 = 0{\cdot}532 \frac{n_e e^2}{\sqrt{m_e kT}} \frac{1}{\sum_k n_k Q_k} . \quad (3.3.55)$$

If we substitute the numerical values for the physical constants, we obtain

$$\sigma_0 = 0{\cdot}385 \,.\, 10^{-7} \,.\, \frac{n_e}{\sqrt{T}} \,.\, \frac{1}{\sum_k n_k Q_k} \,. \tag{3.3.56}$$

The conductivity of the gas is expressed in [S/m], the temperature in [deg . K], the density in [cm^{-3}], and the effective cross-sections in [cm^2]. Eq. (3.3.54) indicates conductivity values only 12% higher than the Chapman-Cowling equation; this inaccuracy is small with regard to the knowledge of the effective cross-section values.

The expression $\sum_k n_k Q_k$ can be rewritten:

$$\sum_k n_k Q_k = \sum_n n_n Q_n + n'_A Q_A + n_e Q_i \,. \tag{3.3.57}$$

Index n refers to the various species of neutral gas, index A to the alkali metal atoms, and i to the ions

The effective cross-section for the electron-ion collision can be computed according to Sherman

$$Q_i = 0{\cdot}90 \left(\frac{e^2}{kT} \right)^2 . \ln \left[\frac{1{\cdot}5}{\sqrt{2} e^3} \left(\frac{k^3 T^3}{\pi n_e} \right)^{1/2} \right] \tag{3.3.58}$$

or after the substitution of numerical values for the constants

$$Q_i = \frac{2{\cdot}52 \,.\, 10^{-6}}{T^2} \ln \left(0{\cdot}873 \,.\, 10^4 \frac{T^{3/2}}{n_e^{1/2}} \right) \tag{3.3.59}$$

The units are the same as substituted in Eq. (3.3.56).

The values of the effective cross-sections for electron-atom collisions in the region of low electron energies are far from being fully known. An idea of the values involved is given in Table 3.3.1.

Electron density is one of the quantities that occur in the equations for electric conductivity. A method of computing the latter is demonstrated in the following paragraphs.

Table 3.3.1 EFFECTIVE CROSS-SECTION FOR COLLISION WITH ELECTRONS, IONIZATION POTENTIALS, AND RATIOS OF STATISTICAL WEIGHT OF SOME IMPORTANT ATOMS AND MOLECULES

Substance	Q [cm²]	U_i [eV]	$2g^+/g$
Caesium	$4 . 10^{-14}$	3·893	1
Potassium	$4 . 10^{-14}$	4·339	1
Sodium	$3 . 10^{-14}$	5·138	1
Nitric oxide	$(5{\cdot}4$ to $5{\cdot}6) . 10^{-16}$	9·250	—
Oxygen (molec.)	$(3{\cdot}3$ to $3{\cdot}6) . 10^{-16}$	12·075	8/3
Water	$(5$ to $7) . 10^{-15}$	12·590	4
Hydroxide	$2 . 10^{-15}$	13·180	3/2
Hydrogen (atom.)	$1{\cdot}5 . 10^{-15}$	13·595	1
Oxygen (atom.)	$(1{\cdot}6$ to $2{\cdot}1) . 10^{-15}$	13·614	8/9
Carbon dioxide	$(1{\cdot}2$ to $3) . 10^{-15}$	13·790	4
Carbon monoxide	$(7$ to $8) . 10^{-16}$	14·010	4
Hydrogen (molec.)	$1{\cdot}2 . 10^{-15}$	15·427	4
Nitrogen (molec.)	$(6{\cdot}2$ to $6{\cdot}4) . 10^{-16}$	15·600	4
Argon	$2 . 10^{-17}$	15·755	12
Helium	$6{\cdot}8 . 10^{-16}$	24·580	4

Let us assume that the seed material, whose original density is n_A, is the only gas component to be ionized, while the others remain electrically neutral. The ionization reaction has the form

$$A = A^+ + e^-,$$

under the condition that A, A^+ and e^- do not participate in any other reaction.

The electron density is proportional to the initial density of the seeding atoms and to the degree of ionization α according to the relation

$$n_e = \alpha n_A. \tag{3.3.60}$$

After the ionization, the density of neutral seeding atoms is

$$n'_A = n_A - n_e = n_A(1 - \alpha). \tag{3.3.61}$$

The relation between the density and pressure of particles is defined by

$$p = nkT. \tag{3.3.62}$$

The total pressure is the sum of the partial pressures of the individual gas components, hence

$$p_{\text{total}} = p_{\text{n}} + p'_A + p_{\text{e}} + p_{\text{i}} = p_{\text{n}} + p_A(1 + \alpha). \qquad (3.3.63)$$

The degree of ionization in Eq. (3.3.60) for the given component, temperature and pressure is calculated with the aid of Saha's equation; in our case, we must introduce $p_{\text{total}} - p_{\text{n}} = p_A(1 + \alpha)$ for pressure, because we have assumed that only the atoms of the seeding material take part in the ionization:

$$\frac{\alpha^2}{1 - \alpha^2} = \frac{2g_+}{g}\left(\frac{2\pi m_e}{h^2}\right)^{3/2} \frac{(kT)^{5/2}}{p_A(1 + \alpha)} \cdot \exp\left(-\frac{eU_i}{kT}\right). \qquad (3.3.64)$$

From this equation we can derive the equilibrium constant K_1, defined by the partial pressures:

$$K_1 = \frac{\alpha^2}{1 - \alpha} p_A = \frac{p_{\text{i}} p_{\text{e}}}{p'_A}. \qquad (3.3.65)$$

For practical reasons it is preferable to calculate $\log K_1$

$$\ln K_1 = \frac{5}{2} \ln T - 6.49 - \frac{5050 U_i}{T} + \ln \frac{2g_+}{g}. \qquad (3.3.66)$$

The values of the ionization potentials [in eV] and the ratios of the statistical weights of positive ions to neutral components are tabulated in Table 3.3.1. The weight ratio for all alkali metals equals unity.

By substituting α from Eq. (3.3.60) into Eq. (3.3.65) and using Eq. (3.3.62) we obtain for the electron pressure the quadratic equation

$$p_e^2 + K_1 p_e - K_1 p_A = 0, \qquad (3.3.67)$$

hence the electron concentration is

$$n_e = \frac{p_e}{kT} = \frac{3.62 \cdot 10^{21}}{T}\left(-K_1 \pm \sqrt{K_1^2 + 4K_1 p_A}\right). \qquad (3.3.68)$$

We take only the positive sign before the radical, because the pressure must have a positive value. K_1 is calculated from Eq. (3.3.66), and the partial pressure of the seeding material p_A from the known pressure of the carrier gas p_n, the contents of seeding and carrier gas b_A and b_n, respectively [in weight per cent], and from the molecular masses M_n and M_A; the relation is

$$p_A = p_n \frac{b_A M_n}{b_n M_A}. \tag{3.3.69}$$

This completes the description of the procedure. In actual practice, however, the calculation proceeds the opposite way: At the start, we know the kind of the gas, its temperature and the percentage of the seeding material. We successively calculate p_A, K_1, n_e, Q_i and finally σ_0. The computation is tedious, and approximate formulae are frequently preferred. For combustion gases with a seeding addition of potassium we use to advantage an empirical formula derived by Swift--Hooke and Wright:

$$\sigma_0 = \left(\frac{T}{2{,}000}\right)^{10} . p^{-0\cdot 4}, \tag{3.3.70}$$

and conductivity [S/m], temperature [deg. K] and pressure [atm] are substituted. The formula is valid for conductivity in the range of $1 \leqq \sigma_0 \leqq 100$ S/m.

The conductivity of nitrogen with an addition of approximately 1% of potassium may be calculated according to a simplified formula [cf. Ref. 3.3.14]

$$\sigma_0 = \frac{9{\cdot}05 . 10^3}{\sqrt{p}} . \exp\left(-\frac{2{\cdot}6 . 10^4}{T}\right). \tag{3.3.71}$$

The electric conductivity of some gas mixtures is plotted against temperature in Fig. 133.

The electric conductivity varies with the amount of the seeding addition. It is therefore essential to know the optimum ratio of the seeding materials which—all other conditions being equal—produce the highest electric conductivity of the working gas. For constant temperature and pressure and a low degree of ionization, the electron concentration may be calculated from the transformed Saha equation:

$$n_e^2 = n_A . \text{const.} \tag{3.3.72}$$

The value thus found is substituted in the equation for electric conductivity

$$\sigma_0 = C \cdot \frac{n_A^{1/2}}{\sum_k n_{nk}Q_{nk} + n_A Q_A + n_A^{1/2} Q_i \cdot K_1}, \tag{3.3.73}$$

and all constants are included in C. We differentiate with respect to n_A, and from the maximum conductivity value we obtain—after a simple

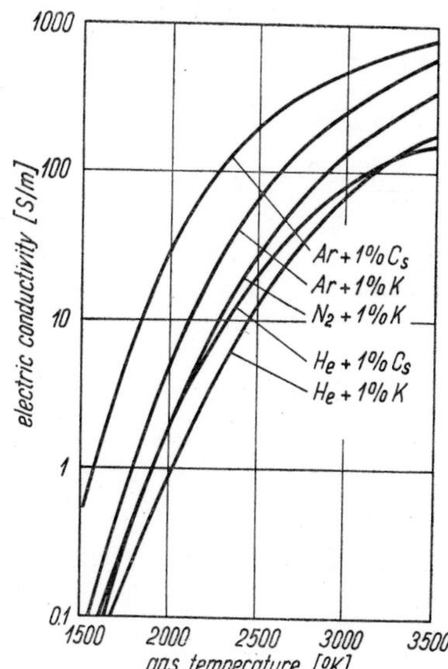

Fig. 133 Electric conductivity of some gas mixtures as a function of gas temperature

manipulation—the ratio of atomic densities of the additive and the carrier gas

$$\left(\frac{n_A}{n_n}\right)_{opt} = \frac{Q_n}{Q_A} \tag{3.3.74}$$

Eq. (3.3.69) enables us to express the optimum ratio of the addition to the carrier gas in weight per cent

$$\left(\frac{b_A}{b_n}\right)_{opt} = \frac{M_A Q_n}{M_n Q_A}. \tag{3.3.75}$$

The optimum percentage by weight of the addition is

$$(b_A)_{opt} = 100 \, . \frac{\dfrac{M_A Q_n}{M_n Q_A}}{1 + \dfrac{M_A Q_n}{M_n Q_A}} ; \tag{3.3.76}$$

numerical values for some working gases are shown in Table 3.3.2.

Table 3.3.2 OPTIMUM AMOUNTS OF SEEDING MATERIALS

Mixture	He + Cs	He + K	He + Na	Ar + Cs	Ar + K	Ar + Na	N_2 + Cs	N_2 + K	N_2 + Na
Amount of addition [% *w/w*]	36·2	14·2	11·5	0·16	0·05	0·04	7·00	2·16	1·54

The indicated method of computing the electric conductivity of a partially ionized gas is primarily subject to the condition that Saha's equation is applicable. That equation has been derived under the assumption that the individual constituents of the gas are in thermodynamical and chemical equilibrium. The working gas in a magnetoplasmodynamic generator need not always satisfy this condition. In some cases we even deliberately aim at achieving a non-equilibrium state. A typical example of great practical importance is a working gas with an elevated electron temperature, which is being studied with the intention of attaining the necessary electric conductivity at lower temperatures. Kerrebrock [Ref. 3.3.15] was the first to devote attention to this problem. He started from the assumption that in a monoatomic gas the collisions of electrons with neutral particles are elastic. In these collisions, the electrons lose no more then an immaterial part of their energy, because their mass is by several orders of magnitude smaller than that of atoms. If a current flows through the plasma, most of the Joulean heat is taken over by the electrons; hence their energy increases. In a polyatomic gas, the collisions between electrons and neutral particles are inelastic and the electrons lose more energy, when in collision.

Therefore only in monoatomic gases can the electron temperature be higher than the temperature of the neutral gas particles.

From this concept Kerrebrock derived a relation describing the degree of ionization and electric conductivity as a function of the temperature of the neutral atoms and of electric-current density. The expression for conductivity consists of two members

$$\ln \sigma = f(T_n) + b \cdot \ln j. \tag{3.3.77}$$

The first member on the right-hand side corresponds to the "conventional conductivity" at gas temperature T_n and the second stands for the increase in conductivity depending on the current that flows through the plasma. At low plasma temperatures the value of b is close to unity. The increase becomes noticeable if the current density attains a value of several $A\ cm^{-2}$. The validity of these conclusions has been checked by experiments and has aroused a lively interest in magneto-plasmodynamic generators working on helium or argon in closed circuits. Contaminants other than the seeding material, which would deprive the electrons of energy, must not be contained in the gas. Experiments have also shown that the working temperature of the gas may be far lower than 1,800°C without impairing the efficiency of the generator by insufficient electric conductivity.

Ionic and radical reactions can also greatly affect electric conductivity. Negative ions are most undesirable, because their formation abstracts electrons which otherwise could have taken a greater part in the conduction of the current through the plasma. Electronegative components are always present in combustion gases, and the effects of oxygen atoms and hydroxyl radicals are particularly strong. Hence, hydroxides of alkali metals, A, are not the best ionizing additives, since they dissociate as indicated below:

$$\left.\begin{array}{l} A\,OH \rightleftarrows A + OH \\ \text{and further} \\ OH + e^- \rightleftarrows OH^- \end{array}\right\}. \tag{3.3.78}$$

Owing to this phenomenon, electric conductivity in the region of maximum temperatures at the generator inlet may be reduced by as much as 10%.

Chlorides of alkali metals, though inexpensive, are also of little advantage as seeding additives. The bond between the alkali-metal and the chloride atoms is so strong that there is virtually no dissociation at the temperatures used; and if there were any, the strong electronegative chlorine atom would produce an unfavourable effect. Therefore carbonates and nitrates seem to be the most promising alkali-metal compounds.

Free halogen atoms may sometimes exert a favourable effect on conductivity. Padley [Ref. 3.3.22] found that in the presence of free halogen atoms and hydrogen the electron density in a flame is increased. In the absence of halogens the ionization reaction is

$$A + X \rightleftarrows A^+ + e^- + X, \tag{3.3.79}$$

where A is an alkali-metal atom and X some other atom. However, in the presence of halogen atoms B, the reaction is apparently

$$A + B \rightleftarrows A^+ + B^-, \tag{3.3.80}$$

and if free hydrogen is also present, then

$$B^- + \mathrm{H} \rightleftarrows \mathrm{H}B + e^-. \tag{3.3.81}$$

Since compounds of the HB type are very stable at high temperatures, this last reaction contributes to an enhanced electron density and conductivity of the plasma. A quantitative estimation of the effect which those reactions produce on the gas conductivity in the plasmodynamic generator is very difficult.

3.3.6 Anisotropy of Electric Conductivity in the Plasma

In the last Chapter, we discussed the calculation of the specific electric conductivity σ_0 of a plasma under the assumption that no external magnetic field acts upon the latter. However, the presence of a magnetic field is indispensable in the magnetoplasmodynamic conversion of energy. Let us demonstrate its effect on the conductivity of the plasma.

The path of charged particles moving in a plane perpendicular to a magnetic field will curve, the more so the stronger the magnetic field; yet the mean free path of the particles practically does not alter. Consequently their velocity in the original direction decreases and a velocity component perpendicular to the original direction appears. Under these conditions the electric conductivity of the plasma loses its isotropic character; it can no longer be considered a scalar quantity, but is a tensor. Let us start from Eq. (3.3.52).

$$\mathbf{E}^* - \frac{\boldsymbol{j} \times \mathbf{B}}{n_e e} = \frac{\boldsymbol{j}}{\sigma_0}$$

and vectorially multiply both sides by the magnetic field strength $\mathbf{B}$. Unlike in our former procedure, when we equated $\mathbf{B} = 0$, we shall now take account of the second left-hand member of the equation and calculate the current density

$$\boldsymbol{j} = \frac{\sigma_0}{1 + \left(\frac{\sigma_0 \mathbf{B}}{n_e e}\right)^2} \left[\mathbf{E}^* + \frac{\sigma_0}{n_e e}(\mathbf{B} \times \mathbf{E}^*)\right]. \tag{3.3.82}$$

We use the expression for the cyclotron frequency of the electron $\omega_e = eB/m_e$ and substitute from Eq. (3.3.53) for σ_0. Then, obviously, we have

$$\frac{\sigma_0 B}{n_e e} = \omega_e \tau_e. \tag{3.3.83}$$

The product $\omega_e \tau_e$, called the Hall coefficient β, expresses the number of electron revolutions in the magnetic field in the time interval between two consecutive collisions.

The equation (3.3.82) is then simplified to

$$\boldsymbol{j} = \frac{\sigma_0}{1 + \beta^2} \left[\mathbf{E}^* + \beta\left(\frac{\mathbf{B}}{B} \times \mathbf{E}^*\right)\right]. \tag{3.3.84}$$

We easily find out that the equation can also be written in matrix form

$$\begin{Vmatrix} j_x \\ j_y \\ j_z \end{Vmatrix} = \frac{\sigma_0}{1 + \beta^2} \cdot \begin{Vmatrix} 1 & -\beta & 0 \\ \beta & 1 & 0 \\ 0 & 0 & 1 \end{Vmatrix} \cdot \begin{Vmatrix} E_x^* \\ E_y^* \\ E_z^* \end{Vmatrix}, \tag{3.3.85}$$

which is very clear and offers much advantage in the calculation of the individual current components. We leave the orientation of the principal vectors as they were before: velocity **v** has the direction of the positive x axis, and magnetic field strength **B** the direction of the positive z axis. From Eq. (3.3.85) we can immediately derive the current components

$$\left.\begin{aligned} j_x &= \frac{\sigma_0}{1+\beta^2}(E_x^* - \beta E_y^*) \\ j_y &= \frac{\sigma_0}{1+\beta^2}(\beta E_x^* + E_y^*). \\ j_z &= 0, \quad \text{because} \quad E_z^* = 0 \end{aligned}\right. \tag{3.3.86}$$

We see that the magnitude of the current components in the directions of the x and y axes depends on the value of the Hall coefficient. Moreover, the conductivity in different directions differs.

The spatial location of the current and electric-field vectors differs from that in the idealized case, when we considered conductivity a scalar quantity. Both vectors lie in a plane perpendicular to the z axis.

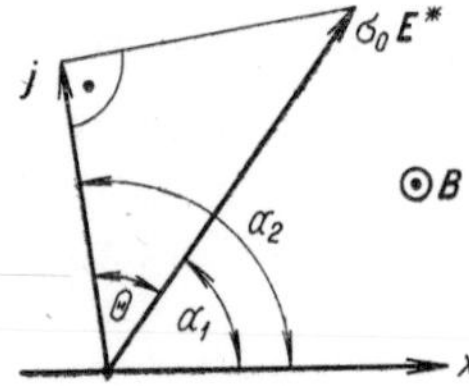

Fig. 134 Electric field E *and current j in magnetic field* B *oriented with respect to x axis*

With the aid of the Lorentz transformation we convert the former mobile reference system of coordinates into a fixed laboratory system

$$\left.\begin{aligned} E_x^* &= E_x \\ E_y^* &= E_y - vB \\ E_z^* &= E_z = 0 \end{aligned}\right\}. \tag{3.3.87}$$

The angle Θ included by **E** and **j** is calculated from the proportion of the absolute values of the vectors

$$\left.\begin{aligned} |\mathbf{E}^*| &= \sqrt{E_x^{*2} + E_y^{*2}} \\ \text{and} \qquad |\mathbf{j}| &= \sqrt{j_x^2 + j_y^2} \end{aligned}\right\} \tag{3.3.88}$$

From Fig. 134 follows that

$$\cos\Theta = \frac{|\boldsymbol{j}|}{\sigma_0|\mathbf{E}^*|} = \frac{1}{\sqrt{1+\beta^2}}, \tag{3.3.89}$$

and computation shows that

$$\tan\Theta = \beta = \omega_e\tau_e. \tag{3.3.90}$$

Let the positive direction of the x axis include the angle α_1 with $\mathbf{E}^*$ and the angle α_2 with the current $\boldsymbol{j}$. Then

$$\begin{aligned} \tan\alpha_1 &= \frac{E_y^*}{E_x^*} = \frac{E_y - vB}{E_x} \\ \tan\alpha_2 &= \frac{j_y}{j_x} = \frac{\beta E_x + E_y - vB}{E_x - \beta E_y + \beta vB}. \end{aligned} \tag{3.3.91}$$

In a generator with continuous electrodes throughout the whole length of the working chamber, $E_x = 0$, and we obtain from Eq. (3.3.91)

$$\begin{aligned} \tan\alpha_1 &= -\infty \\ \tan\alpha_2 &= -\frac{1}{\beta}. \end{aligned} \tag{3.3.92}$$

The field $\mathbf{E}^*$ is thus oriented in the direction of the negative y axis, and the current $\boldsymbol{j}$ is aimed obliquely against the gas flow (Fig. 135).

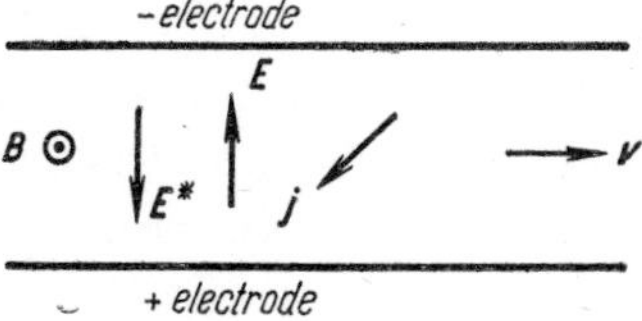

Fig. 135 Orientation of velocity current, magnetic and electric fields with the Hall effect prevailing

If the Hall coefficient is small enough, the directions $\mathbf{E}^*$ and $\boldsymbol{j}$ are almost identical; in very strong magnetic fields, however, the direction of the current is almost identical with the negative y axis.

In deriving these relations, we neglected the current of positive ions. This is justified when the values of Hall's coefficient are low or medium, not higher for instance than 10.

If β is far smaller than 1, the equations (3.3.86) assume the form

$$\begin{aligned} j_x &= \sigma_0 E_x^* \\ j_y &= \sigma_0 E_y^* \end{aligned} \tag{3.3.93}$$

and electric conductivity can be considered a scalar quantity.

The value of the Hall coefficient is affected by magnetic field strength **B**, the chemical composition of the gas, and the quantities of state.

From the relations used in deriving Eq. (3.3.54), we may also derive that

$$\beta = \omega_e \tau_e = \frac{eB}{m_e} \frac{1}{v_e} = \frac{eB}{m_e} \sqrt{\frac{\pi m_e}{8kT}} \cdot \frac{kT}{\sum_k p_k Q_k} =$$

$$= 3{\cdot}91 \,.\, 10^{-16} \frac{BT}{\sum_k p_k Q_k} \,. \tag{3.3.94}$$

When a plasmodynamic generator works on combustion products of about 2,700°K maximum temperature, the Hall coefficient is approximately proportional to $\dfrac{B}{p}$ according to the relation

$$\beta \cong (1 \div 3) \frac{B}{p} \,. \tag{3.3.95}$$

The magnetic field strength is expressed in Tesla units in this formula, and pressure in atmospheres. If the Hall coefficient is to be smaller than 1, magnetic field strength will have to be roughly smaller than 1·5 T.

If the collinearity of the current and electric-field intensity is breached, this affects the output parameters of the generator, in particular its specific electric output power. In a well-designed generator this need not have a negative effect; on the contrary, the high value of magnetic field strength may thus be used to improve the properties of the generator.

3.3.7 Configuration of the Magnetoplasmodynamic Generators

The simplest type of conduction generator—and the one most thoroughly studied so far—has a straight electrode channel (cf. diagram Fig. 130). Generators of this type can by a relatively simple modification be adjusted to the desired magnetic-field value. In the first Chapters of this Division we have already described a generator with a straight channel and continuous electrodes. Let us now supplement this description by discussing the influence which the Hall coefficient exerts upon the specific electric output power.

The electric output power derived from the unit volume of the channel is

$$P = -j_x E_x - j_y E_y. \tag{3.3.96}$$

Since the potential difference is constant throughout the length of the electrodes,

$$E_x^* = E_x = 0. \tag{3.3.97}$$

After substitution from Eq. (3.3.86) into Eq. (3.3.96) we obtain

$$P = \frac{\sigma_0}{1 + \beta^2} v^2 B^2 (1 - K) K. \tag{3.3.98}$$

The specific output power is smaller by a factor of $(1 + \beta^2)$ than when we neglected the Hall effect. The reduction in the specific generator output due to an intensive magnetic field can be partly attenuated by increased gas pressure. The highest possible output P_{max}, corresponding to a magnetic field of infinite intensity, is

$$P_{max} = \frac{n_e^2 e^2}{\sigma_0} v^2 (1 - K) K. \tag{3.3.99}$$

The current flowing along the channel is diminished if we divide the electrodes into short sections. The condition characteristic of the resultant straight channel with segmented electrodes is

$$j_x = 0. \tag{3.3.100}$$

In practice this condition is only partly satisfied, since the electrode pairs are always of a finite length. In the ideal case, the specific output of this type is independent of the Hall effect because

$$P = \sigma_0 v^2 B^2 (1 - K) K. \tag{3.3.101}$$

The optimum operation of a generator with segmented electrodes requires every pair of electrodes to be connected in a separate current circuit with a load resistance adapted to the operational parameters of the particular channel section. As a power generator is about 10 m long, the requisite number of electrode sections is very large. In principle, this disadvantage could be counteracted by connecting all sections in series in a single current circuit according to Fig. 136a. This would also result in a high output voltage, but balancing currents would flow between neighbouring electrodes of various electric potentials. An improvement is achieved – according to Wasserrab – by connecting the sections in such a way that neighbouring electrodes belong to different current circuits (Fig. 136b). The balancing currents

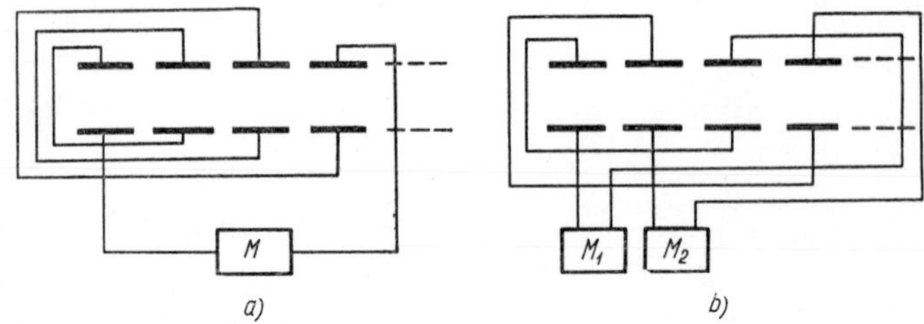

Fig. 136 Connection of electrodes in magnetoplasmodynamic generator. Alternative (b) suppresses the balancing currents, which flow between neighbouring electrodes in the instance indicated in (a) (M designates current converters)

are suppressed and the generator has a limited number of external loads or inverters. This solution does not, however, do away with the overloading of the electrode tips by excessive current densities. In a strong magnetic field, the ends of the current flow lines are shifted towards the electrode tips, where greatly enhanced electron emission and current density may cause overheating and ultimately the fusion of this part of the electrode.

If the electrodes in every section are short-circuited, this forms what is called a Hall-type generator. The external current circuit is connected only to the first and last electrode sections. Electrically, the Hall generator is marked by the condition

$$j_z = 0 \qquad \text{and} \qquad E_y = 0. \tag{3.3.102}$$

The specific electric power of the Hall generator is

$$P = \frac{\beta^2}{1 + \beta^2} \sigma_0 v^2 B^2 (1 - K) K. \tag{3.3.103}$$

The power increases with the value of the Hall coefficient, and at very high values of β, Eq. (3.3.103) changes into Eq. (3.3.101). The Hall coefficient should have a higher value than 10.

The no-load voltage and short-circuit current may be calculated from the relations

$$\begin{aligned} (E_x)_0 &= -\beta v B \\ (j_x)_k &= -\frac{\beta}{1 + \beta^2} \cdot \sigma_0 v B; \end{aligned} \tag{3.3.104}$$

it follows from them that at high values of β the Hall-type generator operates as HT supply.

A survey of the three basic types of straight-channel generators proves that none of them is suitable for values of β in the range between 2 and 8.

An analogue to the Hall generator with straight channel is the cylindrical Hall generator (Fig. 137). Its magnetic field is radial and

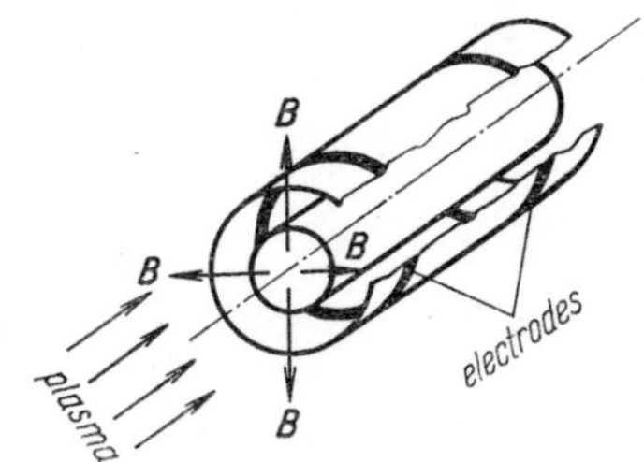

Fig. 137 Cylindrical Hall-type generator

the gas stream flows through a cylindrical gap. The azimuthal current component is suppressed in favour of the axial one. It is worth noting that as early as about 1940 this geometrical arrangement was selected by Karlowitz when he made the first industrial attempt at the construction of a plasmodynamic generator, at the Westinghouse Laboratories. His attempt failed, because the working gas he used in his generator had not the requisite conductivity.

A type whose geometry is similar to that of a steam or gas turbine is known as vortex generator (Fig. 138); it does not utilize the Hall

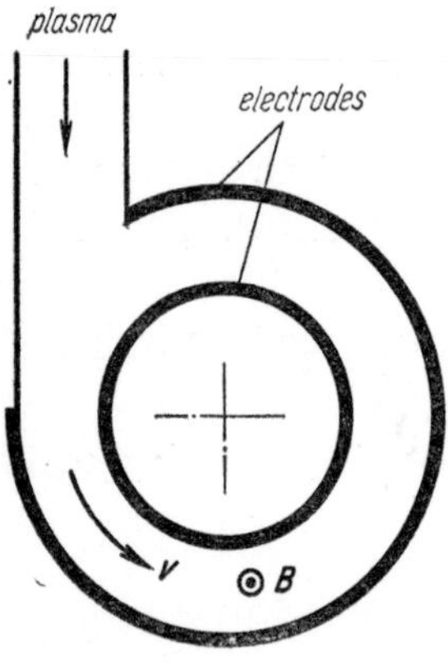

Fig. 138 Vortex generator

current. The condition $\beta < 1$ must be satisfied to prevent the Hall effect from markedly reducing the effective conductivity. The gas enters the chamber tangentially and issues from its centre part. The main advantage of this type is in the small chamber and therefore the relatively small electromagnet generating the external field. This generator promises to be smaller in size and lower in weight and price than any other type. However, because of the two-dimensional flow of the working gas the computation of the vortex generator is far from easy.

3.3.8 A. C. Magnetoplasmodynamic Generators

Certain inherent drawbacks common to all D.C. generators could be avoided by directly producing A.C. This would, in the first place, obviate the need of an inverter plant at the power station, which costs roughly as much as the magnetoplasmodynamic generator itself. In some cases, the mean temperature of the chamber wall could also be reduced, for instance in types which periodically change the conductivity of the plasma by altering its temperature. A third important advantage would consist in doing away with the electrodes; the production of these parts, working under heavy thermal and electric loads, is one of the most difficult technical problems involved in the construction of magnetoplasmodynamic generators. The alternating power generated in the electrodeless generator can be transferred into the external circuit by inductive coupling.

The simplest alternating generators—electrode generators with a variable magnetic field—are of a design similar to that of straight D.C. generators, from which they differ by their variable magnetic field. If the electric output power is to equal that of the D.C. generator, the A.C. generator must also equal the latter in the effective value of magnetic induction. Its specific power is therefore

$$P = \sigma v^2 B_{ef}^2 (1 - K) K. \tag{3.3.105}$$

In the A.C. generator, unlike in the D.C. one, we have to face losses mainly caused by the hysteresis of the magnetic circuit and by eddy currents in the plasma and the adjacent parts of the electrodes. Smy [Ref. 3.3.25] asserts that these losses can be reduced and the efficiency of the generator secured by dividing the electrodes into small sections. The trouble involved in a large number of external circuits can be avoided by connecting the electrode sections to the output coil which is inductively coupled to the coil energizing the magnetic field of the generator. The internal and load resistances are thus in accordance and, in addition, the output voltage of the generator can be varied virtually arbitrarily. The disproportionately high cost of the reactive power source for feeding the electromagnet remains an unsolved problem.

Interest is mainly concentrated on electrodeless generators with a variable electric field. Assuming that electrodes covered by an electrical-insulating layer would form a plasma-electrode condenser, Smy studied an intermediate type between generators with electrodes and without. This design would have the double advantage of protecting the electrode surface against the thermal and chemical action of the plasma and obviating the need for sectioning the electrodes. The calculation proved, however, that satisfactory output parameters of the generators would call for a potential gradient of about 10^5 kV/mm on the "condenser" dielectric, which is by some three orders of magnitude more than the highest dielectric strength obtained in known materials. Since the electric stress in the dielectric is directly proportional to the pressure and indirectly to the gas velocity, the realization of this generator would call for a so far unachievable reduction in pressure and increase in velocity.

The induction generator proper can have either a straight channel or one formed by the cylindrical space between the rotationally symmetrical "rotor" and "stator", if we introduce designations formally resembling those used in conventional electric motors and generators.

Travelling-wave generators [Ref. 3.3.27] are of the former type. Along the linear channel, polyphase coils produce a magnetic field which moves at velocity u in the direction of the flowing gas. When the gas moves at a higher velocity than the magnetic field, decelerating currents are induced in it, and energy is transferred into the external circuit. The characteristic parameter is the slip

$$S = \frac{u - v}{v}. \tag{3.3.106}$$

If the apparatus works as a generator, the slip is negative; it is positive when the gas is accelerated in the magnetic field (i.e. the apparatus works as a compressor). The generator power is at its maximum when the gas flows at double the velocity of the magnetic field. Bernstein based his investigations on experiments with the gas flowing at 2,000 m/s and having an electric conductivity of 1,000 S/m. The wavelength was about 1 m, and the ratio of reactive to active power approximately 30 to 1. The reactive power was found to decrease with conductivity or with increasing slip. The frequency of the current generated was 660 c.p.s, and the power take-off from the gas was about 10% of its input energy.

The layout of axially symmetric induction generators resembles that of normal electric motors or generators. A rotating magnetic field is generated in the cylindrical gap—by either electrical or mechanical methods—and the conductive working gas flows through it on a spiral path. Enhanced compactness and a smaller surface are the advantages of axially symmetric generators over linear designs. One of the authors who studied this type in detail is E. J. Jantovskij [Ref. 3.3.28]; his generator is schematically illustrated in Fig. 139. The conductive combustion gases flow through a helical channel, and the combustion air is supplied through another channel, adjoining it, where it is compressed and preheated. Air and fuel are fed from opposite ends of the

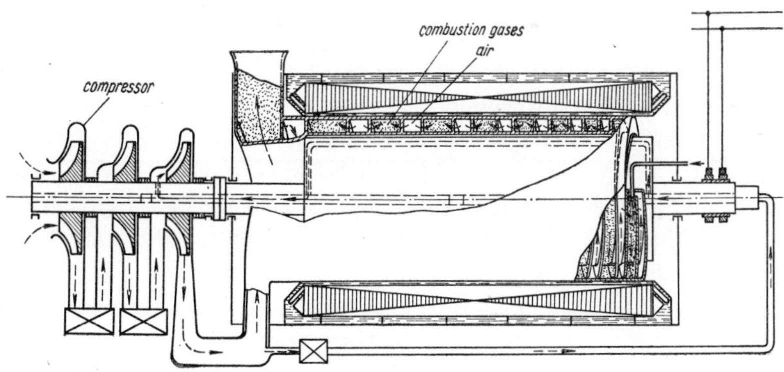

Fig. 139 Induction generator with rotary magnetic field

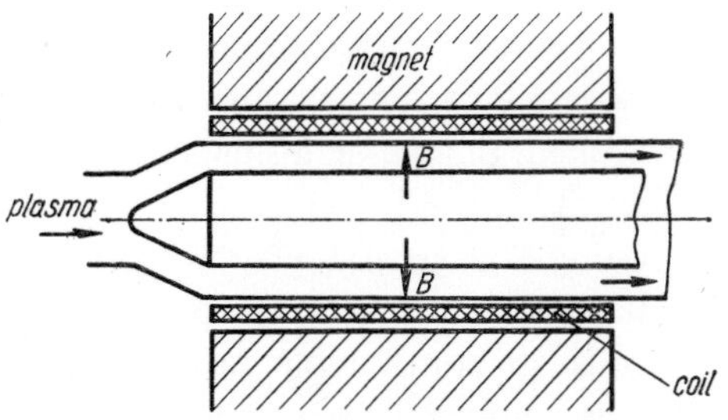

Fig. 140 Induction generator with variable magnetic field or variable plasma conductivity

generator, and the combustion continues throughout the travel of the fuel along its helical path.

In the Harris induction generator (Fig. 140), the conductive gas flows parallel to the centre line of the apparatus. The external alternating magnetic field generates an azimuth alternating current in the plasma, which is inductively coupled with the output coil. The generator can also operate with a constant magnetic field, provided the electric conductivity of the gas is periodically altered.

Alternating generators with a constant magnetic field are based upon the changing electric conductivity or flow velocity of the gas. Periodical variations in the doses of the seeding addition are one way of altering the electric conductivity of the gas, though this method does not seem promising for power generation.

Thring has put forward a system with periodically repeated shock waves. There are suction valves for fuel and air as well as exhaust valves at both ends of a closed tube. The gas mixture is ignited at one end of the tube, and the resulting shock wave traverses the tube, its conductive part interacting with the magnetic field and inducing a current pulse in the external circuit. The residual energy of the combustion gases is utilized to compress the fresh mixture at the other end of the tube; the compressed mixture is then ignited and a new shock wave passes through the tube in the opposite direction. The gas thus alternately moves between both ends of the tube and produces an alternating current.

In this kind of generators with electrodes, the direction of the gas velocity has to be reversed every halfwave in order to produce alterning current only. Failing this, direct current modulated with an alterning component would be generated, and the power station would again need an inverter plant, though on a smaller scale than if equipped with a D.C. generator.

Another interesting A.C. generator is the "parametric generator", working on a principle similar to that of the parametric amplifier. The main component of this apparatus is an insulated tube, with a coil coaxially mounted on its surface. The inductance of this coil changes as a conductive fluid perodically flows through the tube. If the plasma is to exert a strong influence upon the magnetic field of the coil, the period during which the magnetic field diffuses into the plasma must be long in comparison with the time of the latter's passage through the coil. In consequence of this postulate, the magnetic Reynolds number must be high; Woodson and Lewis arrive at the conclusion that it should be higher than 1,000. With the present-day parameters of low-temperature plasma obtained from a high-temperature combustion chamber or by heating of gases in a nuclear reactor, the size of the installation works out too large. It could be reduced if the product of electric conductivity and gas velocity were increased. This method carries promises for plasma of thermonuclear temperature.

3.3.9 Using Magnetoplasmodynamic Generators in Power Plants

A magnetoplasmodynamic generator can be operated in a power plant in either an open or a closed circuit. The basic diagram of a closed

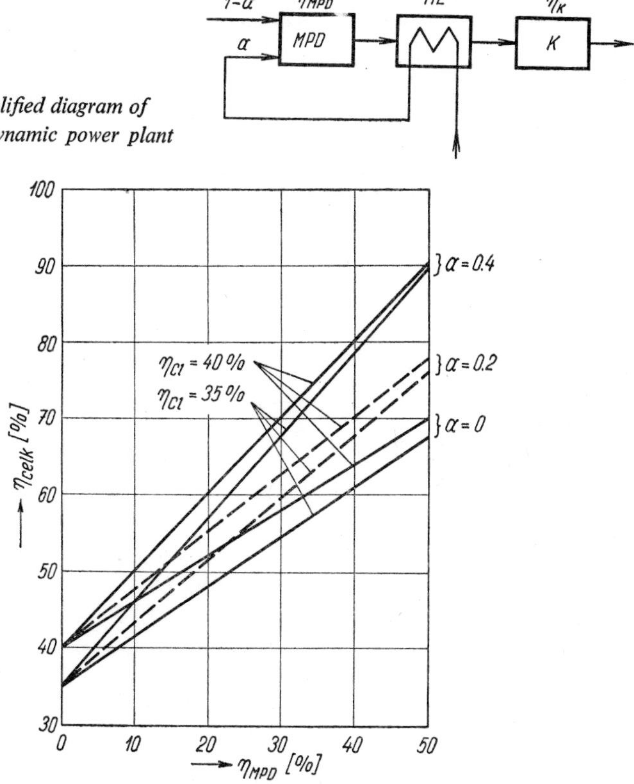

Fig. 141 Simplified diagram of magnetoplasmodynamic power plant

Fig. 142 Total efficiency of magnetoplasmodynamic power plant as a function of generator efficiency η_{MPD} and efficiency of the classical plant stage η_{Cl} depending on the degree of regeneration α

circuit is shown in Fig. 141. Since a relatively small temperature gradient can be tolerated in the generator, the latter is followed by another power-generating stage *K* which can work on combustion gases with the usual parameters.

The heat exchanger *HE*, which preheats the combustion air, plays a very important part in the operation of the power plant. Both the

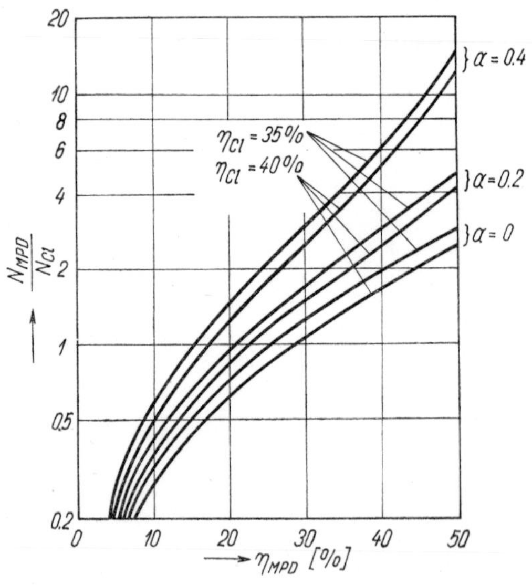

Fig. 143 Ideal N_{MPD}/N_{Cl} ratio for various degrees of regeneration α and in dependence upon the efficiency of the magnetoplasmodynamic generator η_{MPD}

total efficiency of the power plant and, in particular, the power output component generated in its plasmodynamic stage (i.e. combustion chamber plus generator) (Figs. 142 an 143) increase with the heat component "α" supplied to this stage by the combustion air. In the limiting case $\alpha = 1$, i.e. if the entire amount of heat issuing from the generator is regenerated, the classical low-temperature stage *K* can be omitted.

Two kinds of technological problems are involved in the utilization of plasmodynamic generators in electric power plants. In a plant which is to burn fossil fuel, the main problems are the development of the

Table 3.3.3 SOME OF THE FIRST MAGNETOPLASMODYNAMIC GENERATORS

Laboratory	General Electric	Westing-house	Avco Mark I	Avco Mark II	Avco	Avco Mark V
Acting medium	combust. gases	combust. gases	argon	combust. gases	combust. gases	combust. gases
Total temperature [°K]	5,500	3,162	3,260	3,000	3,030	3,000
Gas pressure [atm]	5·5	1·18	up to 2·36	3	—	—
Rate of gas flow [kg/s]	0·5	0·5	0·3	1·8	—	—
Flow velocity [m/s]	1,350	812	688	up to 1,600	—	1,200
Seeding	1% K	2% K	1% K_2CO_3	1% K_2CO_3	K_2CO_3	—
Electric output [kW]	1·5	10·4	12	700	1,350	up to 35 MW
Duration of experiment	9 s	4 min	5 s	45 h	140 h	several minutes or hours
Magnetic field strength [T]	0·86	1·4	1·4	3·3	—	3·5
Electric conduct-ivity of gas [S/m]	350	29·5	65	60·6	—	—

high-temperature regenerative heat exchanger; the chamber through which flows the ionized hot gas; and the recovery of the seeding material. In nuclear power stations, the closed circuit offers better conditions for the optimization of the acting medium. However, the development of a high-temperature reactor for a cooling-gas temperature of 1,500 to 1,800°C is by no means a simple problem.

The development of magnetoplasmodynamic generators has proceeded at a rapid rate in recent years. In 1959, experiments were being performed on a few laboratory generators with outputs of a kilowatt or two and working periods of a few tenths of a second. In 1962, however, a 1,350-kW generator was operated for almost six days without a breakdown. Some data referring to the carliest experimental generators are tabulated in Table 3.3.3. The year 1962 saw the construction of prototype generators with electric outputs of about 20 MW, and the first of these was started up in 1964. These installations are intended to provide the data for the design of power-generating equipment on an industrial scale. However, the first plasmodynamic power stations with an output of about 300 MW must not be expected much earlier than 1970.

The high efficiency and simple design of the generator are the great assets by which the plasmodynamic power plant surpasses the thermal one. As for steam turbines, the concept of turbine efficiency is also introduced for plasmodynamic generators as a measure of the irreversibility of the processes that take place in the generator. Let us write for the stagnation enthalpy of the gas

$$h = i + \frac{v^2}{2}. \tag{3.3.107}$$

The turbine efficiency is expressed in terms of the stagnation enthalpies by the relation

$$\eta_t = \frac{h_1 - h_2}{h_1 - h_2'}. \tag{3.3.108}$$

The indices 1 refer to the inlet cross-section of the generator chamber, and the indices 2 to the exit cross-section. The enthalpy value h_2' is related to isentropic flow.

An important parameter of the power plant is the conversion efficiency of the generator, defined as the ratio of the electric-power output to the stagnation enthalpy h_0 of the gas in the combustion chamber. Before entering the generator chamber, the gas passes through a diffuser in which part of its potential energy is transformed into kinetic energy and its flow velocity is thus enhanced. The enthalpy of the gas on its entry into the generator is $h_1 = \eta_1 h_0$, η_1 being the efficiency of the diffuser. The conversion efficiency is then

$$\eta_{tr} = \eta_t \frac{\eta_1 h_0 - h_2}{h_0}. \tag{3.3.109}$$

If we substitute Eq. (3.3.107) into this relation and assume that the specific heats are constant, then

$$\eta_{tr} = \eta_t \cdot \frac{\dfrac{\varkappa}{\varkappa - 1}\dfrac{R}{M}(\eta_1 T_0 - T_0) - \dfrac{v_2^2}{2}}{\dfrac{\varkappa}{\varkappa - 1}\dfrac{R}{M}T_2}; \qquad (3.3.110)$$

$\varkappa$ stands for the Poisson constant, M is the molecular mass of the gas, and R the gas constant. The conversion efficiency will be highest when the outlet velocity of the gas issuing from the generator is zero, i.e. the entire kinetic energy of the gas has been used up.

The power-plant generators studied at present are sets with power outputs of several hundred megawatts and something like the following characteristic values:

electric conductivity of gas	about 40 S/m
gas velocity	about 800 m/s
current density	in the range of 10^4 A/m^2
electric field intensity	several thousand V/m
specific power	several tens or hundreds of MW per cubic metre of plasma.

Present-day iron-cored electromagnets permit the generation of fields up to 3 T, while superconductive magnets are expected to yield an induction up to 10 T in the air gap. In large generators, approximately 5% of the total generated output goes to energize the magnetic field, but in superconductive magnets the corresponding amount will be less than 1%. The specific mass of the generator is about 15 tons per megawatt, and the built-up space about 4 m^3/MW. The net cross-sectional area of the generator chamber amounts to approximately 1 m^2, and its total length to several tens of metres. The maximum pressure in the chamber is about 15 atm.

To attain the requisite temperature of the combustion gases, the combustion air must be heated to a relatively high temperature. However, this temperature can be drastically reduced if the air is enriched with oxygen, introducing what is called an "oxygen cycle". The enrichment considered most frequently is expressed by the molar ratio $N_2 : O_2 = 2$.

The recovery of the seeding material from an open circuit poses a difficult chemical problem. It is believed that about 90% of the seeding additive has to be recovered from the circuit lest the expense of fresh material disproportionately increases the cost of the power.

A special se tof problems is raised by the combination of the plasmodynamic generator with a nuclear reactor. A monoatomic gas, either helium or argon, will be the best choice for heating in the reactor. The favourable composition of the acting medium will mainly result in a higher terminal voltage, current densities increased by a factor of up to ten, and smaller dimensions of the generator. The closed circuit will permit operation with lower gas pressures. The Hall coefficient will be approximately 10 for helium and even higher for argon, hence the generator must be of a Hall-type design. Ionization due to the radiation of fission products will slightly increase electric conductivity.

The successful advance, hitherto, in the field of plasmodynamic energy conversion is largely due to knowledge and experience drawn from plasma physics. Further scientific and technological progress in the field of partly ionized plasma may be confidently expected to pave the way to more perfect solutions in the plasmodynamic conversion of energy.

3.4 High-Temperature Chemical Synthesis in the Gaseous Phase

3.4.1 Introduction

The rapid advance of plasma technology, in recent years, has suggested new potential applications of electric arcs with extremely high temperatures and heat flows in the gaseous-phase synthesis of chemical compounds. The highest temperatures commonly used in chemical practice do not exceed 2,000°K; only in resistance-, arc- and induction-furnaces (employed for metallurgical purposes and the production and processing of some oxides, carbides, borides and nitrides) are temperatures in the range of 3,000°K, i.e. the maximum employed in present-day chemical technology on an industrial scale. Such temperatures attained in the electric arc were hitherto limited to only two manufacturing processes in the gaseous phase—the direct synthesis of nitrogen oxide for the production of nitric acid, and more recently the production of acetylene by cracking hydrocarbons. These processes could only prevail, though, where electric power was cheap.

The discovery of the electric arc, early in the nineteenth century, and its subsequent improvement stimulated the first attempts at high-temperature chemical reactions, though long before that date some chemical compounds were known to be formed during spark discharges in gaseous media. In 1758, ozone was detected in a spark discharge in air, and in 1785 H. Cavendish prepared nitric oxide by similar discharges. In the later half of the nineteenth century, high-temperature reactions in the electric arc were intentionally brought about, and some of them are being studied up to the present time.

Morren [Ref. 3.4.1] succeeded in 1859 in producing hydrogen cyanide by passing mixtures of nitrogen with either hydrogen or hydrocarbons through an electric arc between carbon electrodes, the reaction being

$$C + 0{\cdot}5\,H_2 + 0{\cdot}5\,N_2 \;=\; HCN. \qquad (3.4.1)$$

Acetylene was another compound prepared in the electric arc. Berthelot (in 1863) carried out the direct synthesis of acetylene from the constituent elements by passing hydrogen through a carbon arc [Ref. 3.4.2]. By absorption in a copper-chloride solution he proved the presence of the gas generated according to the reaction

$$2\,C + H_2 \;=\; C_2H_2. \qquad (3.4.2)$$

When Lesius repeated a similar experiment in 1890 [Ref. 3.4.3], he found that in passing through the arc 10% of the hydrogen is converted to acetylene. Pring and Hutton [Ref. 3.4.4] were the first to determine the temperature dependence of the acetylene concentration in the synthesis of the compound from its elements. They found that 0·2 to 0·4% C_2H_2 is formed at a temperature of 1,800 to 1,900°C, approximately 1% at 2,000°C, and 2–4% at 2,500–2,800°C.

Fischer and Pichler [Ref. 3.4.5] made a very thorough study of the synthesis of acetylene from its elements in the arc. For their experiments they used a 750-cm^3 silicon flask cooled with water or liquid air. In it, two carbon electrodes were attached to water-cooled copper tubes, and a D.C. arc burned between them. The gas mixture employed in their experiments consisted of 94·9% hydrogen, 4·9% nitrogen, and

Table 3.4.1 ANALYSIS OF REACTION GAS IN THE DIRECT SYNTHESIS OF ACETYLENE FROM ITS ELEMENTS [VOLUME PER CENT]

Cooling	water				liquid air			
experiment No.	1	2	3	4	1	2	3	4
pressure [mm Hg]	760	760	260	75	760	755	160	66
acetylene	2·2	2·1	1·5	1·2	8·5	9·6	24·2	93·4
ethylene	0·8	0·7	0·0	0·0	1·8	2·4	0·4	0·0
saturated hydrocarbons	1·3	1·5	1·0	0·9	1·8	2·3	3·0	6·6
hydrogen	95·7	95·9	97·5	97·9	87·7	86·1	72·4	0·0
hydrogen cyanide per 100 cm^3 of gas	1·0	1·3	2·7	—	—	—	—	—

0·2% carbon monoxide. The analyses of the gases after a reaction time of one hour are summarized in Table 3.4.1 for different ways of cooling and different initial pressures. Although Fischer and Pichler proved that a yield of almost 95% can be obtained if liquid air is used for the rapid cooling of the reaction gases, this method has never been exploited on an industrial scale.

After Priestley's and Cavendish's exploits, the possibility of a direct synthesis of atmospheric nitrogen and oxygen according to the reaction

$$0{\cdot}5\, N_2 + 0{\cdot}5\, O_2 \quad = \quad NO \tag{3.4.3}$$

was thoroughly studied, and it became the first of the high-temperature syntheses to be industrially used in the manufacture of nitric acid.

Nuthmann and Hofer [Ref. 3.4.6] carried out the direct synthesis of nitric oxide from air in an A.C. arc and found that in the gas—rapidly cooled after passing through the arc—there is 3·6 to 6·7% by volume of nitric oxide. Nernst [Ref. 3.4.7] carried out more accurate measurements of nitric-oxide concentration in the reaction gas under equilibrium conditions at various temperatures. He passed dry air through a red-hot irridium tube, collected the gas straight from the reaction zone in a silicon tube, rapidly cooled, and analyzed it. He found that the equilibrium mixture contained 0·64 volume per cent of nitric oxide at 2,033°K, and 0·97% at 2,195°K. Later on, the utilization of the direct synthesis of nitric oxide in the Norwegian manufacture of nitric acid was based upon Nernst's theoretical and experimental work in this field.

3.4.2 Contemporary Problems of High-Temperature Synthesis

The employment of high temperatures in chemical manufacturing processes is the subject of very intensive studies, at the present time. This line, although not yet beyond the initial stages of its development, is considered a very promising trend of chemical synthesis, particularly well suited for the preparation of highly endothermic compounds or splitting of thermally very stable substances. Various problems, however, have to be solved before high temperatures can win through in industrial applications. Finding an economical way of producing high temperatures is not the least of them.

At the present time, no other means but a D.C. or A.C. arc or else a plasma torch are at our disposal for setting off chemical reactions at high temperatures. Cheaper electricity and a higher efficiency of arc plasma as a source of thermal reaction energy are the fundamental conditions for a wider utilization of the electric arc in chemical manufacturing processes.

Another crucial problem consists in the fact that virtually not even the basic data of high-temperature reaction kinetics are known, so far, and only for a few substances do we know the thermodynamic state functions at such temperatures. In the study of high-temperature reactions, the values of free energies, enthalpies and entropies of compounds are no less vital than when chemical reactions are studied at normal temperatures. To supplement these values is the basic condition for the scientifically founded investigation and industrial utilization of high-temperature reactions.

Rapid cooling of reaction mixtures is also an unsolved problem; it includes the "freezing" of the equilibrium composition (metastable state) of the reaction products achieved at high temperature. Heat exchangers do not cool the reaction mixture rapidly enough; owing to slow cooling, the reaction equilibrium shifts back in favour of the initial substances at the expense of the desired products, or in favour of undesirable secondary reactions, and substantial losses ensue in either case. Provided the pressure in the reaction chamber is at least 20 to 30 atm, expansion in a Laval nozzle results in extremely rapid cooling of the reaction products, at rates ranging up to 10^6 °K/s. This cooling method is applicable where the reaction is not unfavourably affected by the pressure. Cooling the reaction mixture by the addition of inert substances with high vaporization heat does not, as a rule, produce a sufficiently rapid temperature drop. Moreover, this method complicates the purification of the reaction products, reduces their concentration, and cancels out much of the heat energy supplied by the electric arc.

High temperatures and heat flows involve difficulties in the selection of materials suitable for high-temperature reactors. Problems of a similar kind had to be solved in the advance of rocketry, and experience gathered in the operation of thrust chambers can therefore be utilized in the construction of reactors for high-temperature synthesis. The service life of equipment can, even at high temperatures, be lengthened by protecting the surfaces of basic structural materials with some oxides, carbides, nitrides, borides or silicides.

Regenerative cooling has proved its worth in thrust chambers; the temperature of the material is reduced as required, and the heat removed from it is used to preheat the reacting substances. By capacitative cooling, i.e. by creating a cooler boundary layer, we can—apart from reducing the temperature of the reactor wall—also form a protective layer against erosion and aggressive effects of the reacting substances. By ascertaining the physical and chemical properties of heat-resistant materials in various high-temperature media and becoming familiar with their production, it will, no doubt, be possible to solve the problems involved in the construction of reactors for high-temperature synthesis. This conviction is supported by the successful direct synthesis of nitric oxide from its elements and the industrial production of acetylene by the cracking of hydrocarbons in an arc furnace.

At the present stage, the economical exploitation of high-temperature synthesis is limited by comparatively low yields for a high consumption of electrical energy, and by the high price of the latter—higher in most countries than the cost of thermal energy obtained by the combustion of fossil fuels. The reasons why the new techniques do not easily make headway against the well-proven and time-honoured methods are then mainly economic. Once the basic problems of high-temperature synthesis have been solved, the factors decisive for its full industrial utilization will be the price of the energy required for the generation of high temperatures and, in addition, the raw material reserves of the chemical industry. The dwindling reserves of coal and oil will force mankind to utilize them with the utmost economy. Under the realistic assumption that new methods of power generation will provide a sufficiency of inexpensive electrical or thermal energy, the way to new manufacturing techniques in the chemical industry will be open.

3.4.3 Direct Synthesis of Nitric Oxide

The synthesis of nitric oxide from its elements is a highly endothermic reversible reaction described by the formula

$$N_2 + O_2 \rightleftarrows 2\,NO \qquad \Delta H_f^0 = 43{\cdot}8 \text{ kcal} \tag{3.4.4}$$

According to the Guldberg-Waage law, the equilibrium constant of this reaction—its components expressed by the partial pressures—is given by the relation

$$K_p = \frac{p_{NO}^2}{p_{N_2} \cdot p_{O_2}}; \tag{3.4.5}$$

the temperature dependence of the equilibrium constant as formulated by Nernst is

$$\log K_p = \frac{9\,397}{T} - 1{\cdot}75 \log T + 0{\cdot}006\,8T -$$

$$- 2{\cdot}97 \,.\, 10^{-6} \,.\, T^2 + 0{\cdot}442 \,.\, 10^{-9} \,.\, T^3 - 1{\cdot}6 \tag{3.4.6}$$

The equilibrium state of this reaction does not depend on pressure, because the number of moles entering into the reaction and resulting from it is equal. Its temperature dependence, however, is very marked, and the quantity of nitric oxide rises with the temperature. Hence the equilibrium of the reaction will shift to the right as the temperature rises, and simultaneously the rate at which equilibrium is established will be accelerated. According to Waeser [Ref. 3.4.9] equilibrium is established after a minimum of

30 h	at 1,500°C
5 s	at 2,100°C
0·01 s	at 2,500°C
$3{\cdot}5 \,.\, 10^{-5}$ s	at 2,900°C

Table 3.4.2 indicates the amount [volume per cent] of nitric oxide in the equilibrium state of air (approximately 79% N_2 and 21% O_2) at various temperatures.

Table 3.4.2 EQUILIBRIUM CONCENTRATION OF NITRIC OXIDE IN AIR AT VARIOUS TEMPERATURES

deg. K	2,000	2,500	3,000	3,500	4,000	4,500
vol. % NO	0·708	2·286	4·594	7·316	10·177	12·940

The direct synthesis of nitric oxide from its elements is only possible on the condition that the mixture of nitrogen and oxygen is heated to a high temperature and—after a short reaction time—very rapidly cooled down. Gradual cooling would cause the oxide to re-decompose to nitrogen and oxygen, and the resultant nitric-oxide concentration would correspond to the equilibrium state at the final temperature. If the reaction mixture is cooled at a sufficiently rapid rate, the equilibrium is "frozen", and a metastable state arises at which the nitric-oxide concentration is far higher than corresponds to the equilibrium at that temperature.

Ch. Birkenland found, in 1903, that a high-intensity magnetic field is capable of "stretching" an arc to the shape of a disk, forming thus an "artificial sun". He utilized this achievement to form a high-temperature reactive zone for the oxidation of atmospheric nitrogen. Together with S. Eydem he started in Notodden the first nitric-acid factory operating on the electric-arc method. They had three arc furnaces, of 500 to 600 kW output and approximately 75 m^3/min air throughput each.

One of the arc furnaces used for the oxidation of atmospheric nitrogen is shown in Fig. 144. Ducts (1) conveyed the air into the reaction chamber where an A.C. arc burned between the electrodes (4). A strong electromagnet stretched the arc so that it formed a fiery disk of 1·8 m diameter. The reaction gases escaped through apertures (5) into water-cooled heat exchangers. The furnace was lined with fire clay and lasted about a month when in full operation. The copper electrodes, 1·5 cm in diameter, had to be exchanged approximately every six weeks.

As the air passed through the reaction chamber, whose temperature was about 3,500°K, approximately 3 to 4% of nitric oxide was synthesized. The actual oxide concentration in the issuing gas was far lower, though, and mostly did not exceed 2%. Consequently it was a rather difficult job to recover the nitric oxide from such a diluted gas. For rapid cooling, the reaction gases were passed through water-cooled

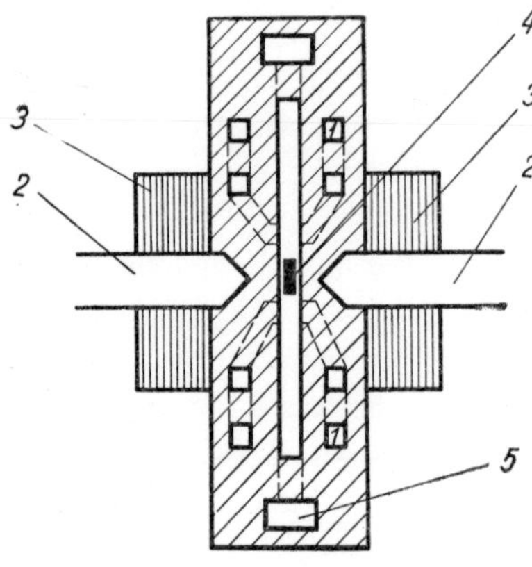

Fig. 144 Arc furnace for oxidation of atmospheric nitrogen

(1) air supply ducts (2) magnet poles (3) coils (4) electrodes (5) outlet for reaction gases

heat exchangers, where part of the heat was utilized to generate steam. The final cooling was carried out in aluminium coolers, where the nitric oxide was simultaneously oxidized to nitrogen dioxide.

Cooled down to 50°C, the gas was passed on into a system of towers where the absorption of the nitric dioxide by water or diluted nitric acid yielded nitric acid and nitric oxide:

$$3\,NO_2 + H_2O = 2\,HNO_3 + NO. \qquad (3.4.7)$$

The nitric oxide was recirculated, oxidized to nitric dioxide, and in this form returned to the absorption towers. The last of these towers was sprayed with milk of lime so as to extract the remaining oxides from the exhaust gases.

In full operation, the three Notodden furances yielded a daily output of 1,500 kg of anhydrous nitric acid, the consumption of electricity amounting to 11 – 17 kWh per kg. These high demands on electric power as well as the difficult production control caused this industrial method, important as it was at its time, to be superseded by an indirect synthesis whose power balance is far more advantageous. According to this method, ammonia synthesized from atmospheric nitrogen is oxidized to nitric oxide by a catalyzed reaction:

$$4\,NH_3 + 5\,O_2 = 4\,NO + 6\,H_2O \tag{3.4.8}$$

Nitric-acid production by direct high-temperature synthesis of atmospheric nitrogen was stopped more than 30 years ago; in recent years, however, it is again being thoroughly studied. Research is primarily concerned with enhancing the thermal effect of the arc in heating the gas, and finding a convenient method of cooling the reaction mixture. Structural materials for reactors and electrode materials are other subjects studied. Since pressure does not unfavourably affect the reaction, a plasma torch working at 20 to 30 atm seems indicated, and a Laval nozzle could be used to cool the reaction mixture very rapidly [Ref. 3.4.10]. Such a reactor is schematically represented in Fig. 145. In order to reduce the electrode wear, pure nitrogen instead of air is

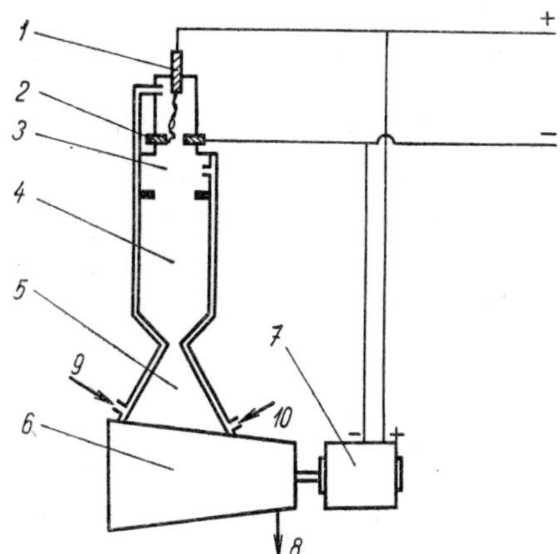

Fig. 145 Plasma reactor for direct nitrogen-oxide synthesis
(1) anode (2) cathode (3) mixing chamber (4) reaction chamber (5) Laval nozzle (6) turbine (7) generator (8) inlet for reaction mixture (9) liquid-nitrogen supply (10) liquid-oxygen supply

heated in the arc, and the nitrogen plasma is subsequently mixed with oxygen to an approximately equimolar composition in a mixing chamber. The heat in the reaction chamber is supposed to be 3,300 to 3,500°K, and after a reaction time of 0·001 s, the mixture is vigorously cooled to approximately 1,400°K by expansion in a Laval nozzle. A gas turbine with sweat-cooled vanes, placed at the end of the Laval nozzle, utilizes part of the kinetic energy of the rapidly flowing gases to drive a D.C. generator. The calculated nitric oxide concentration of the gas issuing from the turbine is approximately 6·5 per cent of volume, the calculation being based on an equimolar mixture of oxygen and nitrogen at a temperature of 3,500°K in the reaction chamber.

The progress of rocket engineering has shown that the electric arc is not the only means of attaining the high temperatures required for the direct synthesis of nitric oxide from its elements. Thrust chambers (i.e. rocket-type combustion chambers) working at 3,000 to 4,000°K and 20 to 40 atm have also been designed for the preparation of nitric oxide. The feasibility of direct nitric-oxide synthesis in the thrust chamber has been experimentally proved by Weissenberg and Winternitz [Ref. 3.4.11]. They burned ethane with an excess of oxygen-enriched air, the amount of the added oxygen being such as to produce an equimolar mixture of nitrogen and oxygen in the combustion gases. The reaction time in the combustion chamber at a pressure of 20 atm and a temperature of 2,800 to 3,100°K was about one thousandth of a second. By expansion in a Laval nozzle, the combustion gases were cooled down to 1,600°K at the rate of approximately 10^6°K per second. A suitable shape of the Laval nozzle will yield an even lower gas temperature after the expansion. The economy of the process is improved if the kinetic energy of the expanding combustion gas is utilized in a turbine. The expanded combustion gas contained 2·7% nitric oxide, which is about 90% of the theoretical yield at 3,000°K under the given conditions.

The oxidizing medium and high temperatures make great qualitative demands on the structural materials of the combustion chambers and turbines. Suitable (e.g. regenerative) cooling systems combined with heat-resisting materials and porous cooling of the gas-turbine vanes [Ref. 3.4.12] will enable the engineering industries to introduce the routine production of such installations.

In shock tubes, reaction mixtures can be cooled at an even higher rate. For the time being, however, this device is not applicable in direct nitric-oxide synthesis, because it works periodically.

To sum up: various problems have to be solved before the direct synthesis of nitric oxide becomes industrially feasible. In the first place there is the most economical utilization of the heat source (arc, combustion); next, suitable structural materials for high temperatures, pressures and oxidizing media have to be developed, and the maximum utilization of the thermal and kinetic energies of the gases or combustion products must be secured.

3.4.4 Production of Acetylene by Cracking of Saturated Hydrocarbons in the Electric Arc

At the present time, the only manufacturing process in the chemical industry to use temperatures higher than 3,000°K is the production of acetylene by the cracking of saturated hydrocarbons (particularly methane) in an electric arc.

Acetylene is now mostly produced from calcium carbide prepared in electric furnaces from calcium monoxide and high-quality coal coke or anthracite coke. The high concentration of the resultant acetylene is the advantage of this process, whereas the valuable raw materials required and the high consumption of electricity are its drawbacks. The pyrolysis of methane or low hydrocarbons is another method of manufacturing acetylene. The heat required for the highly endothermal reaction—cracking the hydrocarbon to acetylene and hydrogen—is supplied either indirectly (regenerative method) or by the direct combustion of part of the methane with oxygen (partial oxidation). The combustion of some of the methane supplies the heat for both the heating of the gas to reaction temperature and the pyrolysis proper. Its drawback, however, is the high consumption of hydrocarbons and the low acetylene content of the reaction gas—8 to 10%.

Various authors studied the cracking of hydrocarbons in an electric arc, in the thirties of this century [Refs. 3.4.13, 3.4.14, 3.4.15 3.4.16]. Soviet research chemists devoted particular attention to the effect of the initial gas on the consumption of electrical energy; they obtained very satisfactory acetylene yields at low power expenditure.

In 1940, Chemische Werke Hüls, an affiliation of I. G. Farben, started an acetylene plant in which hydrocarbons were split in an electric arc by a method developed by Baumann. The raw materials were natural gas, waste gases from coal hydrogenation, the methane fraction of coke gases, or gaseous products of petroleum distillation. In all of these materials, the main component was methane (70 to 90%), mixed with lesser amounts of higher hydrocarbons, olefins and nitrogen. The thermal decomposition of methane, the basic constituent of all these materials, proceeds by the following mechanism [Ref. 3.4.17]

$$\begin{array}{ccll} & & 2\,CH_4 & \\ & & \downarrow & \\ CH_3\,.\,CH_3 & \rightleftarrows & 2\,CH_3 & +\ H_2 \\ & & \downarrow & \\ CH_2 : CH_2 & \rightleftarrows & 2\,\dot{C}H_3 : & +\ H_2 \\ & & \downarrow & \\ CH \vdots CH & \rightleftarrows & 2\,CH \vdots & +\ H_2 \\ & & \downarrow & \\ & & 2\,C & \end{array}$$

also

$$n(CH :) \rightarrow C_nH_n,$$

and

$$n(CH :) \rightarrow C_nH_{n-m} + 0{\cdot}5m\,H_2.$$

The thermal stability of saturated hydrocarbons decreases with increasing molecular mass. The first signs of decomposition in methane appear at 683°C, in ethane at 485°C, in propane at 460°C, and in butane at 435°C.

As methane is split in an electric arc, the most important endothermal reactions are as follows:

thermal dehydrogenation of methane and formation of acetylene:

$$2\,CH_4 = C_2H_2 + 3\,H_2 \quad + 91\ \text{kcal}; \tag{3.4.9}$$

total split-up of methane into carbon and hydrogen

$$2\,CH_4 = 2\,C + 4\,H_2 \quad + 41\ \text{kcal}; \tag{3.4.10}$$

dehydrocyclization of methane and formation of benzene

$$6\,CH_4 = C_6H_6 + 9\,H_2 \quad + 126\ \text{kcal}. \tag{3.4.11}$$

High temperature and low pressure are the best conditions for reaction (3.4.11), and the shortest possible reaction time suppresses the undesirable reaction described in (3.4.12); low temperatures and a longer reaction time favour reaction (3.4.13).

The formation of acetylene by the thermal decomposition of methane is explained by the theory of free radicals. At high temperature (3,000 to 5,000°K) methane splits up into the methane radical CH_2: and a hydrogen molecule:

$$CH_4 = CH_2: + H_2, \qquad (3.4.12)$$

and the subsequent reaction of the methene with methane forms ethane:

$$CH_4 + CH_2: = C_2H_6. \qquad (3.4.13)$$

This theory was first of all confirmed by the considerable amount of ethylene C_2H_4 and ethane C_2H_6 found when methane was heated in a silicon tube. The existence of free radicals during the splitting-up process was proved by spectral methods and by the chemical combination of the radicals in volatile organo-metal compounds; $CH_3^{\cdot}$ radicals were identified in tellurium and mercury compounds, and $C_2H_5^{\cdot}$ and CH_2: in tellurium or iodine compounds. More complex radicals were not found.

The Hüls plant was equipped with 17 manufacturing units, 15 of them in operation. Each unit had two arc furnaces (Fig. 146)—one of them in operation and the other being cleaned or repaired.

Admitted under a pressure of 1·4 to 1·5 atm into the head of the arc furnace, the gaseous starting material is set into a vortical motion by the tangential arrangement of the intake ports and passes into the arc that burns in a tube 1 m long and of 9·5 cm inner diameter. The furnaces have an input of 7,000 kW each, and its increase to 10,000 to 14,000 kW was under consideration. The furnace is fed with 850 to 900 A D.C. at 7,800 V. The gas velocity in the reaction chamber is approximately 1,000 m/s and the duration of the contact between the gas and the arc plasma 0·001 s. The temperature in the arc core is estimated at 3,000 to 5,000°C. Every furnace processes about 2,800 m^3 of gas—a mixture of fresh and recycle gas in equal proportions—per hour; the split-up increases this volume to 4,200 m^3.

The D.C. arc burns between a water-cooled copper cathode placed at the top of the furnace and an earthed copper anode which seals the arc tube—a thick-walled water-cooled steel tube. Though cooled, the reactor tube is rapidly consumed by combustion and has to

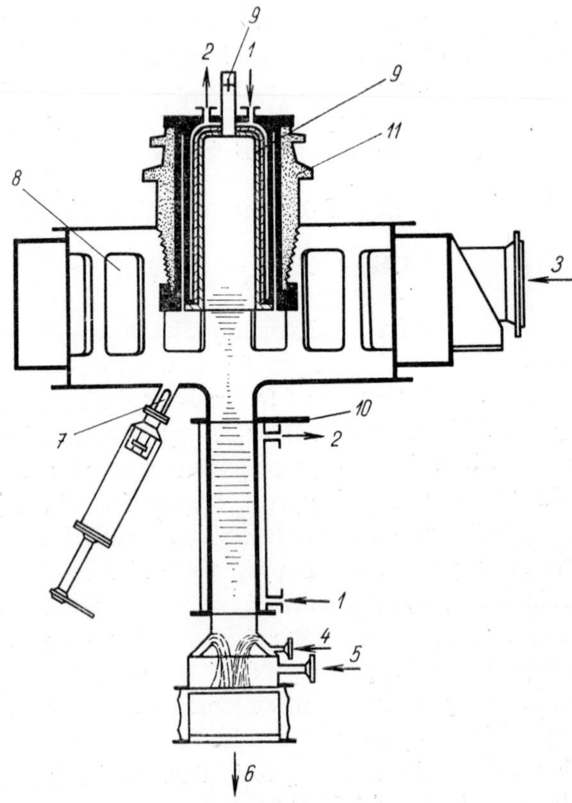

Fig. 146 Arc furnace for decomposition of hydrocarbons

(1) cooling-water inlets (2) cooling-water outlet (3) gas supply (4) cooling water (5) nitrogen supply (6) outlet for reaction gases (7) pilot electrode (8) tangential intakes (9) cathode (10) grounded anode (11) porcelain isolator

be exchanged after 250 or 300 hours of operation. In the expanded bottom part of the furnace the reaction gases are cooled from about 1,600°C to 150 or 200°C.

The average daily output is some 15,000 kg of acetylene from every furnace, and roughly 200 metric tons for the entire plant at full operation.

The analysis of the cracked gas mixture and the consumption of electric power for breaking it down depends primarily on the composition of the initial gas. The processing of hydrogenation gases, which also contain higher saturated hydrocarbons, requires an average of 9 kWh per m^3 of crude acetylene, and the processing of methane (natural gas) about 12 kWh. The theoretical consumption of electricity is about 4·7 kWh per cubic metre of acetylene. From the point of view of power consumption it is therefore advantageous to manufacture acetylene from gases containing higher hydrocarbons, which are thermally less stable than methane. Such initial gases also yield a higher acetylene content in the reaction gases [Ref. 3.4.19].

Table 3.4.3 ANALYSES OF GASES IN THE ELECTRIC-ARC PRODUCTION OF ACETYLENE [PER CENT OF WEIGHT]

	hydrogenation gases		natural gas	
	before	after	before	after
methane and homologues	83·7	24·9	80·2	27·8
hydrogen	9·0	50·8	2·5	46·0
acetylene	0·0	16·8	1·5	13·3
olefins	0·0	3·9	1·4	0·9
carbon monoxide	1·7	0·9	3·0	2·9
carbon dioxide	0·2	0·0	0·3	0·0
oxygen	0·4	0·2	0·3	0·2
nitrogen	5·0	2·5	10·2	8·9

After cooling by a water spray, the reaction gases of every unit are passed through four cyclones where 60 to 70% of the soot are removed; the remainder is eliminated by spraying with water and filtering. The soot-free gas is cooled to 20°C to remove the entrained water and then passed to the washers. High-boiling substances are removed in the first of them, hydrogen cyanide is washed out with water in the second, and sulphur is removed in the third by passing the gas over ferrous oxide. The crude acetylene is then compressed to 19 atm and absorbed by water. Unabsorbed gases, such as methane, ethane, ethylene and hydrogen, are separated by distillation in a low-temperature Linde unit. Ethylene and hydrogen are separated and the saturated hydrocarbons are re-cycled into the arc furnace.

After separation and concentration of the reaction gases, the yield from 100 kg of initial gas is

45 kg of acetylene (96 to 97% pure)
9 kg of ethylene (98% pure)
5 kg of soot
140 m^3 of hydrogen (97% pure).

The content of contaminants in the reaction gas after the de-composition in the electric arc [Ref. 3.4.20] is tabulated below.

Table 3.4.4 CONTENT OF CONTAMINANTS IN REACTION GASES AFTER DECOMPOSITION IN THE ELECTRIC ARC [g/m^3 OF GAS]

	hydrogenation gases	natural gas
hydrogen cyanide	1—3	2—5
naphthalene	1—3	0·12—2·0
benzene	1—6	1·5—8·8
diacetylene	15—30	15—30
hydrogen sulphide	traces	traces
soot	20—25	11

Instead of gaseous hydrocarbons, liquid ones may also be used for making acetylene. Tatarinov cracked liquid hydrocarbons (crude oil, benzene, paraffin, mazut, etc.) in an electric arc [Ref. 3.4.21] and obtained reaction gases containing [per cent of weight] an average of

30 acetylene
10 ethylene
40 hydrogen

15 methane
5 unsaturated hydrocarbons.

Similar results were obtained by Dobryanski [Ref. 3.4.22] and Morozov [Ref. 3.4.23]. Analogous experiments were also made in Germany. A mixture of city gas and petrol vapours (the latter of 65 to 95°C boiling point) was passed through the electric arc and then cooled to 120°C with oil (boiling point 220 to 360°C). The resultant gas had the following analysis [weight per cent]:

0·4 carbon dioxide
18·0 acetylene
7·0 ethylene
0·5 oxygen
6·3 carbon monoxide
54·2 hydrogen
9·8 methane
4·0 nitrogen.

In addition to these gaseous products, the decomposition also yielded soot. The expenditure of electricity amounted to 8 kWh per 1 kg of acetylene or its homologues.

The utilization of the electric arc for chemical manufacturing processes is often dismissed as uneconomical. Reviewing the Hüls method, Cagaš and others [Ref. 3.4.24] emphasized that the economic characteristics of this method can be improved. Their experimental results demonstrated that yields of 11 to 13% acetylene and about 1% of ethylene in the reaction gas are obtainable when A.C. is used instead of D.C. (They worked on natural gas containing 91 to 95% of methane, 3 to 6% of higher saturated hydrocarbons, and 1 to 3% of "non-reactive" gases, mainly nitrogen.) Electric-power consumption was 11 to 12 kWh per kg of acetylene.

The substitution of an A.C. arc for D.C. reduces the prime costs of the industrial feeding circuit by about two thirds, which results in total capital costs cut by approximately 15%. Moreover, the analysis of the thermal balance proves that as much as 50 to 60% of the heat energy supplied to the arc goes to heat the gas to the reaction temperature, and only 40 or 45% of the total energy is used for the chemical reaction. As the reaction gas issuing from the arc furnace is cooled from 1,600°C to 150 or 200°C, its heat energy is virtually wasted.

Cagaš asserts [Ref. 3.4.24] that a two- or more-phase process will greatly improve the exploitation of the thermal energy of the reaction gases. He recommends to spray the reaction gas not with water, but which higher liquid hydrocarbons, which have a lower thermal stability and lower decomposition temperatures than their gaseous homologues. They are heated and partly decomposed by the thermal energy drawn from the gas, and this secondary decomposition – brought about with no additional energy supplied – produces unsaturated hydrocarbons and thus enhances the efficiency of the process. The cooling agent used in Cagaš's experiments was paraffinic low-octane petrol, which was enriched with unsaturated and aromatic hydrocarbons in the process (cf. Table 3.4.5).

Table 3.4.5 COMPOSITION OF PETROL BEFORE AND AFTER USE IN A TWO-STAGE PROCESS [PER CENT OF MASS]

	before use	after use
parafins	60·0	52·0
olefins	0·7	6·7
aromatized compounds	8·7	15·3
naphthenes	30·5	26·0

The two-stage method was found to require less electric power – 10 to 11 kWh/kg of acetylene, and 7 to 8 kWh/kg of acetylene and ethylene.

The experiments performed by Cagaš and his collaborators have shown that the introduction of new techniques is capable of reducing the cost of acetylene synthesis in the electric arc to a level where this method becomes fully competitive and can make the best use of all hydrocarbon fractions contained in the waste gases of chemical industries.

3.4.5 Direct Synthesis of Organo-Silicon Compounds

Organo-silicon compounds are at present prepared by reactions of silicon tetrachloride with Grignard's reagent, in which either alkyls and aryls or their halogen derivatives are formed (tetramethyl silane, tetraphenyl silane, diethyldichloro silane, etc.):

$$SiCl_4 + 4\,CH_3MgCl = (CH_2)_4\,Si + 4\,MgCl_2 \qquad (5.4.14)$$

$$SiCl_4 + 4\,C_6H_5Mg\,Br = (C_6H_5)_4\,Si + 2\,MgCl_2 + 2\,MgBr_2 \qquad (5.4.15)$$

$$SiCl_4 + 2\,C_2H_5MgCl = (C_2H_5)_2\,SiCl_2 + 2\,MgCl_2 \qquad (3.4.16)$$

The preparation of alkyl silanes by Grignard's reagent is too expensive because it requires large amounts of ether and magnesium. A direct method of preparing alkylated and arylated silanes has therefore been developed; it is based on the direct reaction of alkyl halogenides or halogen derivatives of aromatic hydrocarbons with metallic silicon at temperatures between 250 and 400°C in the presence of a copper catalyst.

$$2\,CH_3Cl + Si = (CH_3)_2\,SiCl_2 \qquad (3.4.17)$$

$$2\,CH_3Cl + Si = CH_3SiCl_3 + C_2H_6 \qquad (3.4.28)$$

These methods of preparing organo-silicon compounds yield either fully substituted silanes or their halogenous derivatives. The preparation of incompletely substituted silanes containing no halogen is more difficult.

In 1953, a British patent was granted [Ref. 3.4.25] for the direct synthesis of alkylated, arylated or unsubstituted silanes by the reaction of gaseous hydrocarbons with silicon activated in the plasma jet of an electric arc. The principle of the high-temperature preparation of silanes consists in passing the silicon or the hydrocarbon gas or both of them through the electric arc and separating the reaction mixture into its constituents after cooling it down.

This method, though not yet used on an industrial scale, points a way to the preparation of other organo-metallic compounds such as tetraethyl lead, dimethyl zinc, dibutyl lead, diethyl zinc, etc., by high-temperature synthesis in the gaseous phase. We shall therefore discuss it in more detail.

A mixture of one hydrocarbon or several of them with silicon powder is repeatedly passed through the electric arc. The products are separated from the recycled reaction gas by dissolving them in a mixture of methanol with ammonium hydroxide in a concentrated aqueous solution. The methanol, ammonium and water is distilled off,

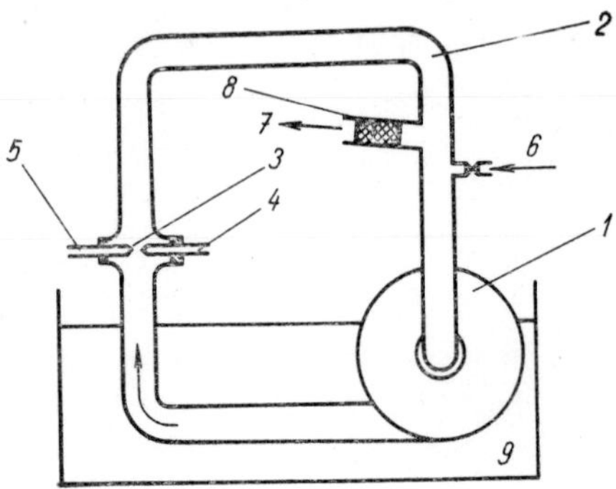

Fig. 147 Laboratory reactor for preparation of unsubstituted and alkylated silanes in an electric arc (Brit. patent 741067)

(1) fan (2) reaction tube (3) reaction chamber (4) carbon electrode (5) carbon electrode (6) gas inlet (7) reaction-gas outlet (8) filter (9) cooling tank

and the oily residue can be burned to silicon dioxide. Such results are obtained by using methane, hydrogen and carbon electrodes. (Depending on the hydrocarbon employed, alkylated or arylated silanes of the general formula R_nSiH_{4-n} are obtained, R being the alkyl or aryl group.) If pure hydrogen and silicon are processed and the electrodes are of copper, silanes such as SiH_4, Si_2H_6 and others are found in the reaction mixture. The reaction gases contain acetylene, ethylene, some unconverted initial hydrocarbons, hydrogen, soot and several per cent of alkyl silanes, mainly dimethyl silane.

The laboratory reactor employed for the direct high-temperature synthesis of silanes in an electric arc is represented in Fig. 147. A fan (*1*) circulates 10 g of pulverized silicon suspended in hydrogen through the reaction chamber (*3*) where the arc burns between the electrodes (*4*) and (*5*). The carbon electrodes, 5 mm in diameter, are gripped in cooled holders. The arc voltage – A.C. or D.C. – is about 200 V, and the current 100 A. A stop cock (*6*) permits replenishment with hydrogen and the reaction products—cleared of silica and soot on filter (*8*)—, are passed on to be absorbed by methanol and ammonium hydroxide. A cooling tank (*9*) regulates the temperature of the gas after its passage through the arc.

The oily substance formed after a reaction time of 3 hours was found to contain 0·7 g of silicon, which is 7% of the amount introduced.

The direct synthesis of silicon with hydrocarbons was carried out in the reactor illustrated in Fig. 148. This apparatus has a cooled copper cathode (*1*) and a water-cooled tubular iron anode (*2*), approximately 1 m long and 70 mm in inner diameter. The initial gas, supplied into reaction chamber (*3*) through feed tube (*8*), is heated to a high temperature and reacts with the silicon dripping into the chamber from

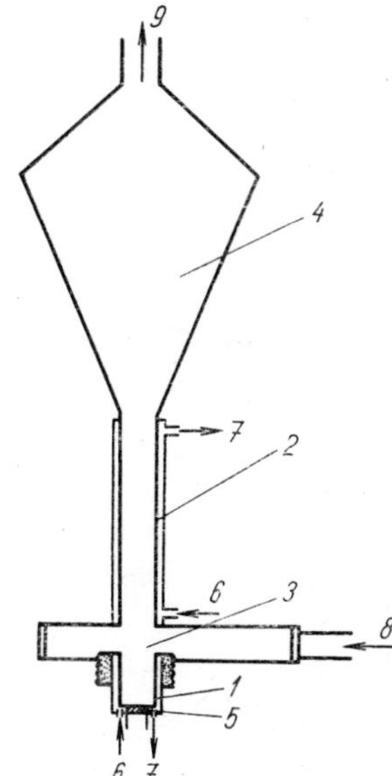

Fig. 148 Laboratory reactor for preparation of alkyl silanes in an electric arc (Brit. patent 741067)
(1) cathode (2) anode (3) reaction chamber (4) powdered-silicon feed hopper (5) slag hole (6) cooling-water supply (7) cooling-water outlet (8) gas supply (9) reaction-gas outlet

hopper (*4*). The hopper is 8 m high, its walls are inclined at 15°, and its capacity is about 1,000 kg of silicon (ferrosilicon of 98% Si and 2% Fe). Dropping into the arc, the pulverized ferrosilicon is vaporized.

The slag of unconverted ferrosilicon is drained through port (*5*). The gaseous reaction products rise from the chamber through the anode, are cooled by their passage through the ferrosilicon powder, and escape through aperture (*9*).

The analysis of the initial natural gas [per cent of volume] was 92% methane, 1·1% olefins, 6·3%nitrogen, 0·3% carbon dioxide, and 0·3% oxygen; its velocity on flowing through the anode was approximately 1,000 m/s. The reaction gas consisted of 13% acetylene, 5% ethylene, 30% unconverted hydrocarbons, 50% hydrogen and 5% nitrogen. The yield of alkylated and arylated silanes (particularly dimethyl silane) isolated from the reaction gases was very low, approximately 400 g of silanes per hour. The arc had an output of 7 kW, which had to be increased to 10 kW when operating under increased pressure (about 5 atm).

No marked improvement in yields was obtained in experiments with coke gas cleaned of carbon monoxide. The initial gas (40% hydrogen, 20% methane, 19% ethylene, 21% nitrogen) with a fine ferrosilicon powder suspended in it was circulated through the electric arc, and after a reaction time of 8 hours the yield was about 1 kg of alkyl silanes, including about 30% dimethyl silane.

The direct electric-arc synthesis of alkylated and arylated silanes from silicon and hydrocarbons or hydrogen gives very low yields. Considering a wider application of this method would therefore be premature. Direct silane synthesis combined with acetylene production in the electric arc seems to hold greater promise.

3.4.6 **Plasma Torches**

The plasma jet generated in the torch is at present the best medium for studying chemical processes and determining the fundamental physical and chemical properties of substances at high temperature.

The highest temperatures applicable in industry can be attained in plasma torches. Depending on the initial gas, plasma consisting of nitrogen, hydrogen or helium atoms and ions is obtained, and according to recent reports oxygen can also be heated to temperatures permitting its transition into the plasma state. The high temperature and intensive heat flow of plasma attract the attention of research chemists as a means of carrying out high-temperature syntheses in the electric arc, studying physico-chemical processes that take place at high temperatures,

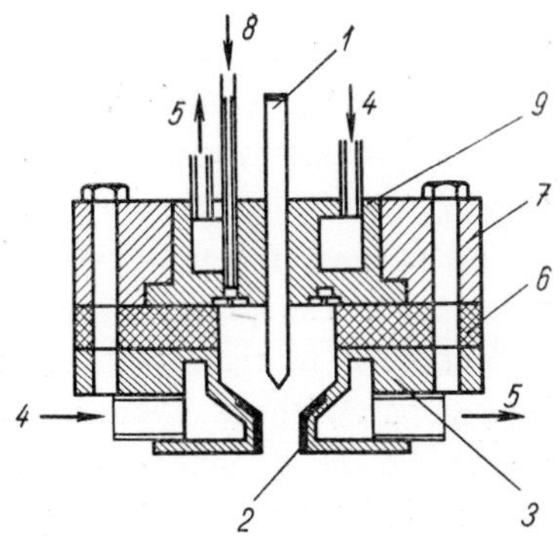

Fig. 149 Nitrogen plasma generator
(1) thoriated-tungsten electrode (2) tungsten lining of anode (3) copper anode (4) cooling-water inlet (5) cooling-water outlet (6) teflon insulation (7) bakelite housing (8) nitrogen inlet (9) cathode body (copper)

and determining the thermodynamic functions of state of substances heated to high temperatures. The plasma state of matter opens up wide vistas of further advance in research.

So far, we do not yet know all the potentialities of this new field of chemical synthesis. At the present stage, the synthesis of highly endothermal compounds, the direct synthesis of high-melting compounds from gaseous phases, and the decomposition of thermally stable compounds seem the most rewarding applications of the plasma jet.

The applicability of a nitrogen-plasma torch in the preparation of endothermal nitrogen compounds and some nitrides has been checked on a laboratory scale by Stokes and Knipe [Ref. 3.4.26].

The torch (Fig. 149) was fed by D.C. (600 A) from a welding set. The original copper electrodes cooled with distilled water had a service life of merely 15 minutes. The cathode material was therefore replaced by tungsten alloyed with 2% of thorium, and the copper anode was given an annular lining of pure tungsten. Before starting, the walls of the cooling chambers in both the cathode and anode parts were rinsed with nitric acid in order to remove contaminants which might impair the heat transfer. The cathode and anode parts were separated by a 1/4-in. (6.35-mm) PTFE insulation. The nitrogen was introduced into the generator chamber through an annular distributor.

The gaseous or solid charge is injected into the plasma by a batcher fixed to the bottom part of the generator. The entire generator is attached to the reactor in which cooling with solid carbon dioxide or liquid air or nitrogen condenses the reaction products. The generator works under moderate overpressure so as to prevent the access of air into the reaction chamber.

Endothermic compounds synthesized in this apparatus were cyanogen and nitrogen dioxide.

Cyanogen was prepared by spraying graphite into the nitrogen-plasma jet. The plasma was sufficiently hot to vaporize the graphite and cause it to react with the nitrogen in the gaseous phase. In the bottom part of the reaction chamber the unconverted graphite condensed, and so did the gaseous products after cooling with solid carbon dioxide. The molecular mass of the product obtained was 50 (while that of cyanogen is 52); however, the temperature dependence of the vapour tension, determined by experiment, proved that the sample was pure cyanogen. The electric input into the plasma ranged between 10 and 14 kW, and nitrogen was fed into the generator at the rate of 9·5 l/min. Only 2% of the total graphite introduced into the plasma was converted to cyanogen.

Nitrogen dioxide was synthesized by blowing oxygen into the plasma jet. The electric input was 10 to 13 kW, and the gas throughput 9·51 l of nitrogen and 8·25 l oxygen per minute. The blue substance obtained by cooling the reaction gases with solid carbon dioxide turned into a green liquid, when melted. Part of this substance was added to water, in the expectation that nitrous and nitric acids would be formed if the product of the reaction was nitrogen tetroxide

$$N_2O_4 + H_2O = HNO_3 + HNO_2. \qquad (3.4.19)$$

The presence of both nitric and nitrous ions in the solution was actually confirmed by the analysis. However, the conversion level was low, about 2% referred to oxygen.

The synthesis of metal nitrides in the gaseous phase was also tried out in the same apparatus. The injection of titanium and magnesium powder yielded titanium and magnesium nitrides, respectively, which condensed on the cooled elements in the reaction chamber. The electric input was 9·5 to 10·5 kW and the nitrogen feed 2·5 l/min. Titanium was fed at the rate of 1·72 g/min, and the condensed nitride formed lustrous golden crystals. Unconverted titanium was removed by crushing and washing of the product with 10% hydrochloric acid. The average yield was about 30%. In the preparation of magnesium nitride, the input into the plasma was 12 to 15 kW, and the colour of the product was a light yellow. The conversion level ranged about 40%. The purification was rather difficult, since magnesium nitride is very rapidly decomposed in the presence of water, liberating ammonia in the process.

The work done by Stokes and Knipe indicates that plasma has a wide scope of potential applications in chemical reactions and further studies in this field are fully justified. The results must not be considered the maximum that can be obtained, but rather a mere proof that syntheses of this kind are feasible. Accelerated processes and improved equipment will certainly enhance the yields.

3.5 Prospects of Plasma in Technical Applications

Among prospective technical applications of plasma – still remote, but very significant in terms of national economy – the conversion of heat into electrical energy holds first place. It has been discussed in Div. 3.3.

For a long time now industry has been utilizing chemical reactions taking place in the plasma. Div. 3.4 treats this application. Little is known about the kinetics of some chemical reactions in high-temperature plasma. There are no high-output plasma torches, so far, which – operating continuously over sufficiently long periods – convert electrical into thermal energy with good efficiency. This is one of the reasons why we are not better informed about the pattern of chemical reactions in high-temperature plasma.

Such plasma torches of minor outputs as are available at present prove useful in many technical applications. The utilization of plasma torches for welding purposes is a subject of intensive studies and trials. Linde's PT-3 plasma welding torch is already in trial operation. [Ref. 3.5.1, 3.5.2]. The main problem in employing plasma for welding is the stabilization of the arc at very low gas-flow velocities, since flowing at higher velocities the stabilizing gas would blow away the weld puddle. The throughflow of stabilizing gas in the PT-3 torch ranges from 56·6 to 283·1 dm^3h^{-1}. Since this amount is insufficient to shield the bath from oxidation, shielding gas is supplied round the nozzle, in a similar way as in the Linde surfacing torch, at the rate of approximately 850 dm^3h^{-1}. Another special feature of the PT-3 torch is a water-cooled copper electrode which can be substituted for the tungsten electrode in order to weld with indirect polarity (the electrode acting as anode). This is very useful in welding aluminium, titanium or zirconium. Arc-welding of titanium and zirconium with an argon-shielded tungsten electrode is impermissible for certain purposes, because tungsten contaminates the weld.

The welding rate attainable with plasma torches is 50 to 300% higher than is obtained in argon-shielded arc-welding with a non-consumable electrode.

A transferred arc is used in welding with plasma; argon or helium is the stabilizing gas employed on aluminium, and a mixture of argon with hydrogen on stainless steel. A single rectifier with 70 V no-load voltage is a sufficient current source when welding with pure argon. If the stabilizing gas contains more than 7% hydrogen, two such sources must be connected in series. The maximum welding current is 450 A.

Research is aimed at the welding of thin aluminium, copper (pure and alloyed with nickel), titanium, zirconium and stainless steel. Extremely thin aluminium foil (down to 0·02 mm) and stainless steel and titanium (down to 0·15 mm) are also being welded.

Automatic controls for mechanized plasma welding are to be introduced in the near future, but hand-operated plasma welding torches are also being developed.

Processing of tough and difficult-to-work materials is another suggested application of plasma torches. The torch can either be used for heating before tooling or machining, or the workpiece can be roughly shaped by fusing and finished mechanically.

The first alternative is actually an improvement on the hot-turning method which has been recommended for years [Ref. 3.5.3]. Apart from concentrating the heat on a well-defined working zone, the plasma torch has a readily controlled heat output and a high rate of heating. The arrangement of the plasma torch and the cutting tool in experimental hot turning [Ref. 3.5.4] is shown in Fig. 150.

Machining by a plasma jet is schematically illustrated in Fig. 151. The material (a), set up in a lathe, is roughed by plasma (b) and simultaneously turned with a cutting tool (c) or machined with a grinding wheel (d) (the alternative is dashed). The machining time for tough high-alloy steels can be drastically reduced by this method [Ref. 3.5.5]. As a rule, torches with a transferred arc are employed, and the workpiece acts as anode. The plasma torch can be employed for smoothing or finishing, in addition to roughing. In this operation, however, the heat expenditure per unit volume of metal removed is about double, because the heat of the plasma jet—tangentially aimed at the workpiece—is

Fig. 150 Arrangement of plasma torch and cutting tool for hot turning (Courtesy of M.v. Ardenne Forschungsinstitut, Dresden)

relatively poorly utilized [Ref. 3.5.6]. Cutting rounded threads was also tried. A screw was cut on a 2-in. (51-mm) rod held in a lathe. The rod

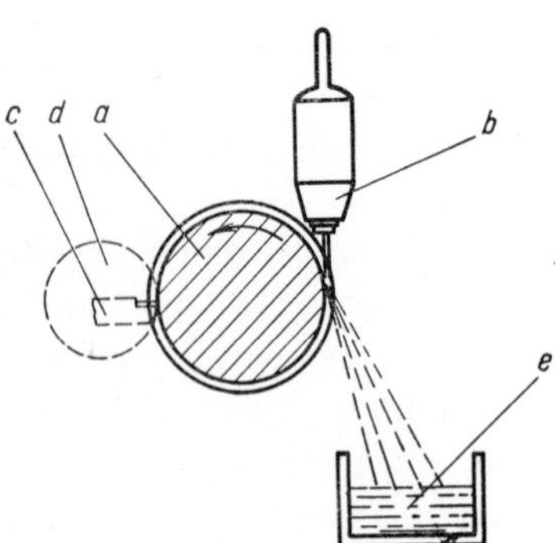

Fig. 151 Schematic illustration of plasma machining with subsequent finish turning or grinding

was rotated at 40 r.p.m. and the plasma torch moved at the rate of a foot per minute. The threads were $\frac{1}{4}$ in. deep, and the pitch was $\frac{1}{2}$ in. (12.7 mm).

Grooving by plasma torches is already being practised. Die blocks of intricate shapes can be roughed by this method. As plasma torches gain in popularity, further special applications will surely emerge.

In metallurgy, plasma torches will open the way for new manufacturing procedures. The advantages inherent in this tool are the easy control of high-temperature processes, the clean flame whose composition can be changed at will, the high gas velocity securing high heat transfer values, and the concentration of the heat energy on small areas. Experimental heats, with melts in conventional furnaces heated by a plasma output of 1,000 kW from an argon-stabilized arc, yielded steel of a quality such as is usually obtained in vacuum melting.

The chlorination of ores finds fairly extensive application, at present. This process exploits the fact that metal chlorides widely differ in their boiling points, which enables the metallurgist to isolate the individual metals from difficult-to-float complex ores by fractionated distillation. Chlorinating experiments with chlorine added into a plasma jet proved that this method can accelerate the process to a mere fraction of the time needed in the classical procedure.

Oxides of aluminium, titanium, silicon, magnesium, etc. which do not readily lend themselves to reduction by traditional methods are more easily reduced by carbon treatment in the gaseous phase. For the time being, no way has been found to cool the gaseous phase rapidly enough to prevent its re-oxidation. Attempts to cool the gases by a water-cooled copper plate or by blowing cold hydrogen into a plasma were only partly successful: the yield of pure metal was between 40 and 70%. Further studies aimed at the reduction of oxides from the gaseous form at reduced pressures continue.

Possibilities of producing carbides and nitrides in plasma have been discussed in Div. 3.4.

In metallurgy as in chemistry, the prerequisite for the application of new manufacturing methods is the solution of design problems posed by high-output plasma torches for continuous operation.

In the USA the production of high temperatures for metallurgical purposes is in the development stage. The combustion products of

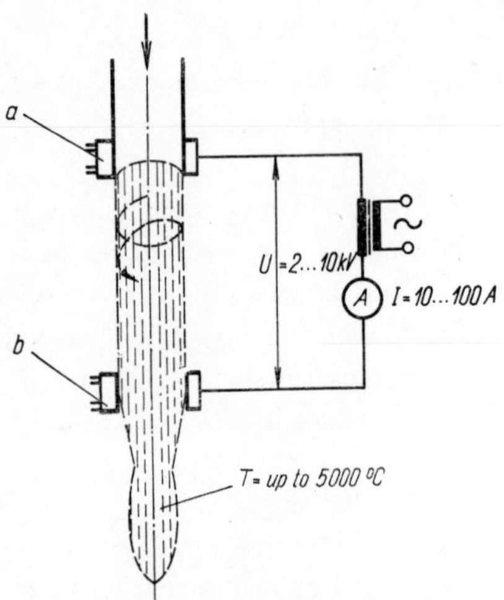

Fig. 152 Supplementary electric-arc heating of combustion gases

fossil fuels are exposed to supplementary heating by an electric arc. The original temperature of the combustion gases is about 1,900°C, but they can be heated to 5,000°C by electricity. The principle is evident from Fig. 152. A water-cooled copper ring (*a*), acting as an electrode, is mounted on the outlet of the burner. The second electrode consists of another water-cooled copper ring (*b*) placed coaxially with the flame at a proper distance from electrode (*a*). The voltage across these electrodes is 2 to 10 kV. The circuit can be simplified by leading the voltage directly into the metal to be melted. This is an analogue to the transferred arc in the plasma torch. Combustion products heated to 1,900°C have a very low degree of ionization and hence of conductivity. In order to initiate the electric heating, the conductivity must therefore be enhanced by the addition of a substance with low ionization energy. This purpose is served by potassium, which ionizes at a relatively low temperature and provides the requisite amount of charge carriers for the initiation. As soon as the current flows through it, the flame is heated by Joulean heat.

However, care must be taken to keep the conditions in proportions which prevent the discharge from contracting into an arc column. The diffuse nature of the discharge can be maintained by the turbulent flow of the combustion products. This problem, however, is far more complex and intricate than it would seem at first sight.—The fact that the device can be operated on alternating current is certainly an advantage. With the voltage ranging between 2 and 10 kV, the current in the experimental devices was 10 to 100 A.

Plasma torches are widely used as laboratory equipment. In rocket engineering they are employed for the production of extremely high flow velocities in hyperthermal tunnels. Plasma torches with a high excess pressure of gases in the electrode chamber are used to

Fig. 153 Thermal Dynamics 5,000-kW plasma torch for hyperthermal tunnels, operating at gauge pressures up to 100 atm

this purpose (cf. Fig. 153 representing a Thermal-Dynamics plasma gun of 5,000 kW output, operating with a gauge pressure of 100 atm). The plasma expands in a Laval nozzle and is accelerated far beyond the velocity of sound. The velocity is enhanced by expansion into a negative-pressure chamber. From this chamber the gases are exhausted by high-performance vacuum pumps through a heat exchanger. The plasma tunnels are used to model the re-entry conditions of rocket and ballistic-missile heads, for high-temperature material studies, etc.

In laboratory practice, small plasma torches with an argon-stabilized arc and reversed polarity,—i.e. the nozzle acting as the cathode—have proved a convenient aid for the spectral analysis of difficult-to-vaporize samples such as slags [Refs. 3.5.8, 3.5.9]. The gauge pressure in the electrode chamber exceeds 0·4 atm, and the orifice of the cathode-nozzle is 1·6 mm in diameter.

Radiation furnaces employed in special laboratories use a carbon anode heated by an electric arc as a source of radiation (cf. Sec. 2.1.5.5). Both the output of the freely burning arc and the surface of the anode—the actual source of radiation—are rather limited. The anode should be heated to a temperature close to the vaporization point of graphite (~ 3,900°C). If the heat supplied per unit area of the anode is too low, it is radiated at a lower anode temperature. By means of a plasma torch we can very considerably increase the heat flow striking the anode, which permits us to enlarge its radiating surface. Consequently the optical system produces a larger high-energy image in the focal plane, and thus a larger area with relatively uniform heat flow. In some applications—e.g. in experimental singeing of animal skins in research aimed at the curing of burns [Ref. 3.5.10]—uniform burns on larger surfaces are desirable.

The modelling of the re-entry conditions for rocket heads calls for the highest purity of the air-plasma stream, and so do thermodynamic measurements. However, in plasma torches with rod-shaped cathodes—no matter whether these are made of carbon or tungsten—the evaporation of the cathode material and hence the contamination of the arc is unavoidable.

A far purer plasma is obtained by using magnetic stabilization and a rotating arc. As the magnetic field acts upon the charged particles that move in the discharge path of the arc, it generates forces which cause the cathode and anode spots to travel quickly along the electrode surface, and this reduces the evaporation of the electrode material.

Even the form of the discharge changes in a sufficiently strong electric field: Owing to the rapid rotation, the arc body extends and forms a continuous ring between the concentric electrodes. This kind of discharge is termed a diffuse arc.

The principle of such an installation is illustrated in Fig. 154. Wiring the central electrode as the anode, and the the wall of the arc

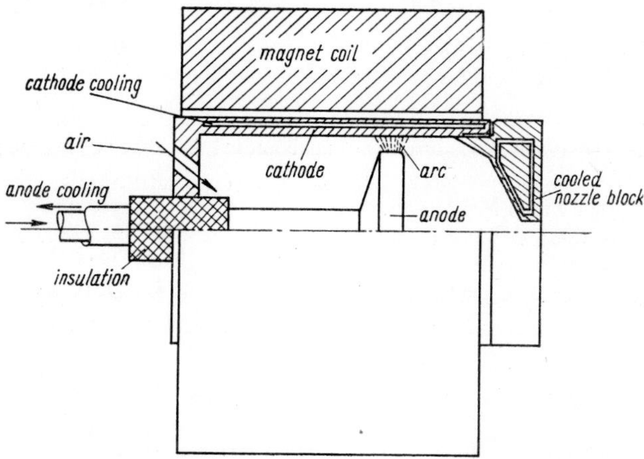

Fig. 154 Arc air heater with stabilized arc rotated by magnetic field

chamber containing it as the cathode, has been found the more convenient alternative: The cathode spot moves more rapidly and therefore runs ahead of the anode spot. Moreover, the larger circumference of the cathode reduces its current density per unit area, and the cathode is also more efficiently cooled. Both electrodes are made of copper.

This system produces a very pure air-plasma stream, and also increases the specific enthalpy of the air, as its pressure and flow quantity is reduced. So far, however, no such apparatus fully simulates the re-entry conditions for rocket heads, because air streams of extremely high enthalpy prove unobtainable: the reduced flow quantity rapidly reduces the efficiency of the arc heater and limits its output.

References

3.1.1 Miklóssy, K.: Lecture read at National Welding Conference, Prague, 4. 10. 1962

3.1.2 Linde Bulletin No. 51—405

3.1.3 Rosenthal, D.: The Theory of Moving Sources of Heat and its Applications to Metal Treatments. Oxford, Pergamon Press, 1946

3.1.4 Rykalin, N. N.: Berechnung der Wärmevorgänge beim Schweissen. (Computing the Thermal Phenomena of Welding Practice) Berlin, Verlag Technik, 1957

3.1.5 Wells, A. A.: *Weldg. Journal* **32** (1952). May, p. 263

3.1.6 Wells, A. A.: *Brit. Weldg. J.* **2** (1955), No. 9., p. 392

3.1.7 Wells, A. A.: *Brit. Weldg. J.* **8** (1961) No. 3, p. 79

3.1.8 Tykva, J.: *Strojírenství* **13** (1963), No. 8

3.1.9 Minutes of Comparative Cutting Tests, Bridličná, 8. 11. 63, Weldg. Res. Inst. Bratislava

3.1.10 O'Brian, R. L.—Wickham, R. J.: *Weldg. J.* **42** (1963) No. 2, p. 107

3.1.11 Matejec, M.: *Research Report Weldg. Res. Inst.* Bratislava No. R VI-3-4-1

3.1.12 Miklóssy, K.: *R.I.E.E. Research Report* Z-1097

3.1.13 Matejec, M.: *Zváranie* XII (1963) No. 1, pp. 2—6

3.1.14 Heinze, B.: *Die Technik* **17** (1962) p. 404

3.1.15 Browning, J. A.: Private communication

3.1.16 Browning, J. A.: *Weldg. J.* **41** (1962) pp. 453—455

3.1.17 Rychovki, D. G.: *Avtomaticheskaya svarka* (1962) No. 4, pp. 44—53

3.1.18 von Ardenne, M. et al.: *Die Technik* **18** (1963), pp. 645—651

3.1.19 Bízek, V.: Material Research Institute Prague, Report 5855

3.1.20 Kulhánek, L.: Research Institute f. Weldg. Technology, Report No. 51-021

3.1.21 Miklóssy, K.: Candidate's thesis, Czech Technical University, Prague, 1964

3.1.22 Málek, B.—Věchet, B.: Measurements of radiation and noise produced by operation of plasma torches, Prague, Institute of Hygiene and Epidemology

3.1.23 Speicher, W.: *Journal of Environmental Health* **102** (1962) p. 278

3.2.1 SCHOOP, M. U.: *Patent specifications* DRP 2,585,005, Sweden 49,270, France 406,387 and 407,170, GB 5712

3.2.2 SCHOOP, M. U.—DAESCHLE, CH.: Handbuch der Metallspritztechnik (Handbook of Metal Spraying) Zürich, Rascher & Cie. 1935

3.2.3 AULT, N. N.: *J. Amer. Ceram. Soc.* **40** (1957) No. 3, pp. 69—74

3.2.4 MAYER, H.: *Werkstoffe und Korrosion* **11** (1960), pp. 601—616

3.2.5 LEVY, H. J.: *SAE Journal* **67** (1959), No. 4, pp. 84—87

3.2.6 Oblouková metalizace (Electric-Arc Metallization) Prague, SNTL, 1961

3.2.7 KRETZSCHMAR, E.: Metall, Keramik- und Plastspritzen (Spraying Metals, Ceramics and Plastics). Berlin, Verlag Technik, 1963

3.2.8 MAYER, H.: *Ber. Dtsch. Keram. Ges.* **39** (1962) No. 2, pp. 115—124

3.2.9 MASH, N. E.—WEARE, WALTER: *J. of Metals* **13** (1961), pp. 473

3.2.10 NADLER, R. M.: Plasma Flame-Spraying Equipment Development *ASME Paper* No. 59-A 236

3.2.11 LEVINSTEIN, M. A.—EISENLOHR, A.—KRAMER, B. E.: *Weldg. J.* **40** (1961), pp. 8—13

3.2.12 New Coatings from the Plasma Arc. *Materials in Design Engrg.* **54** (1961) No. 6, pp. 127—128

3.2.13 VESELÝ, V.: *Schweisstechnik* **13** (1963) pp. 393—395

3.2.14 STETSON, A. R.—HAUCK, C. A.: *J. of Metals* **13** (1961) p. 479

3.2.15 *Design Engrg.* **54** (1961) No. 6, pp. 129—132

3.2.16 GERHOLD, E. A.: *Brit. Weldg. J.* **7** (1960), pp. 327—330

3.2.17 SINGLETON, R. H.—BOLIN, E. L.—CARL, F. W.: *J. of Metals* **13** (1961) pp. 483—486

3.2.18 *Weldg. Engrg.* **47** (1961) No. 7

3.2.19 Plasmadyne Corporation Prospectus: S-series plasma systems

3.2.20 VON ARDENNE, M.: *Die Technik* **18** (1963) pp. 263—274

3.2.21 Plasmadyne Corporation Prospectus No. 547—7027

3.2.22 ZUCHOWSKI, R. S.—CULBERTSON, R. P.: *Weldg. J.* **41** (1962) pp. 548—555

3.2.23 ZUCHOWSKI, R. S.—GARRABRANT, E.: *Weldg. J.* **43** (1964) pp. 13—26

3.2.24 LÖBL, K.: Navařování (Building up by Welding) Prague, SNTL 1961

3.3.1 PAUTHENIER, M.: *Revue Generale de l'Electricité,* **65** (1939) pp. 583—589

3.3.2 German patent specification 66,127 (1934) AEG

3.3.3 German patent specification 329,422 (1917) G. Braun

3.3.4 A. S. SEMENOV: Soviet patent specification 54,934 (1937)

3.3.5 B. M. MOLCHANYUK: Soviet patent specification 117,379 (1958)

3.3.6 BRDIČKA, M.: Mechanika kontinua (Continuum Mechanics). Prague, NČSAV, 1939

3.3.7 NEURINGER, J. L.: *J. Fluid. Mech.* **7** (1960) pp. 287—301

3.3.8 ROSA, R. J.: *Phys. Fluid.* **4** (1961) pp. 182—194

3.3.9 SARYCHEV, V. M.: *Zhur. prikl. mekh. tekh. fiz.* (1962) No. 2, pp. 3—6

3.3.10 NOWACKI, P. J.: *Res. report of Polish Nuclear Research Institute,* Warsaw, 1961

3.3.11 Dahlberg, E.: *Trans. R. Ins. Technology* No. 208 (1963)
3.3.12 Crown, J. C.: *Proc. Fourth Biennial Gas Dynamics Symposium*
3.3.13 Chapman, S.—Cowling T. G.: The Mathematical Theory of Non-Unifom Gases. Cambridge University Press, 1952
3.3.14 Aravin, G. S.—Shevelev, V. P.: Termicheskaya ionizatsia i elektroprovodimost nekotorikh smesey i produktov sgorania. *PMTF* (1962) No. 2, pp. 20—31.
3.3.15 Second Symposium on Engineering Aspects of Magnetohydrodynamics 1961
3.3.16 McGrath, J. A.—Siddall, R. G.—Thring, M. W.: Advances in Magnetohydrodynamics. Oxford, Pergamon Press, 1963
3.3.17 Coe, W. B.—Eisen, C. L.: *Electrical Engrg.* (1960) pp. 957—1004
3.3.18 Grekov, L. I.—Favorski, O. N.: *IANSSSR — Energetika i avtomatika* (1961), No. 4, pp. 46—54
3.3.19 Sherman, A.: *Voprosy raketnoy tekhniki* (1961), No. 4, p. 21
3.3.20 Plazma v magnitnom pole i priamoe preobrazovanie teplovoi energii v elektricheskuyu (Plasma in the Magnetic Field and the Direct Conversion of Thermal into Electrical Energy — Collected papers), Atomizdat, 1961
3.3.21 Ionnie, plazmennie i dugovie raketnie dvigateli (Ionic, Plasma and Electric-Arc Rocket Engines — Collected Papers) Atomizdat, 1961
3.3.22 Padley, P. J. et al.: *Trans. Faraday Soc.* **57** (1961) No. 465, pp. 1552—1562
3.3.23 Harris, L. P.—Cobine, J. D.: *ASME Paper* 60-Wa-329 (1960)
3.3.24 Wasserrab, T.: Patent specification GFR 1,116,801 (1959)
3.3.25 Smy, P. R.: *J. Appl. Phys.* **32** (1961) pp. 1946—1951
3.2.26 Clark, R. B. et al.: *Brit. J. Appl. Phys.* **14** (1963) pp. 10—15
3.2.27 Nernstein et al.: An Electrodeless MHD Generator. *Second Symposium on Engineering Aspects of Magnetohydrodynamics,* 1961
3.2.28 Jantovski, E. J.: *Energetika i avtomatika* (1961) No. 6
3.2.29 Jakovlev, V. S.: *Avtomatika* (1962) No. 4, pp. 74—78
3.2.30 Lindley, B. C.: *Symposium on MPD Electric Power Generation,* Newcastle, 1962
3.3.31 Strakhovich, K. I.—Sokovishin, J. A.: *Elektrichestvo* (1963) No. 9, pp. 15—22
3.3.32 Shih-I-Pai: *Physical Review* **105** (1957) No. 5, pp. 1424—1426
3.4.1 Remy, H.: Ànorganická chemie. Prague, SNTL, 1962
3.4.2 Berthelot, M.: *Ann. chim. et phys.* **67**
3.4.3 Lepsius, B.: *Ber.* **23** (1890) p. 1638
3.4.4 Pring, J. N.—Hutton, R. S.: *Trans. Chem. Soc.* **89**, 1906, p. 1591
3.4.5 Fischer, F.—Pichler, H.: *Brennstoff-Chemie* **19** (1938) p. 377
3.4.6 Muthmann, W.: *Berl. Ber.* (1903) p. 438
3.4.7 Nernst, W.: *Nachr. Kgl. Ges. d. Wissenchaftn.* (1904) p. 261
3.4.8 Stokes, C. S. et al.: *Ind. Eng. Chem.* **52** (1960) No. 1, p. 75
3.4.9 Waeser, B.: Die Luftstickstoffindustrie (The Atmospheric-Nitrogen Industry) Leipzig, 1922

3.4.10 Phillips, R. C.—Fergusson, R. A.: High-Temperature Technology. *Proceedings of International Symposium*

3.4.11 Weisenberg, I. J.—Winternitz, P. F.: *Sixth Symposium on Combustion,* Reinhold (1957)

3.4.12 Mac Leod, G. J.: *M.E. Thesis, California Inst. of Techn.* 1952

3.4.13 Egloff, G.—Schaad, R. E.—Lowry, C. D.: *J. Phys. Chem.* **34** (1930), p. 1617

3.4.14 Bozhko, N. P.: Poluchenie atsetilen-soderzhashchikh gazov iz metana v elektricheskoy duge. Moscow, 1940

3.4.15 Koller, L. K.: *Khimia tverdogo topliva,* 7 (1936) and 8 (1937)

3.4.16 Vasilev, S. S.: *Zhur. an. khim.*, No. 6, 7, 8 (1936), No. 9 (1937)

3.4.17 Nash, Stanley: J. *Chem. Soc. Ind.*, **48**, 1 (1929)

3.4.18 Bažant, V.—Šorm, F.: Organická technologie (Organic Technology) I, Prague, 1952

3.4.19 Bránský, E.: Acetylén (Acetylene) Prague, 1958

3.4.20 Fedorenko, N. P.: Metody poluchenia atsetilena (Methods of Producing Acetylene) Moscow, 1959

3.4.21 Tatarinov, V.: Soviet patent 40, 362 (1934)

3.4.22 Dobriański, A. F.: *Zhur. prikl. khim.* 10 (1947) No. 20

3.4.23 Morozov, I. N.: *Nauch.-inf. bul. Leningr. inst.* 12 (1939) No. 10

3.4.24 Cagaš, F.—Štaud, M.—Lazarev, A.: *Chem. prům.* **9** (1959) pp. 185, 225, 242

3.4.25 British patent 741,067 (1953)

3.4.26 Stokes, C. S.—Knipe, W. W.: *Ind. Eng. Chem.* **52** (1960) No. 4, p. 287

3.5.1 Plasma Arc: *The Iron Age,* Jan. 10 (1963) pp. 49—51

3.5.2 Plasma Arc: *The Iron Age* Jan. 10 (1963) p. 19

3.5.3 Tour, S.—Fletcher, L. S.: *The Iron Age* July 21 (1949)

3.5.4 von Ardenne, M.: *Die Technik,* **18** (1963) No. 4, pp. 263—274

3.5.5 Browning, J. A.: *Metalworking Production* Sept. 12 (1962)

3.5.6 O'Brien, R. L.—Wickham, R. J.: *Welding Journal* **42** (1963) No. 2, pp. 107—111

3.5.7 Tyler, P. M.: *Journal of Metals* **13** (1961), January, pp. 51—54

3.5.8 *Engineer's Digest* **20** (1959) No. 9, p. 351

3.5.9 Korolov, F. A.—Kvaratchkeli, I. K.: *Optika i spektroskopia* X (1961) pp. No. 3, p. 398

3.5.10 Butler, C. P.: *Proceedings of International Symposium on High Temperature Technology.* N. York, McGraw-Hill, 1959, p. 7

Index